Photons

Klaus Hentschel

Photons

The History and Mental Models
of Light Quanta

 Springer

Klaus Hentschel
Section for History of Science
and Technology, History Department
University of Stuttgart
Stuttgart, Baden-Württemberg
Germany

Translated by Ann M. Hentschel, Translator, Stuttgart, Baden-Württemberg, Germany

ISBN 978-3-030-07001-4   ISBN 978-3-319-95252-9   (eBook)
https://doi.org/10.1007/978-3-319-95252-9

Printed on acid-free paper

This Springer imprint is published by the registered company Springer International Publishing AG
part of Springer Nature
The registered company address is: Gewerbestrasse 11, 6330 Cham, Switzerland

Preface and Acknowledgements

"Light quanta" or "photons,"[1] derived from the Greek word for light, $\phi\omega\varsigma$ ("phos"), are omnipresent in modern science and technology. These terms appear all the time in the modern media, in science, and technology, as well as in art and popular reporting. "Quanta of light" and "photons" are used ubiquitously.[2] Applications such as the laser and its derivative devices, such as the CD and DVD player or the barcode scanner, define our daily lives. "Photonics" centers are cropping up everywhere for scientists and engineers in materials research and technology to collaborate on conceiving and developing new applications of quantum optics. Even so, this frequent usage of the term "photons" in research and production and the incidental mentionings of such a concept do not necessarily imply that it is generally known what these "photons" actually are. The mental models for this concept used by various groups could hardly be more different. The only reason why they can understand each other more or less is that generally no questions are asked about what exactly this concept might mean.

The history of the concept is mostly given short shrift with a reference to Einstein's first proposal in 1905 and its subsequent rapid rise to current ubiquity. Half-truths combined with a formidable number of historical myths are what this book intends to correct. We shall see that the history of such a complex concept as the "light quantum" is composed of many different layers, many of which formed well before 1905 while other additional ones formed much later on, during the twentieth century. Altogether, this book identifies and analyzes twelve different semantic layers that must be viewed together, in order to be able to understand properly the meaning of the concept of "light quanta" and the historical usages of

---

[1] The German expression "Lichtquant" first appears in Einstein (1905a) p. 144. The "photon" concept is mostly attributed to G. N. Lewis, who introduced it 1926 in a "Letter to the editor [of *Nature*]: The conservation of photons." However, it had already been in use in many other contexts within the physiology of sight and in biochemistry—albeit without having any lasting effect—and Lewis had probably never encountered it. On the history of this term in these other contexts, see Helge Kragh (2014a–b) as well as here Sects. 2.3, 3.7, and 4.8.

[2] See p. 36 for some quantitative instances.

such terms as "light energy quanta," "elementary quanta," "energy projectiles," "light corpuscles," or simply "quanta," as Einstein already used.[3] One of the oldest of these layers, the notion of light as being similar to a particle, even reaches back to antiquity. We shall see later on that the more precise form to develop out of this notion was radically reshaped many times over the course of history. We now know that the naive idea of a point-like particle is anything but suitable in the case of light quanta, because experimental situations exist in which light quanta seem to have considerable extension. At the beginning of the nineteenth century, in order to be able to explain such phenomena as interference, Thomas Young and Augustin Fresnel postulated their wave model of light as an alternative to Newton's corpuscular model.

According to the present ontology, light can appear either as a particle or as a wave, depending on the experimental situation (wave–particle duality). Here, too (as we shall see in Sect. 3.8), Einstein made decisive contributions in less well-known papers from around 1909. But we are going to encounter a number of other actors besides, who illuminated other facets. We shall see that experimenters such as Johannes Stark, of all people, who later gained notoriety as an anti-Semite, were among the earliest adherents of Einstein's postulates. These, in turn, were so controversial that Einstein himself only presented them from behind the skirts of a "heuristic hypothesis." Ironically again, Einstein's greatest benefactors and promoters (such as Max Planck or Hendrik Antoon Lorentz) started out as the most stubborn opponents of these postulates.[4] The American physicist Robert Millikan, no friend of the Einsteinian style of model building, who even expressly came forward to refute by experiment what he viewed as speculations by Einstein, was of all people, the person to confirm in 1916 Einstein's prediction of a strict proportionality between energy and frequency (see p. 62). We shall see that the broad front ranged up against Einstein's light quantum only slowly began to crumble from 1923 on, when the results of A. H. Compton's experiments on X-ray scattering off electrons became known, which confirmed empirically this conception of a point-like interaction between quasi-particulate X-rays.

Thus, an innovative concept such as Einstein's light quanta was initially far less clear and self-evident for those contemporaries than we would now expect. History is more convoluted than more tautly drawn retrospective allows it to appear. It also took 45 years for a reaction to Einstein's famous paper of 1915 on the spontaneous and induced emission of light quanta to lead to the development of the laser, based upon this idea.[5] A very decisive step toward deepening our understanding of the light quantum came with a paper by the Polish physicist Ladislas Natanson in 1911. It pointed out the need to radically reinterpret Planck's findings statistically, thereby anticipating by a decade one of the most important core assumptions of quantum

---

[3] For more specific citations and the usage frequency of these six approximately synonymous terms in the oeuvre of Albert Einstein, see Table 2.1.

[4] See Sects. 4.3–4.7.

[5] See here Sect. 4.9 and Lemmerich (1987), Bromberg (1991), Bertolotti (1999), Hecht (2005).

statistics—namely, the inability to distinguish between quanta.[6] But more than a decade had to elapse before what is known as Bose–Einstein statistics could emerge out of these insights by a few shrewd pioneers.[7] Even when a completely unknown Indian physicist by the name of Satyendra Nath Bose (1894–1974) applied to Einstein directly in 1924 to arrange for the publication of an article of his in *Zeitschrift für Physik*, Einstein needed some time and many false starts to recognize what other profound statistical implications were contained in Bose's paper besides his elegant new derivation of Planck's radiation density ("only under the condition that the smallest elementary cell in phase space has the volume h^3").[8] He forwarded the paper to the journal editor with the following comment that later also appeared at the foot of it: "Bose's derivation of Planck's formula signifies, in my opinion, an important advance."[9] Following the appearance of the concept of intrinsic angular momentum (spin), the quantum statistics for "bosons"—i.e., all particles with integer spin, and hence also including Einstein's light quanta—emerged.

The next important stage in the history of the concept was attained with the formulation of what is referred to as quantum electrodynamics (QED). It reinterpreted the photon in electromagnetic interactions as a virtual massless exchange particle bouncing back and forth between the interacting charges like a ping-pong ball. The exciting thing about these last layers of quantum statistics, of QED, as well as of the later quantum mechanical insights about the entangled states of correlated photon pairs, is that many of the formerly held basic characteristics of light quanta, in particular their particulate nature, localizability, and individualizability, are completely dissolved at this final stage.

The gradually effectuated semantic enrichment or accretion of layers of meaning, analyzed in detail in this book using the light quantum as its test case, is not a purely cumulative process but nonlinear: individual layers can also reshape themselves or even completely disappear again. The history of the terms and mental models of the light quantum or photon is paradigmatic of the complexity as well as of the intellectual tension of such concept formation and development processes. Many published commentaries on individual episodes of this historical event,

---

[6] Natanson demonstrated that only when one presupposes indistinguishability does Planck's formula for the mean energy E of a resonator result, after introducing the entropy formula $S = k \ln(W)$ and the second derivative. Distinguishable light quanta yield the Boltzmann distribution. Cf. Sect. 3.11 and additionally Kastler (1983), Monaldi (2009), Borelli, Saunders as well as Huggett and Imbo in Greenberger, Hentschel and Weinert (2009) pp. 299ff., 311ff., 611ff. and further references given there.

[7] See Delbrück (1980), Bergia (1987) and Stachel (2000). As Darrigol (1991) p. 239 describes this tragicomical history of quantum statistics: the "erroneous or opportunistic transposition of [combinatorial] formulas resulted in what we now call the Bose–Einstein statistics."

[8] Quoted from a letter by S. Bose to A. Einstein enclosing the draft of Bose's paper, dated June 4, 1924, reprinted in the *Collected Papers of Albert Einstein* (in the following abbreviated CPAE), vol. 14, doc. 261, p. 399 and in facsimile in https://en.wikipedia.org/wiki/Satyendra_Nath_Bose (last accessed on March 17, 2016). On Einstein's conceptualization at the time cf. Perez and Sauer (2011).

[9] See A. Einstein in a "remark by the translator" at the end of Bose (1924) p. 181.

especially in physics textbooks (cf. Chap. 6), by comparison, are noticeably under-complex, sometimes unduly simplistic, diluted or even pseudo-historically distorted. Thus, it shouldn't come as a surprise that an entire book be necessary for a thorough study of these processes. Nor am I alone in making this assessment: "many of the portrayals of the photoelectric effect suffer from inclusion of quasi-history and a partially wrong portrayal of the concepts themselves. [...] The situation vis-à-vis the concept of the photon is much more complex than can be portrayed in a short summary [...], and a thorough discussion of the various aspects would surely require a large volume."[10]

Precisely this is what the present book intends; and it ought to be of interest to many different groups of readers: obviously first and foremost to my own professional colleagues in the history of science, but also to physicists, astronomers, chemists, biologist, and other scientists and technologists with specialties in which photons now play a part, as well as to anyone else familiar with the basics in high-school physics and mathematics who wants to find out what this strange concept of a light quantum is and where it comes from. I did my best to present even the most complex topics, experiments, theories, and conceptual models in easily comprehensible form. My hope is to motivate cognitive scientists, linguists, historians of concepts, and historians of ideas to make something of this book from the point of view of method, as it attempts to clear aisles in those directions as well.

Acknowledgements

Although this book is the first historiographic integrative account of the photon, there are a number of excellent texts on specific aspects covered in individual chapters or sections, from which I was able to profit during my decade-long research on this book. Many are specifically cited in the notes, but I would like to point them out here as well, in approximately chronological topical order:

- Russell McCormmach 1968/69, Simon Schaffer 1979 and Jean Eisenstaedt 1996, 2005, 2012 on Newtonian emission theory
- John Heilbron, Alan Needell and Dieter Hoffmann on Max Planck
- Hans Kangro 1970 on experiments on radiation density
- Martin Klein 1964 and Thomas S. Kuhn 1978 on Planck's and Wien's, Einstein's and Ehrenfest's stepwise theory-formation
- Samuel Goudsmit 1971, Dirac 1974/75 and Sin-Itiro Tomonaga 1974/97 as eye-witnesses of the early history of spin and the theorem of spin statistics

---

[10] Klassen (2011) pp. 5–6. On the critique of "quasi-history" see, e.g., Holton (1973), Whitaker (1979), Simonsohn (1979, 1981), Jones (1991), Kragh (1992), Franklin (2016), Norton (1916), Passon and Grebe-Ellis (2016), and further references there.

- Roger Stuewer 1975a, 2014 and Allan Franklin 2013, 2016 on Millikan's measurement of h, on the Compton effect and other experiments on the light quantum within the framework of the old quantum theory
- Bruce Wheaton 1983 on early experiments and theories about the wave-particle duality
- Silvan Schweber 1994 on the history of quantum electrodynamics
- Alexei Kojevnikov 2002 on theories and experiment about fluctuations
- Lisa Bromberg 1991, 2006 and Jeff Hecht 2005 on the history of lasers as well as on the context of quantum optics particularly in the U.S.A.
- Roger Stuewer 1998, 2014, Steven George Brush 2007, Helge Kragh and Allan Franklin 2013, 2016 on critical assessments of pseudo-historical accounts by individual actors, such as Millikan and Compton, whose widely disseminated and still much-read descriptions of their own historical roles are hardly more than strategic posturing and "potted history"
- Gerhard Simonsohn, Stephen Klassen, Oliver Passon et al. for critical analyses of influential school and college textbooks
- Harry Paul 1985, 1986, Marlan Scully 1997, Boschi et al. 1998, Serge Haroche 2006, 2012, Raymond Chiao 2008, Anton Zeilinger et al. 2005, 2012, Yin 2017, Popkin 2017 on recent experiments on non-locality and entangled photons
- Indiarana da Silva and Olival Freire Jr. 2013 on experiments by Hanbury Brown and Twiss as well as John F. Clauser
- Kärin Nickelsen 2013, 2016 on controversies over quantum yield in photosynthesis

The abovementioned studies by these highly esteemed colleagues in the history of science, by nuclear physicists and quantum optics specialists, are finely detailed and micro-historically precise. Yet, they each merely cover small mesoscopic excerpts of the overall course of the centuries-long sequence of experiments regarded here, along with their intricately interlocked approaches to concept and theory formation.[11] None of the available historical studies on physics, its concepts or ideas, had the stamina to examine this centuries-long development while grappling with issues in the cognitive sciences. This is the ambition here.

The centennial of Einstein's *annus mirabilis* 1905 provided me with the opportunity to present preliminary talks on this topic at the 2005 international physics convention in Berlin as well as at an international conference on Einstein and Bose in Dhaka, organized by the *Bangladesh Academy of Sciences*. Initial publications about Einstein's approach to the light quantum soon followed.[12]

A written version of a series of public lectures at a symposium covering the topic of the formation and development of scientific concepts in conceptual history, delivered in Heidelberg at the end of May 2014, appeared the following year in

---

[11] For far more on the interplay between scientific instrument design, experimental practice, and theory formation, see Galison (1987) as well as my *Habilitation* thesis: Hentschel (1998b).

[12] See Klaus Hentschel (2005, 2005/07, 2007b, 2009a, 2015). On layer 10 (indistinguishability of photons—quantum statistics) see furthermore Hentschel and Waniek (2011).

Berichte zur Wissenschaftsgeschichte. These five contributions were trial borings into the formidable massif of the theme and constitute the core of the present monograph. It takes up each one of these twelve semantic veins (p. 39) and delves further into it, while integrating them into histories of science, concepts, and ideas combined with issues from cognitive psychology and linguistics.

The physicist and physics historian Dr. Peter Kasten (Göttingen) and Dr. Bernd Kröger (Tübingen) were so kind as to read through this book manuscript with a critical eye. I thank these two former auditors of my lectures at the Universities of Göttingen and Stuttgart for many useful comments and corrections. My thanks also go to members of the audience at my lectures on the topic (in Dhaka, Bangladesh 2005, as well as in Berlin 2005, Stuttgart 2009, Heidelberg 2014 and Dresden 2016). My students and colleagues at Göttingen, Bern, and Stuttgart participated in stimulating discussions for which I am grateful. The anonymous referees and members of the editorial team at Springer-Spektrum Publishers also deserve special mention, in particular, Dr. Lisa Edelhäuser, Ms. Stella Schmoll (both Heidelberg), Dr. Matthias Delbrück (Dossenheim), and Dr. Claus Ascheron for their active involvement in the German book project, Viju Falgon, Vani Gopi, Renu Boopalan in its English version.

For permissions to reprint illustrations, I duly acknowledge the *American Association for the Advancement of Science*, the *American Physical Society*, *Cambridge University Library*, *CCC Publications*, the *Nobel Foundation* in Stockholm, Sweden, *Nokia-Bell Labs Archives*, *AIP Publishing LLC*, *Cambridge University Press*, *Princeton University Press*, *Taylor & Francis* in London, *Springer-Verlag*, Heidelberg and *Springer Nature* in London; further Prof. Ferenc Krausz at the *Max Planck Institute for Quantum Optics* in Garching near Munich, as well as Prof. Olivier Darrigol (CNRS, Paris) and Dr. Rainer Reuter at the teaching unit of Science Education through Earth Observations for High Schools (Oldenburg). My wife, Ann M. Hentschel, a professional translator, kindly translated the whole book and helped me in the proofing stage.

Stuttgart, Germany   Klaus Hentschel

Contents

Chapter 1
Introduction

Why could it be useful—indeed 'important'—to a modern reader to think about the complex history of a concept like the photon instead of just concentrating on today and tomorrow? The reason is that the dense stratification of those twelve older layers of meaning, which have fused together into this concept, is still a live issue right now.[1] For a deeper understanding of what we mean by light quanta, it is highly instructive to study the history behind the concept and the cognitive obstacles that feature in it, faced by some of the most brilliant physicists. Einstein, in any case, never was able to come fully to grips with his own conceptual creation. In 1951 he wrote to a lifelong friend and confident: "All those 50 years of careful pondering have not brought me closer to the answer to the question: 'What are light quanta?' Today any old scamp believes he knows, but he's deluding himself."[2] And even Willis Lamb, like Einstein a theoretical physicist and Nobel laureate for his influential research on quantum optics, announced as late as 1995: "there is no such thing as a photon. Only a comedy of errors and historical accidents led to its popularity among physicists and optical scientists. I admit that the word is short and convenient. Its use is also habit forming."[3] These persistent, conspicuous and deep problems with which some of the greatest minds in the history of physics struggled should not be taken lightly or so easily dismissed. Participation in these profound, at times, heated debates counts, on one hand, among the most fascinating episodes in the history of physics. On the

---

[1] See p. 6 for an explication of the modeling of this process, suggested here only metaphorically, which Grattan Guiness denoted as a "convolution" and Reinhart Koselleck described using geological imagery.

[2] A. Einstein in a letter to his former colleague at the Bernese patent office, Michele Besso, 12 Dec. 1951, in: Speziali (1972) p. 453. Unless otherwise specified, all translated quotations in this book are ours.

[3] Lamb (1995) p.77; analogously Jones (1994), cf. Sulcs (2003) pp. 367ff. on the "Lamb–Jones opinion".

© Springer International Publishing AG, part of Springer Nature 2018
K. Hentschel, *Photons*, https://doi.org/10.1007/978-3-319-95252-9_1

other hand, it opens deep insights into the way in which our conceptual apparatus operates, into the genesis of new concepts and new mental models.[4]

Framed within these problems in history and cognitive psychology, the present monograph maps new territory, departing from a stimulating example. The case chosen here is so suitable for our enterprise because the emergent phase circumscribing and defining the concept persisted not for a matter of a few months or years, but for many decades—indeed, in the case of some layers, for centuries, even. Thus we have here complex phases that otherwise proceed in very rapid succession extenuated as if in slow motion. It therefore suits close analysis. I see parallels with earlier texts in the history of ideas, such as, the one by Max Jammer (1915–2010) on the concepts of mass and space or by Norwood Russell Hanson (1924–1967) on the history of the positron. More recent monographs have treated the history of the electron in similar complexity.[5] Theodore Arabatzis went a step further by writing his study from 2006 as a kind of biography of a scientific entity.—I definitely disagree, because I consider the biography metaphor misleading with reference to an inanimate object. Historically grounded studies already exist for some other fundamental concepts of modern physics, such as mass, the field, or even for more specific entities such as the electron. But this is the first book devoted to the history of the light quantum or photon that combines historical analysis with the cognitive perspective.[6] Obviously, one should not consider this book a general history of optics, very many of which already exist for all grades, to which reference is merely made.[7]

1.1 Methodology of This Study

History of concepts generally inquires about the historically changeable meaning of specific terms.[8] The frequent additions and alterations to which all natural languages are subject generates high interest in this in all national languages going back to early

---

[4]On 'mental models' and for further examples of the use of this concept from cognitive psychology, see Gentner and Stevens (1983), Collins and Gentner (1987).

[5]See Hanson (1963), Jammer (1961/74, 1974) or Davis and Falconer (1997), Dahl (1997). For further perspectives on the electron: Buchwald and Warwick (2001), Arabatzis (2006).

[6]Books such as the one by Zajonc (1993) remain on a far too popular level, whereas pamphlets such as the one by Fred Bortz (2004) in the "Library of subatomic particles" offer only a disappointingly brief introduction strewn with false myths that, as a consequence, is deplorably wrong in many passages. The best overview comes from the pen of an expert in quantum optics, Paul (1985).

[7]Among the popular books, I prefer the one by Park (1997) on the "Fire in the Eye." Among the more learned treatises, I recommend Mach (1921), Weinmann (1980), Darrigol (2012) and Smith (2014); concerning Darrigol, it is with the limitations spelled out in Hentschel (2012/14). My provisos for Mach, Smith and other classics of this field are: only up to early modern times, and the 19th century, respectively. An anthology of primary sources is offered by Roditschew and Frankfurt (1977). On the ontology of light rays, also in comparison with heat rays and radiation at other wavelengths of the electromagnetic spectrum and other types of particle rays (such as α- or β-radiation), see Hentschel (2007a) and further sources cited there.

[8]For good historiographic surveys in German, see Meier (1971), Richter (1987).

modern times, ever since the relinquishment of Latin as the *lingua franca*. Reference works such as *Grimms Wörterbuch*, the *Oxford English Dictionary* and other lexica for the main Western languages are to this day convenient resources for getting a quick idea of the first usages and shades of meaning of all commonly used words in the pertinent idioms. More specific reference works have developed since 1900 not only to define as precisely as possible central terms for individual subject areas, such as, in particular, philosophy and historiography, but also to work out the etymology of those terms. In the German-speaking world, for instance, pioneers in this direction include Rudolf Eisler's *Wörterbuch der philosophischen Begriffe* (1909), followed since 1955 by the *Archiv für Begriffsgeschichte*, founded by the medical historian Erich Rothacker in order to supplement the fourth edition of Rudolf Eisler's dictionary (1927–30), which had become outdated, as well as to compile his reference work *Bausteine zu einem historischen Wörterbuch der Philosophie*. Under the direction of the historian Joachim Ritter, it was later published under the title *Historisches Wörterbuch der Philosophie* in 1971. Likewise, from 1979 Otto Brunner's, Werner Conze's and Reinhart Koselleck's *Geschichtliche Grundbegriffe: Historisches Wörterbuch zur politisch-sozialen Sprache in Deutschland*. From 1980, the four-volume *Enzyklopädie Philosophie und Wissenschaftstheorie* by Jürgen Mittelstraß and collaborators followed (completed 1996). However, all of these reference works were limited to concepts in the core disciplines of philosophy and history and just contain a few fundamental scientific concepts; and these are rather treated from the perspectives of philology, philosophy and general history.

The **history of ideas** is an interdisciplinary clustering stimulated by the histories of philosophy, literature and art, the natural and social sciences, religions and political thought.[9] Forerunners in the history of philosophy included works by the cultural philosopher Ernst Cassirer (1874–1945) on the problem of knowledge in the philosophy and science of modern times (*Erkenntnisproblem in der Philosophie und Wissenschaft der neueren Zeit*, 4 vols., 1906–57) and by the historian of philosophy Edwin Burtt (1894–1989), who retraced the theological motives behind Newton's physics, in his classical work from 1924/25 *Metaphysical Foundations of Modern Physical Science*. Another canonical text was Arthur Lovejoy's *The Great Chain of Being* (1st ed. 1936), which staked out the method that would henceforth determine the field:

1. Distilling out so-called 'unit ideas,' i.e., elementary "types of categories, thoughts concerning particular aspects of common experience, implicit or explicit presuppositions, sacred formulas and catchwords, specific philosophical theorems, or the larger hypotheses, generalizations or methodological assumptions of various sciences." (Lovejoy 1936, p. 533)
2. Tracking down and pursuing these 'unit ideas' and their constellations in all fields of knowledge, coupled with searching for links between science, philosophy and religion as well as influences between one cultural area and another

---

[9]On major works, see Mandelbaum (1965), Kelley (1990, 2002), Grafton (2006).

3. Transgressing national and linguistic boundaries as well as temporal periods, by diachronic analyses through the centuries
4. Focusing on widely disseminated and influential ideas represented preferably by large portions of the "educated class, though it may be a whole generation, or many generations" (Lovejoy 1936, p. 19)
5. Taking an interest in the formation of new ideas as well as the replacement, fusion and diffusion of ideas (Lovejoy 1936, p. 20).

Examples of such 'unit ideas' include the guiding notions: natural laws, advancement, or civil rights, or else, Lovejoy's great chain of natural beings, which influenced the classification scales of natural history well into the eighteenth century. The idea of light quanta or photons could also be conceived as a 'unit idea' of science in Lovejoy's meaning.

With the work of Arthur O. Lovejoy (1873–1962) in Baltimore and Alexandre Koyré (1892–1964) at Harvard and Paris, along with the appearance of the *Journal of the History of Ideas* (founded in 1940 by Lovejoy together with Philip Wiener (1905–1992), who also published the first *Dictionary of the History of Ideas*), this field formulated since the 1930s the claim to broad integration of historical trends, motives, forces, attitudes and moods, freely hovering between and above conventional disciplinary boundaries.

Intellectual history appeared around 1960, a kind of fusion between the history of ideas, the social history of ideas and cultural history. Among its early promoters, Anthony Grafton demanded that the matter covered by the history of ideas be extended to include *all* textual and cultural products of human thought. The argument is that the classical history of ideas only looks at the tip of the iceberg. Intellectual history, on the other hand, takes into account less well received contemporary groups and individuals as well as the social and intellectual conditions of the emergence, duration and demise of ideas.[10]

Conceptual historians and historians of ideas typically come from the philological faculties, from philosophy or general history, which explains their preference for general cultural guiding concepts—for instance, peace, justice, unity or polarity, resistance or revolution. Following this reasoning, this means that many scientific concepts have yet to be satisfactorily analyzed—this also applies to the concept of light quanta and the entire field of related verbal terms, such as 'elementary quanta,' 'energy projectiles,' 'light corpuscles,' 'bullets of light,' 'photons,' etc. (See Sect. 2.5 for an overview and specific references.)

Such an analysis of the gradual formation and multiple changes in meaning of the term 'light quantum' could just as well be performed on many other concepts that are semantically more or less closely related. For instance, 'particles,' 'mass' or 'velocity' experienced a similar variety of deformations. Approaches along these lines have hitherto always either belonged *solely* to the history of ideas or *solely*

---

[10]See again Grafton (2006), further Greene (1957), Mandelbaum (1965), Kelley (1990, 2002) and, e.g., Horst and auf der (1998) on the nature concept.

to etymological history of concepts.[11] A *combined* history of the idea and physical concept is what will be attempted here that relates philological and cognitive psychological approaches with the history of science.

1.2 Terminological Distinctions Between Term, Concept and Mental Model

At first glance, this book's approach appears to be full of confusing variety: Why this double labeling of 'concept' and 'term'? Don't these two terms together really suffice to stake out our field sufficiently? Let us start by setting this straight and laying down definitions. In the following, the expression 'term' is understood as a concrete, linguistically fixed denotation for a defined phenomenal area, object or process. 'Concept' is a clearly outlined notion to which many very different terms may possibly be attached. A 'mental model' is the representation of an examined object, for instance: an object or process in the consciousness. In particular, it also includes more detailed notions about its properties, operation or handling schemes, causal entanglements and temporal courses.

We do, in fact, need this triad of terms in order to adequately grasp the genesis of the complex concept 'light quantum.' Pure histories of terms or expressions are not enough. We need to have the full arsenal available in historical analysis of terms, ideas and mental models. One obvious reason is that the development of terms limps behind the formation of scientific hypotheses. Fully shaped stabilized scientific terms, such as, 'light quantum' or 'photon' only appear at a relatively late stage in their development. Conceptual efforts and groping attempts at finding the right term and hammering out hypotheses and models occur long beforehand, which we definitely must also incorporate into the picture. In order to understand what is going on beneath the surface of language in the minds of those introducing new terms at some point, we must also include the mental models connected to those terms.[12] Mental models are not directly observable since they are in the mind of a researcher, but we can study their effect on the spoken word, the way they are used and connected. We thus infer them from statements, often only side-remarks, about how a hidden object is imagined to look, or how a hidden process is presumed to function. Certainly this inference remains a conjecture, but the historical actors analyzed in Chap. 4 of this book left enough textual or visual traces to allow a relatively unambiguous reconstruction of their mental models.

---

[11] For the classics in the history of ideas in physics, see Hanson (1963), Jammer (1966, 1974); on the etymology and history of concepts: Walker and Slack (1970), Caso (1980), Müller and Schmieder (2008) and Kragh (2014a, b).

[12] On mental models, cf. Gentner and Stevens (1983), Collins and Gentner (1987).

1.3   Concept Formation as Layered Semantic Accretion

The genesis of terms and concept formation, in my view, are complex 'nonlinear' processes of accretion, the gradual addition of many layers of meaning that extend variously far back in history and also endure variously long and have weightings within the total package that occasionally change variously. Old layers traceable back to early modern times lie right beside newer layers of quantum theory that only became formulable after 1900. In this "non-simultaneity of the simultaneous" (a bon mot by Ernst Bloch), just as in the turn of phrase of superposed layers of time, my approach undoubtedly resembles the conceptual history in semantics by Reinhart Koselleck (1923–2006). However, this historian in Bielefeld actually only draws metaphorical parallels to geology in introductory passages,[13] but never offers his readers any structure diagrams depicting merging, superpositions and layering of different semantic planes. I do try to draw an analogy by this metaphor and draft clearly legible schematic graphs reflecting the patterns of those processes of superpositioning and concentration. History is not a series of 'point events' but rather a web of lines of development: on one hand, strands of research (cf. Fig. 1.1), on the other hand, semantic layers as well, that continuously accumulate more layers in processes of accretion. Analogously to geological superpositioning, the old layers occasionally get compressed beyond recognition in the process, are bent or folded. These nonlinear processes continually generate new meanings out of already shaped terms, concepts and mental models.

The geological metaphors of layered superpositioning or accretion initially suggest a strictly cumulative picture of knowledge development. Nevertheless, my emphasis here is on 'nonlinear' features in multiple respects:

(i) This process is not at all even or steady—on the contrary, there are phases of dramatic change as well as stabile plateaus.

(ii) This is not plain accumulation, cumulative growth, but a process of complex cognitive interactions between old and new layers of meaning, which may also experience shifts in meaning and cut offs. (In the following we shall discuss one example, namely, the supposed point-shape of light quanta.)

(iii) There are incidents of readopted classical models attributable if not to Newtonians then to Newton himself.[14]

---

[13]E.g., Koselleck (2000) p. 9, as well as in the cover design of his hardcover book jacket, but not in the main text, which (as is typical of general historians) is a lead desert without any attempt at graphical visualization. The same applies to Koselleck (2010) and to Brunner et al. (1979).

[14]For instance, the article by Einstein (1924/25) in *Berliner Tageblatt* about Compton's experiments 1922/23: "Newton's corpuscular theory of light is coming back to life"; or by Sommerfeld (1919c) p. 59: "A ray in which energy and momentum are localized in a point shape does not essentially differ from a corpuscular ray; we have revived Newton's corpuscles."

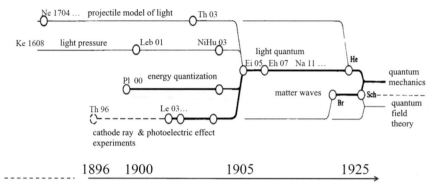

Fig. 1.1 Research strands along the way to the light-quantum hypothesis. This diagram cannot be more than a schematic and greatly simplified illustration of the complex superpositions and increasingly interconnected research strands that had previously been independent. During periods in which many strands are involved, such as here around 1905 and 1925, nonlinear, if not 'turbulent' phases form. Abbreviation key: Ke: Kepler, Ne: Newton, Leb: Lebedew, NiHu: Nichols and Hull, Le: Lenard, Th: J.J. Thomson, Pl: Planck, Ei: Einstein, Eh: Ehrenfest, Na: Natanson, Br: Louis de Broglie, He: Heisenberg and Sch: Schrödinger. Author's modification of the time line in Hund (1984) p. 20

The designation "convolutions" (as folds of meaning) is perhaps more appropriate than the geological metaphor of semantic superpositions. Ivor Grattan Guiness (∗1941) coined it in what he actually intended as a response to the never-ending debate about evolution vs. revolution, but it also fits the formation of terms and concepts.[15]

---

[15] See Grattan-Guiness (1990).

Chapter 2
Planck's and Einstein's Pathways to Quantization

When the young Max Planck[1] (1858–1947), offspring of a family of theologians and scholars, was considering which field of study best suited him, the mathematical physicist Philipp von Joly (1809–1874) at Munich advised against his choosing physics in 1874, because he believed that theoretical physics could no longer offer good prospects and experimental physics would only be insignificantly chasing the next possible fraction of the decimal point in its measurements of the natural constants.[2] Fortunately, Planck did not follow this advice and commenced studies in mathematics and science at the *Ludwig-Maximilians-Universität* in Munich. In 1877 he transferred to the *Friedrich-Wilhelm-Universität* in Berlin, to attend lectures in theoretical physics by Gustav Robert Kirchhoff (1824–1887) and Hermann von Helmholtz. Planck was among the first physicists to specialize exclusively in the theory and not also conduct experiments. His dissertation 'On the Second Law of Mechanical Heat Theory' was submitted there in 1879.[3] One year later Planck was already ready to present his second thesis on 'Equilibrium States of Isotropic Bodies at Various Temperatures,' to apply for his *Habilitation*, a degree permitting teaching at the academic level. Topics in thermodynamics and statistical mechanics, as were being pursued by Ludwig Boltzmann in Vienna, continued to preoccupy him,[4] along with other topics in physical chemistry. Planck adopted the mathematical techniques of statistical mechanics employed by Boltzmann, the Viennese pioneer of gas theory,

---

[1] Planck was born in Kiel, but his family subsequently moved to Munich in 1867, where he went to school and later to university. On Planck's life and work see, e.g., Heilbron (1986), Hentschel and Tobies (1999/2003), Hoffmann (2010).

[2] For depictions of the mood among physicists around 1900 see, e.g., Badash (1972), Brush (1987) and Sibum (2008). The mood among elder physicists around 1920 are described by McCormmach (1982).

[3] See Hentschel and Tobies, eds., 2nd ed. (2003) pp. 254ff. for the favorable but hardly enthusiastic evaluation of it.

[4] Cf. Badino in Joas et al. (2008), Badino (2009, 2015) Sects. 3.2 and 4.5, as well as further literature cited there.

© Springer International Publishing AG, part of Springer Nature 2018
K. Hentschel, *Photons*, https://doi.org/10.1007/978-3-319-95252-9_2

along with his interest in finding a physical explanation for the concept of entropy that Rudolf Clausius (1822–1888) had introduced in 1865 to gauge the lack of order in a system. The first law of thermodynamics is the law of energy conservation. The second law of thermodynamics, according to Clausius, assumed the simple form that the entropy of a closed system always strives toward a maximum. Statistically speaking, it steadily increases and never decreases of its own accord.[5] Throughout his life, Planck regarded these two fundamental principles of thermodynamics, which are valid without exception and entirely generally formulable, as his theoretical ideal. His research in thermodynamics attempted to apply them to all the other subfields of physics he was working on as well.

Planck gathered his first teaching experience in 1880 as a freshly graduated private lecturer at Munich. In 1885 he received an appointment as extraordinary professor at Kiel and 1889 succeeded his former academic teacher Gustav Robert Kirchhoff in Berlin. Planck's style of argumentation emulated Kirchhoff's,[6] the first professor of theoretical physics at the University of Berlin. In stark contrast to the thinking in visual models typical of British physicists, Kirchhoff and Planck avoided detailed models of matter, preferring rather to work with very general assumptions independent of any modeling. Planck's quantum of action h as well as his 'resonators'—precisely *not* concrete atoms or molecules but general systems capable of oscillating—and his strongly idealized 'black bodies' are all examples of this conceptual style of general abstraction that was later also characteristic of Einstein.[7] When Planck decided in 1899 to postulate another natural constant h as a unit of action in addition to the Boltzmann constant k, he justified this step—in conformity with his general style—as follows: "By utilizing both constants k and h, the possibility is given to posit units of length, mass, time and temperature that necessarily retain their meaning independently of specific bodies or substances, for all time and all, even extraterrestrial and nonhuman, cultures, and which therefore can be described as 'natural units of mass'."[8] The figure 6.88510^{-27} erg s that Planck provided for h was, incidentally, just 4% above the current value.[9] Planck's lifelong search

---

[5]On the context of these theories see, e.g., Brush (1970, 1976) as well as the primary sources discussed there.

[6]On Kirchhoff's style see my contribution in Hentschel and Zhu (2017).

[7]Planck was occasionally faulted for this abstractness. Niels Bohr, for example, declared during a talk before the Danish physical society on 20 Dec. 1913: "no one has ever seen a Planck's resonator, nor indeed even measured its frequency of oscillation; we can observe only the period of oscillation of the radiation which is emitted" (Engl. transl. from 1922, cit. in Kragh (2014c) p. 16). A visualization aid (obviously not in accord with Planck's personal preference!) can be found in Giulini (2011) p. 118.

[8]Planck (1897) part V, pp. 479–480, presented at the meeting of the Prussian Academy of Sciences on 18 May 1899.

[9]Caution: the foregoing papers denote what was later called the quantum of action (*Wirkungsquantum*) as b. From 1900 on, it is referred to as h and numerically set at 6.5510^{-27} erg s. For the sake of clarity, I have conformed the nomenclature here throughout as h.

for absolutes, for invariants also explains his early interest in Einstein's theory of relativity. The invariance of the velocity of light c and the square of the four-vectors, led him to interpret it as a theory of absolutes and support it energetically.[10]

2.1 Planck and Energy Quantization 1900

Upon following a call to Berlin, it did not take Planck long to come into contact with the Imperial Bureau of Standards situated in the suburb of Charlottenburg, the *Physikalisch-Technische Reichsanstalt* (in the following abbreviated as PTR). Kirchhoff had been its founding director up to his sudden death in 1887. The tasks performed at the PTR were not only practice-oriented, in the areas of metrology and standardization, but also proper research. It was one of the largest and most well-equipped research institutions for precision experiments in all subfields of physics.[11] These tasks included, in particular, high-precision measurements of temperature and heat intensity as well as the intensity of luminous emissions from different types of lamps. On the practical side of things, manufacturers and users of competing gas lamps and electrical lighting were the main interested parties. They wanted to know how much of the chemical or electrical energy supplied to these lamps was converted into visible light and how that energy was distributed over the total emission spectrum as well as how much of that energy dissipated as thermal radiation in ranges imperceptible to human vision. On the other hand, these practice-oriented contexts also touched on theoretical issues that Kirchhoff's pupils and successors in Berlin examined. Since 1860 Gustav Robert Kirchhoff had been very generally studying light radiation and reabsorption in matter and had been able to show that the emission and absorption coefficients E and A are always equally large. That explained, for example, why any luminous gas that emits specific spectral lines can also absorb incident radiation of exactly that wavelength. It also explained why the positions of their bright emission lines and dark Fraunhofer lines in the solar spectrum coincide so perfectly. Only then did it become possible to conclude the presence of all the elements in the solar atmosphere from the positions of those dark spectral lines, which are bright lines at exactly the same places in terrestrial spectra.[12] Kirchhoff excluded any dependence on material or form by adding another idealization: so

---

[10]Ultimately, Planck and his pupil Max von Laue were the driving force behind Einstein's appointment in 1914 to an extremely comfortable position in Berlin as director of an institute that was founded for him personally, the *Kaiser-Wilhelm-Institut für Physik*, and that until 1937 actually only existed on paper. See Kirsten and Treder (1979).

[11]On the history of the PTR and the measurements performed there, see Cahan (1989), Kangro (1970) and Hoffmann in Büttner et al. (2000) and the primary literature cited there.

[12]On these basics in the theory and practice of spectroscopy, see Kirchhoff (1860) and Hentschel (2002) and the primary sources cited there. The subtle shifts in position of these spectral lines through the relative motion of the light source or the receiver or else through gravitational red shift are discussed with primary references in Hentschel (1960b).

typical of his style, he limited his observations to ideal 'black bodies,' which he described as follows:

> When a cavity is entirely surrounded by bodies at the same temperature that are impenetrable to rays, then every beam of radiation in the interior of that space must, with regard to its quality and intensity, be constituted as if it had emanated from a perfectly black body at the same temperature and must therefore be independent of the form and nature of those bodies, having been determined by the temperature alone. One sees the validity of this assumption when one considers that a beam that has the same form and the opposite direction to the selected one is entirely absorbed after undergoing the enumerable successive reflections inside the imagined bodies. Accordingly, the same luminosity always occurs in the interior of an opaque glowing body at a particular temperature, irrespective of how it is otherwise composed.[13]

Attached to this idealization was the guarantee that the density of the radiation energy $\rho(\nu, T)$ would be independent of the material. But it also offered the possibility to transfer the concept of temperature away from the cavity walls onto the radiation in its vicinity, taking into account the thermal equilibrium between matter and radiation. It then made sense to speak of the temperature or entropy of radiation. This *per—constructionem*—guaranteed material independence, and the equivalence between the emissive power E and the absorptive power A at any place in the spectrum, which follow out of Kirchhoff's proposition, did not predict anything yet about how the density of the radiation energy $\rho(\nu, T)$ depended on the temperature T of the luminous body and the frequency ν. Kirchhoff pronounced that it was a primary goal of theoretical physics to determine these functional dependencies at thermodynamic equilibrium. Experimentally, a 'black body' approximating the thermodynamic ideal of perfect absorptive properties could be realized at the PTR shortly before 1900. The cavity's inner walls were specially treated with a powdering of platinum dust.[14]

Kirchhoff's successor Planck also adopted this problem when experimental physicists at the PTR confronted him with it in Berlin. Josef Stefan (1835–1893) in Vienna had already been able to demonstrate in 1879 that the total amount of radiation emitted by a luminous body at temperature T increases by the fourth power of T. Therefore, one must heat up an iron bar relatively intensely before it gradually starts to glow but only in the deep red range of the spectrum, whereas a very highly heated iron bar glows white. That is, it emits a generally much more intense energy spectrum shifted toward higher frequencies. Planck's colleague Wilhelm Wien (1864–1928) derived in 1893–94 the distribution law named after him out of electrodynamic and thermodynamic premises, according to which the spectral energy density $\rho(\nu, T)$ is, to good approximation, proportional to the third power of the frequency ν and could additionally only depend on a dimensionless function $f(\nu, T)$:

$$\rho(\nu, T) = \alpha \nu^3 f(\nu/T).$$

---

[13] Kirchhoff (1860) p. 300.

[14] See Lummer and Pringsheim (1897, 1899a, c, 1900) and Kangro (1970) pp. 149ff., Hoffmann and Lemmerich (2000) and Hoffmann in Büttner et al. (2000).

The problem defined by Kirchhoff one generation before was thus reduced to the question of what form this dimensionless function $f(\nu, T)$ should take for the idealized 'black body' at radiation equilibrium. Einstein described this situation in historical retrospect, with characteristic irony:

> It would be edifying if the brain matter sacrificed by theoretical physicists on the altar of this universal function f could be put on the scales; and there is no end in sight to this cruel sacrifice! What's more: classical mechanics also fell victim to it, and one still cannot tell whether Maxwell's electrodynamic equations will survive the crisis that this function f has brought about.[15]

Wilhelm Wien, who was co-editer of *Annalen der Physik* at the time, had been one of the first to make a concrete suggestion regarding the form this function $f(\nu, T)$ could take[16]:

$$\rho(\nu, T) = \alpha \nu^3 e^{b\nu/T}.$$

For a number of years Planck believed that this formula was correct. He attempted repeatedly to derive it out of fundamental electrodynamic and thermodynamic theorems, but it refused to work.[17] In 1900 Planck learned from Berlin experimenters that this formula agreed with their laboratory results to good or very good approximation only for large ν. It evidently completely failed for small ν. Another formula fit extremely well for the low-energy end of the spectrum, that is, toward the red, and even more so in the infrared spectral range. Lord Rayleigh and William Jeans in England had derived it from Maxwell's electrodynamics and from statistical mechanics[18]:

$$\rho(\nu, T) = \frac{8\pi\nu^2}{c^3} k_B T.$$

Planck heard about this conflict between the two fit formulas when Heinrich Rubens was visiting him at his home in the Grunewald suburb of Berlin on 7 October 1900. A few hours later he was able to produce an interpolation formula, which approaches the Rayleigh-Jeans limit for lower frequencies ν and approaches the Wien limit for high ν, with a smooth transition in between[19]:

$$\rho(\nu, T) = \frac{8\pi\nu^2}{c^3} \frac{h\nu}{e^{h\nu/k_B T} - 1}.$$

In this formula k_B is the Boltzmann constant of statistical mechanics and h is the quantum of action that Planck had already introduced into the discussion in 1899 and

---

[15]Einstein (1913), p. 1078 (CPAE, vol. 4, doc. 23, transl. ed., p. 273).

[16]See Wien (1896). Wien's b corresponds to h/k in our current nomenclature.

[17]On these efforts by Planck 1897–99, see Kangro (1970) pp. 93ff., Kuhn (1978) pp. 114ff. and Gearhart in Hoffmann (2010) along with the primary literature cited there.

[18]See Kangro (1970) pp. 189ff., Kuhn (1978) pp. 144ff., Giulini (2011) and Chiao and Garrison (2008) pp. 5–8.

[19]See Planck (1900a, 1943) and Kangro (1970) for a comparison with experiments from the time.

was later named after him. Further precision measurements conducted at the PTR proved its merit. The examinations by Rubens and Kurlbaum in the long-wave range, and by Lummer and Pringsheim in the short-wave range demonstrated a surprisingly good empirical match.[20]

Knowing what we do now about Planck's theoretical ideals, it is clear that he could not accept this situation. In his view theoretical physics must achieve more than simply supply empirically useful and lucky-strike fit formulas. Planck started an intense search for a satisfactory and reasonable way to derive the formula from more general considerations. In December 1900 he finally succeeded,[21] but it came at a high price. He used a statistical method by Boltzmann that he and his assistant Zermelo had already heftily criticized, to calculate the entropy S from the number of macroscopically indistinguishable microscopic 'complexions' K, that is, from distributions of the total available energy onto the individual resonators. The most probable macroscopic state is the one to which the most (macroscopically indistinguishable) complexions K correspond, as microstates. In order to be able to apply this method, Planck had to divide the energy up into finite packets, to be able to perform the combinatorics in the manner of Boltzmann. Classical physics had always treated energy as continuous. Unlike Boltzmann in 1877, however, he could not let the magnitude of these energy packets ϵ approach zero at the end of his calculation. The result was that the energy remained chopped up into finite packets, that is, 'quantized.' How dire this situation must have been for Planck to make him venture this formal step toward quantized energy, is revealed in his own words in a letter to the American experimental physicist Robert Williams Wood (1868–1958) from 1931:

> In a word, I could call the whole deed an act of desperation. For I am, by nature, peaceable and not inclined to dubious adventures. But I had been wrestling with the problem of the equilibrium between radiation and matter for 6 years [since 1894], without success. I knew that this problem was of fundamental importance to physics. I was familiar with the formula describing the energy distribution in a normal spectrum; consequently, a theoretical interpretation had to be found at any price, no matter how high. Classical physics was not good enough, that was clear to me. Because, according to it, over time the energy in matter must convert entirely into radiation. In order for it not to do so, we need a new constant [Planck's quantum of action h] that assures that the energy not disintegrate. [...] one finds that this dissipation of energy as radiation can be prevented by the assumption that energy be compelled from the outset to stay together in specific quanta. That was a purely formal assumption and I did not really consider it much, just that I must, under all conditions, cost what it may, force a positive result.[22]

---

[20]For the experimental research on black-body radiation conducted around 1900, see Lummer and Pringsheim (1897–1900), Kurlbaum and Lummer (1898, 1901), Rubens and Nichols (1896), Rubens (1917), Rubens and Kurlbaum (1900, 1901); further Kangro (1970, 1970/71) and the primary references there.

[21]On the following, see Planck (1900b) and, e.g., Kuhn (1978), Kangro (2009, 2015) Chap. 4, Darrigol (1992), Gearhart (2002) and Duncan (2012) pp. 8–14 on the derivation of this formula from radiation entropy and its derivatives for temperature.

[22]Planck to Robert Williams Wood, 7 Oct. 1931, cited from Hermann (1969/71) p. 31 or (1969/71b) p. 23.

The long interval between this retrospective statement and the developments around 1900 prompts one initially to ask how reliable this assertion might be. I do think, though, that recollections of such an obviously so agonizing dilemma for him don't fade, and for that reason I rather attach considerable plausibility to this report, especially considering that it also suits perfectly what we otherwise already know about Planck's intellectual world view and personality. Why would such an austere arguer and author, as Planck was, stress this situation in historical retrospect so much, if it had not in fact seemed so dramatic to him then? Such a thing is not so easily forgotten, either. Einstein characterized Planck's way of going about this in a later talk:

> To solve the problem of radiation, Planck concluded that one would have to introduce a new physical quantity: the famous quantity h, in order to arrive at a reasonable formula for radiation. But this calculational figure has a very real meaning in nature, in the sense that radiation forms or vanishes only in the magnitude $h\nu$. When a bell rings, it sounds loudly when it is struck firmly, and more quietly, the weaker it is struck; it receives a greater or lesser amount of energy. In radiative processes, this is not so to the same degree; rather, energy cannot be introduced into a luminous structure in arbitrarily small amounts, never less than one quantum and always only integral multiples of this quantum are taken up or released again by a structure able to radiate.[23]

Planck's derivation in 1900 was intrinsically discordant, if not downright schizophrenic. In the argument for the energy density ρ of the field as a function of frequency, he applied a formula following out of classical electrodynamics: $u_\nu = 8\pi\nu^2/c^3$. That is, the oscillation mode of the radiation field was presumed to be strictly continuous. In combinatorially computing the number of 'complexions' K, i.e., the number of microstates corresponding to a given macrostate of fixed energy and temperature, he followed the model of Boltzmann (1877) in setting $K = (N + P - v)!/N!P!$ Therefore, in this step the absorption and release of energy by the resonators was presumed to be discontinuous. In order to be able to compute the number of these complexions combinatorially at all, it was compellingly necessary that the portions of energy for distribution among the resonators be finite. However, as Planck put it, this remained "merely a formal assumption," even though—different from Boltzmann 1877—later having the boundary transition $h \to 0$ was no longer possible. How inconsistent Planck's reasoning and formulations about this quantization were is revealed not least in the following remark he made in a paper dated December 1900 about cutting up the total energy E into tiny energy packets ϵ.—Darrigol and other historians view this as the finishing touch to their evidence that Planck did not discover quantum theory already in 1900: "By division of E by ϵ, we obtain the number P of energy elements, which are to be distributed among the N resonators. If the quotient is not an integer, then one takes for P an integer near by.[24] This sentence is struck in the revised reprint of Planck's paper in the *Annalen der Physik* for January 1901! Did its editors, Planck and Wien, already notice this inconsistency in the condition of strict quantization? We cannot tell.

---

[23]Einstein (1927) p. 546.
[24]Planck (1900b) p. 240.

Whereas critics of the standard historiography, such as Kuhn or Darrigol, appear to suppose that such a radical new theory can suddenly develop from one day to the next, I see the genesis of quantum theory rather as a stepwise process. At the end of 1900 Planck was just in the act of climbing up one of those high steps. It is not really surprising that there should be full clarity about an advance as radical as is quantization only retrospectively, after an interval of some ten years. In a later lecture before members of the *Mathematisch-Physikalischen Arbeitsgemeinschaft* at the University of Berlin, Einstein summed this up as follows:

> 27 years ago, Planck had deposited a huge gadfly with his theory of radiation [...], although at first it was just a small one, so, many people did not notice it.[25]

What Einstein is adeptly holding back here again is that Planck himself did not grasp the full consequences of his own assumption at first and was shocked at himself when he eventually did.[26]

2.2 Einstein's Line of Thought Prior to His 1905 Paper

As we have seen in the foregoing—to borrow Friedrich Hund's words—Planck's quantum hypothesis was "somewhat of a premature birth. Essential properties of nature to which the quantity h could be attached still had been scarcely researched or classified. While Planck's radiation formula rapidly gained acceptance as empirically valid, initially his theory was not explored further."[27] The first person to study more closely not only Planck's interpolation formula but also his derivation from December 1900 and the theoretical assumptions underpinning them was Albert Einstein (1879–1955), at that time a still unknown examiner at the Bernese patent office.[28] Although Einstein had never officially studied under either Kirchhoff, Planck, Clausius or Boltzmann, he had very thoroughly read the books and papers by these four theoreticians and had also intensely discussed them with some student friends of his in their informal 'Olympia Academy.'[29] This intense study of thermodynamics and statistical mechanics is also evidenced by Einstein's first scientific papers which he, as a freshly engaged third-class patent expert, submitted to *Annalen der Physik*. At that time this leading German journal was being edited by Max Planck

---

[25]Einstein (1927) p. 546: This is taken from a paraphrasing of Einstein's talk, but this particular passage is rendered in quotations as essentially verbatim.

[26]See again Planck's letter to Robert W. Wood (first quoted in Hermann (1969/71) p. 31, quoted above on p. 14).

[27]Hund (1984) p. 29.

[28]On Einstein's oeuvre and vita see, e.g., Schilpp (1959), Pais (1982), Kirsten and Treder (1979), Home and Whitaker (2007), Janssen and Lehner (2014) and further literature cited there.

[29]See Solovine (1956), Speziali (1972), Pyenson (1985) and Howard and Stachel (2000) Howard on the intellectual context of the young Einstein, and Norton (2016) on his heuristics.

and Wilhelm Wien, who probably had also reviewed Einstein's submissions.[30] These debutant publications were followed shortly thereafter by twelve papers that were to change the course of physics decisively in 1905, which was to become Einstein's *annus mirabilis*. The first of these was the kettledrum roll which concerns us here: "On a Heuristic Point of View Concerning the Production and Transformation of Light." However, the other papers were also highly significant—especially considering that they were closely interconnected with each other not only by their times of appearance but also by their substance (cf. Fig. 2.1).

What exactly was so "very revolutionary" about Einstein's paper from March 1905? Foremost: the introduction of light quanta. Quantization was explicitly not limited to the resonators (as had been the case with Planck) or to the interaction between matter and the field. Quantization was required of the energy of the electromagnetic field, too, now: "when a light ray is spreading from a point, the energy is not distributed continuously over ever-increasing spaces, but consists of a finite number of energy quanta that are localized in points in space, move without dividing, and can be absorbed or generated only as a whole."[31]

Why was this thought "revolutionary"? The idea of light being like a corpuscle is very much older (cf. Sect. 3.1). The novelty was quantizing its energy—i.e., the requirement that in certain circumstances the light's energy cannot vary continuously anymore. It agreed neither with continuous classical mechanics nor with Maxwell's electrodynamics generally. Einstein fully realized this, as he told Lorentz: "neither molecular mechanics, nor Maxwell-Lorentz's electrodynamics can be brought into agreement with the radiation formula."[32] Needless to say, his counterpart was anything but convinced of this hypothesis (cf. Sect. 4.3 for Lorentz's own mental model).

Einstein knew that his suggestion was bold and that he needed to be cautious, therefore the conjunctive mood to his wording from March 1905. He definitely did not say: Light quanta of energy $E = h \cdot \nu$ exist. His formulation was much more tentative: It would be possible to conceive monochromatic radiation of frequency ν at the Wien limit *as if* it were composed of mutually independent energy quanta. Then, consequently, the matter–field interaction would be emission and absorption of such quantized energy packets. This idea reappears in 1913 in Bohr's model of the atom, but such a visual model is not part of Einstein's postulate in 1905.

What was Einstein's reasoning in 1905 in support of the existence of such light-energy quanta, or of their plausible assumption, at least? He used a two-track derivation so typical of his way of working (see Box 1). Einstein first analyzed a

---

[30]Their evaluation reports unfortunately have not survived but all of Einstein's papers from this period are available by now, variously annotated or commented on. See *The Collected Papers of Albert Einstein* (CPAE), 15 volumes of which have appeared so far.

[31]This is Einstein's definition of the light quantum in March 1905, CPAE vol. 2, doc. 14, p. 133, (transl. ed., p. 87). Cf., e.g., Pais (1982) p. 373. Fölsing (1993) even considered this to be the "most revolutionary" statement in 20th-century physics as a whole. See also Rigden (2005) p. 18 for five general criteria permitting a scientific article to become 'revolutionary': it must be (i) comprehensive ("a big idea"), (ii) provocative or contradictory to current conceptions, (iii) initially almost unanimously repudiated, (iv) later hesitantly accepted, and (v) finally widely accepted.

[32]Einstein's synopsis four years later in a letter to H.A. Lorentz, 30 March 1909, CPAE, vol. 5, doc. 146, transl. ed., p. 105.

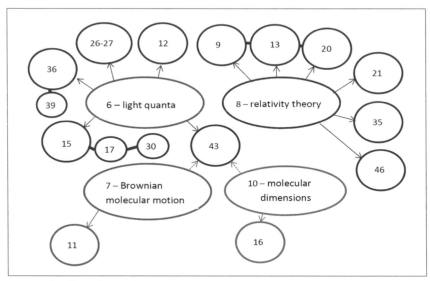

6. AdP **17**, 132 (1905) [17 pp.] (CPAE 2, 149) *On a Heuristic Point of View concerning the Production and Transformation of Light*
7. AdP **17**, 549 (1905) [12 pp.] (CPAE 2, 223 *On the Movement of Small Particles suspended in Stationary Liquids required by the Molecular-Kinetic Theory of Heat*
8. AdP **17**, 891 (1905) [31 pp.] (CPAE 2, 275) *On the Electrodynamics of Moving Bodies*
9. AdP **18**, 639 (1905) [3 pp.] (CPAE 2, 311) *Does the Inertia of a Body depend upon its Energy Content?*
10. AdP **19**, 289 (1906) [18 pp.] (CPAE 2, 346) *A New Determination of Molecular Dimensions*
11. AdP **19**, 371 (1906) [11 pp.] (CPAE 2, 333) *On the Theory of Brownian Motion*
12. AdP **20**, 199 (1906) [8 pp.] (CPAE 2, 349) *On the Theory of Light Production and Light Absorption*
13. AdP **20**, 627 (1906) [7 pp.] (CPAE 2, 359) *The Principle of Conservation of Motion of the Center of Gravity and the Inertia of Energy*
15. AdP **22**, 180 (1907) [11 pp.] (CPAE 2, 378) *Planck's Theory of Radiation and the Theory of Specific Heat*
16. AdP **22**, 569 (1907) [4 pp.] (CPAE 2, 392) *On the Limit of Validity of the Law of Thermodynamic Equilibrium and on the Possibility of a New Determination of the Elementary Quanta*
17. AdP **22**, 800 (1907) [1 p.] (CPAE 2, 404) *Correction to my Paper: `Planck's Theory of Radiation, etc.'*
20. AdP **23**, 371 (1907) [14 pp.] (CPAE 2, 413) *On the Inertia of Energy required by the Relativity Principle*
21. A. Einstein & J. Laub, AdP **26**, 532 (1908) [9 pp.] (CPAE 2, 508) *On the Fundamental Electromagnetic Equations for Moving Bodies*
26-27. A. Einstein & L. Hopf, AdP **33**, 1096 (1910) [9 pp.] & AdP **33**, 1105 (1910) [11 pp.] (CPAE 3, 258 & 269) *On a Theorem of the Probability Calculus and Its Application in the Theory of Radiation / Statistical Investigation of a Resonator's Motion in a Radiation Field*
30. AdP **34**, 170 (1911) [5 pp.] (CPAE 3, 408) *A Relationship between Elastic Behavior and Specific Heat in Solids with a Monatomic Molecule*
35. AdP **35**, 898 (1911) [11 pp.] (CPAE, 3, 485) *On the Influence of Gravitation on the Propagation of Light*
36. AdP **37**, 832 (1912) [7 pp.] (CPAE 4, 114) *Thermodynamic Proof of the Law of Photochemical Equivalence*
39. AdP **38**, 881 (1912) [4 pp.] (CPAE 4, 165) *Supplement to my Paper: 'Thermodynamic Proof of the Law of Photochemical Equivalence'*
40. AdP **38**, 888 (1912) [1 p.] (CPAE 4, 171) *Response to a Comment by J. Stark: `On an Application of Planck's Fundamental Law...'*
43. A. Einstein & O. Stern, AdP **40**, 551 (1913) [10 pp.] (CPAE 4, 274) *Arguments for the Assumption of Molecular Agitation at Absolute Zero*
46. AdP **49**, 769 (1916) [54 pp.] (CPAE 6, 283) *The Foundation of the General Theory of Relativity*

Fig. 2.1 Einstein's interrelated papers from his *annus mirabilis* 1905 and 1906–16 in the *Annalen der Physik* (abbreviated AdP), reprinted in *The Collected Papers of Albert Einstein* (CPAE), enumeration from http://users.physik.fu-berlin.de/~kleinert/papers/einstein-in-adp-papers.htm (access 10.3.2016)

given system by two different theoretical techniques as far as each would take him and then, in a second step, required that the same physical quantities apply to the two resulting expressions to make them consistent with each other. Given n of these light quanta—or more generally, these corpuscular local systems—in a volume V_0. Then one can ask how probable it is for all the point-shaped systems n to be inside the smaller partial volume V within the beginning volume V_0. The smaller V is, compared to V_0, the more improbable it is. Solutions to this kind of problem can be found by general probability theory (the left-hand side of Box 1) as well as by Wien's and Planck's radiation theory (the right-hand side of Box 1). Since they are supposed to be compatible, both these expressions must be equal to each other, which is only the case when $E = h \cdot \nu$—*quod erat demonstrandum.*

Box 1: Einstein compares ideal Boltzmann gas with radiation at Wien's limit

The probability W that at one arbitrary instant all independently moving points n within a given volume V_0 happen to be in the same smaller volume V is calculated by two different methods. Likewise for the resulting entropy of the system. The left-hand column shows the calculation according to kinetic gas theory, by analogy to an ideal Boltzmann gas; the right-hand column, according to Planck's radiation theory at the Wien limit.

Boltzmann gas	Wien limit
$S - S_0 = \frac{R}{N} \cdot \ln(W)$	$S = V \cdot \int_0^{\inf} \phi(u, v)dv$
$W = (\frac{V}{V_0})^n$	$u_{\text{Wien}} = \alpha \nu^3 \cdot e^{-(\beta\nu)/T}$
$S(V, T) - S(V_0, T) =$	$1/T = d\phi/du$
$\frac{R}{N} \ln[(\frac{V}{V_0})^n](*)$	$= \frac{R}{N} \ln[(\frac{V}{V_0})^{NE/R\beta\nu}](**)$
$(*), (**) \Rightarrow n = \frac{NE}{N\beta\nu}$	$= E/(k\beta\nu) = E/(h \cdot \nu),$

The final equations substitute $R/N = k$ and $\beta = h/k$. By comparing the two expressions, which can only agree when $E = h \cdot \nu$, Einstein derived his light-quantum hypothesis: "Monochromatic radiation of low density (within the range of validity of Wien's radiation formula) behaves thermodynamically as if it consisted of mutually independent energy quanta $[n]$ of magnitude $[= h \cdot \nu]$."

What did this "heuristic point of view" achieve? The first step down this long path. This new vantage point allowed the prediction of many experimental effects as well as explanations for known empirical findings. Einstein mentioned the following topics in 1905:

1. Stokes's rule on photoluminescence (1852): The re-emission frequency is always smaller than the initiator frequency. Einstein realized that this previously incomprehensible rule simply follows from the law for the conservation of energy, $E = h \cdot \nu$.
2. The photoelectric effect (first observed in 1888 by Hallwachs and experimentally analyzed by Lenard in 1902): UV radiation that hits a cathode in a vacuum triggers the emission of cathode rays. Lenard attributed the energy of the released radiation to an energy within those particles prior to the emission. Einstein thought it originated from the absorbed UV light quanta, less the work of emission W_A. This explains why $E = h \cdot \nu - W_A$ and not $E \sim$ square of the amplitude, as would be expected in the classical theory.

3. The short-wave limit of x-ray bremsstrahlung $\nu < E_{max}$: This, too, follows from the energy conservation law. The energy released in the form of x-rays at frequency ν from charged particles that are suddenly slowed down, cannot exceed the maximal energy of the slowed-down particles.
4. Spectral density of black-body radiation. Einstein discussed, further, Planck's considerations about black-body radiation and demonstrated the compatibility of his light-quantum hypothesis with Planck's formula for the energy density of black-body radiation from 1900. It had already been confirmed experimentally, but this provided a deeper theoretical underpinning.[33]

The harmony between all of these empirically verified predictions or conclusive new explanations of known effects that had hitherto been difficult to understand, if at all, yielded a "consilience of inductions" as defined by the English philosopher of science William Whewell (1794–1866), which is one of the strongest indicators of the correctness of a proposition.

Einstein continued to write papers about radiation theory and the problem of specific heats until 1909, which can only be skirted here. With falling temperatures there is a growing discrepancy with Dulong and Petit's law of constant specific heat per gram-equivalent and degree of freedom of a substance. Einstein was able to derive this from the low-temperature limit of Planck's energy distribution. His model easily allowed one to anticipate that the specific heat would drop strongly at very low temperatures.[34] One of the sources of Einstein's data on specific heat was, incidentally, his doctoral advisor Heinrich Friedrich Weber (1843–1912). The success of the light-quantum hypothesis proved not to be ephemeral. It became a fruitful concept in many areas of physics.

2.3   Einstein and Planck: A Comparison

Einstein did not shy away from calling his light quantum idea "revolutionary," at least in private letters to his same-aged friends. Max Planck, his elder by 21 years, acted much more cautiously. He wove his theoretical web mainly around the goals of stabilization, improvement and, if at all, minimal modification. It was a matter of 'limiting the damage' to classical physics, which he wanted to preserve as best as possible, as the following quotations evidence:

> For I do not seek the meaning of the elementary quantum of action (light quantum) in the vacuum but in the sites of absorption and emission, and assume that the processes in the vacuum are described exactly by Maxwell's equations. At least I do not yet see any

---

[33]On these experiments, see the literature cited in see footnote 20 above as well as Dorling (1971), Norton (2008), Darrigol (2014) on Einstein's theoretical underpinning.

[34]Einstein (1907a) resp. CPAE, vol. 2, doc. 39; Einstein (1911/12) resp. CPAE, vol. 3, doc. 26, as well as Debye (1911), Pais (1979), Gearhart (2010) and Gearhart in Greenberger, Hentschel and Weiner (2009), pp. 719–721.

compelling reason to abandon this assumption, which seems to me the simplest for the time being.[35]

And in a paper on the theory of thermal radiation from 1910 he advised:

> It appears to me that utmost caution against the new Einsteinian corpuscular theory of light would be warranted ... The theory of light would be thrown back not decades, but centuries, to the time when Christian Huygens dared to take up his fight against Newton's overpowering emission theory ... And all of these accomplishments, which are among the proudest successes of physics, indeed of scientific research overall, are supposed to be sacrificed for the sake of some still quite contestable observations? Heavier artillery would really need to be run out to sway this, by now, very firmly founded edifice of electromagnetic light theory.[36]

Planck was trying to avoid any outright quarrel with the solid foundations of classical physics. It seems to have been downright embarrassing for him that with his initially harmless interpolation proposal in 1900 to explain the distribution of radiative energy measured at the PTR he, of all people would have triggered such a far-reaching development. Consequently, Planck did everything he could, particularly with his so-called second quantum theory between 1907 and 1911, to mend the rupture again.[37]

In 1986 John Heilbron very fittingly characterized Planck as a "conservative revolutionary"—one could also say: a renitent revolutionary, because he neither sought nor wanted to assume this role as discoverer of energy quantization, as has often been ascribed to him in the later historiography.[38]

I place my own interpretation among the central historiographic positions in Fig. 2.2. As I see it, Planck was a figure of transition. Although his thinking and argumentation was still situated wholly within the contexts of the tried-and-true theories, he had inadvertently ventured out onto new terrain. He was partly aware of this but shrank back from consistently exploring it—he had stepped out onto it, nonetheless.

Irrespective of the reading one might choose for Planck's writings from around 1900, Einstein's approach was definite. Quite in contrast to Planck, he offensively set out in search of cracks, pointed them out and took them as his starting point toward something new. Evidently, the no longer extant first draft of his paper, in which he criticized Planck's half-hearted position on energy quantization, was much more aggressively direct. But Einstein's close friend and colleague at the Bernese patent office, Michele Besso (1873–1955) persuaded him to soften the tone a bit and present his hypotheses about the light quantum much more deferentially. Looking back on the good old days, Besso listed in a letter the many things he felt indebted to Einstein for, also mentioning some things he felt he could return in kind: "For my part, I was your public during the years 1904 and '05; in the formulation of your

---

[35]Planck to Einstein, 6 July 1907 (Phys. Abh. II, 292ff.), CPAE, vol. 5, doc. 47, transl. ed., p. 31.

[36]Planck (1910b) pp. 242f.

[37]See here p. 114. This second quantum theory should not be confused with the 'second quantization' in QED, which will be discussed later in Sect. 3.12.

[38]This still is a controversial position among experts, though: see Heilbron (1986) as well as Darrigol and Gearhart in Büttner et al. (2000).

Discontinuity	Indetermination	Continuity
Textbooks Klein 1961-7 Hund 1967 Jost 1995	Kangro 1970 Needell 1980, 1988 Galison 1981	Kuhn 1978
	Rosenfeld 1936 Jammer 1966	Planck 1920, 1943, 1948 Darrigol 1988, 1992

Fig. 2.2 Situating the different historiographic interpretations of Max Planck's contribution in 1900. At the far left, some advocates of the standard thesis that Max Planck broke with the past by introducing quantum theory in 1900; at the far right, defenders of the opposing thesis that Planck was still working along the lines of the classical theories of mechanics and electrodynamics; in the middle, some in-between positions that presume an incomplete transition or Planck's indecisive vacillation between these extremes. Source: Darrigol (2000/01) p. 6. Reprinted by permission

communications on the quantum problem, I robbed you of some of your fame, but for it, procured you Planck as a friend."[39] Hence the conspicuously modest aim in the title of the published version of Einstein's article: a "heuristic point of view."[40] It insinuates another meaning of this multifaceted concept: the opening up of new horizons along the exploratory path toward novel, deeper insights.

Einstein kept looking for arguments that might convince Planck and other skeptics that it was not just the emission and absorption of the resonators that had to be quantized but energy generally and even the radiation field itself. In 1906, he tried it with a general statistical derivation of the mean energy (from resonators but also from arbitrarily oscillating systems) without taking recourse to radiation measurements! The hypothesis Einstein presented was that Planck's derivation of the spectral energy density "makes implicit use of the [...] hypothesis of light quanta."[41]

Instead of avidly claiming the priority of this 'discovery' for himself, as lesser minds would surely have done, Einstein followed the opposite strategy in 1906 and ascribed to Max Planck, a scholar one generation his senior, the role of having "introduced into physics a new hypothetical element: the hypothesis of light quanta." In view of the opposition which Planck himself had mounted right after 1905 against this radical interpretation of the light quantum by Einstein, this move was more than a little unusual. It probably gave Planck quite some headache, because he definitely did not want to assume this role. He did everything he could at the time to try to find another, more conservative interpretation of what he had perpetrated in 1900.

---

[39]M. Besso to A. Einstein, 17 Jan. 1928, in Speziali (1972) pp. 237–238.

[40]Einstein (1905): CPAE vol. 2, doc. 14, transl. ed., p. 86, dated Bern, 17 March 1905, originally appeared in the issue dated 9 June 1905 of *Annalen der Physik*, at that time edited by Wilhelm Wien and Max Planck, the two theoretical physicists upon whose works Einstein was directly departing from.

[41]Einstein (1906b) p. 199 (CPAE 2, doc. 34, transl. ed., pp. 192–199, esp. pp. 192, 196). Davisson (1937) p. 388 formulated it thus: Einstein "outplancked Planck in not only accepting quantization, but in conceiving of light quanta as actual small packets or particles of energy transferable to single electrons *in toto*" (original emphasis).

2.4 Planck's Second Quantum Theory 1909–13

Between 1909 and 1913 Planck worked on what came to be known as his 'second quantum theory.' Einstein interpreted the quantization as quantization of the energy within an electromagnetic field. Planck, however, took the quantization merely as an epiphenomenon caused by the limitations of the resonators' oscillatory modes, which emit the electromagnetic energy. Accordingly, the energy quantization was just a consequence of those releases of energy, artificially produced packet-wise, so to speak, by the resonators, whereas the transmitting fields themselves could have any arbitrary energy. Mathematically speaking, Planck (and later also Debye and Sommerfeld) thus presupposed a quantization of phase space: only when the smallest packet size h—the Planck quantum of action—had been stringed together, could the packet be dispatched by the emitter. He considered the radiation itself as continuous and also the later re-absorption of electromagnetic waves by the absorber.[42] Planck was not alone in this. Some of his fellow physicists at home and abroad were amenable to such an attempted reinterpretation and tried to come up with their own variants of a safely defused 'second quantum theory.'[43] The Ukrainian-born physicist George Gamow (1904–1968), who emigrated to the U.S.A. in 1933, drew a particularly engaging analogy to illustrate how Planck's quantization is not applied to the radiation field itself: according to Planck, "Radiation is like butter, which can be bought or returned to the grocery store only in quarter-pound packages, although the butter as such can exist in any desired amount."[44]

Einstein was completely unimpressed by Planck's arguments and attempt to salvage classical electrodynamics. He wrote to the aspiring young physicist Jakob Laub (1884–1962) on 17 May 1909, to whom he was a kind of mentor:

> I busy myself incessantly with the question of the constitution of radiation, and am conducting a wide-ranging correspondence about this question with H.A. Lorentz and Planck. The former is an amazingly profound and at the same time lovable man. Planck is also very nice in his letters. His only failing is that it is hard for him to follow other people's trains of thought. This might explain how he could have made totally wrong-headed objections against my last radiation paper. But he did not adduce anything against my criticism. [...] This quantum question is so extraordinarily important and difficult that everybody should take the trouble to work on it.[45]

This reinterpretation of energy quantization may well appear, from our perspective today, as a compulsive attempt to hang onto outdated ideas and stubbornly reject the radical novelty of Einstein's interpretation. It did have an interesting ancillary effect, though. The phase-space consideration in Planck's second quantum theory led to new mean values for the energy. The smallest mean energy of a phase-space

---

[42]On Planck's 'second quantum theory,' see Planck (1910b, 1911, 1912, 1913), as well as Needell (1980) and Gearhart in Hoffmann (2010) p. 116 and here p. 114.

[43]E.g., Debye and Sommerfeld (1913) and Millikan (1913) p. 123; cf. here Sect. 4.7.

[44]Gamow (1966), also cited in Weinberg (1977) p. 20; cf. further Paul (1985) p. 57 and here p. 65 for another catchy comparison with continuous soup and discrete spoonfuls.

[45]A. Einstein to J. Laub, 17 May 1909, CPAE, vol. 5, doc. 160, transl. ed., p. 119.

cell of radiation at frequency ν did not yield zero anymore (as with Einstein 1905) but $1/2h\nu$.[46] Thus Planck predicted a nonvanishing zero-point energy—as we now know, this prediction was right, but it had been made on the basis of an entirely wrong starting hypothesis.[47] This advance cost dearly, though, because seen purely mathematically, the frequency integral of the sum of all the zero-point energies diverged.[48]

Nevertheless, Einstein avoided directly criticizing Planck in public. Possibly also at Besso's advice, he occasionally even swung over to somewhat lavish praise in excusing his mentor's hesitation about the light-quantum hypothesis:

> [Planck's] derivation was of unparalleled boldness, but found brilliant [empirical] confirmation. [...] However, it remained unsatisfactory that the electro-magnetic-mechanical analysis, [...] is incompatible with quantum theory, and it is not surprising that Planck himself and all theoreticians who work on this topic incessantly tried to modify the theory such as to base it on noncontradictory foundations.[49]

As we shall see in Sect. 4.3, it would still be a while, certainly not before Einstein's death, for any interpretation of light quanta to be found that was impeccably consistent—it is even doubtful whether we have achieved this today, but we shall discuss this later.

2.5 The Polyphony of Terms *in statu nascendi*

Let us start with an accounting of the various verbal descriptions that Einstein chose to use between 1905 and 1925 for his new concept of 'light quanta' (*Lichtquanten*). Table 2.1 displays an astonishing variety of expressions that he used to denote his object of study.

In 1927 Einstein said in a popular lecture that some experiments supported "that light be projectile-like in character, hence be corpuscular."[50] Faced with this abundance of synonyms, it is rather surprising that—as far as I can see—Albert Einstein never once used the expression 'photon,' despite being employed in an English-speaking setting since 1933 in the *Institute for Advanced Study* at Princeton. I have yet to find any statements by him explaining this refusal to use the term 'photon.'

---

[46]See, e.g., Planck (1911) and Einstein and Stern (1913). On zero-point energy cf. furthermore Whitaker (1985) p. 266 with a derivation, by series expansion, of Planck's spectral energy density, and Kragh (2014c); on the connection with vacuum fluctuations cf. Gerry and Knight (2005) pp. 29–33.

[47]Nowadays zero-point energy is based on the Heisenberg uncertainty relation $\Delta p \cdot \Delta q \geq \hbar$, which Werner Heisenberg formulated a while later, in 1927.

[48]See Planck (1958) vol. 2, p. 249 and Giulini (2011) p. 131.

[49]Einstein (1916a) p. 318 (CPAE, vol. 6, doc. 34, transl. ed., p. 212).

[50]Einstein (1927) p. 546. Of course, he also said that other properties, such as the capacity of light to interfere, "are not explained by the quantum interpretation." See Sect. 3.8 on the wave-particle duality, esp. Sommerfeld (1919) p. 59: "A ray in which energy and momentum are localized in a point shape does not essentially differ from a corpuscular ray; we have revived Newton's corpuscles."

Table 2.1 Einstein's own terminology with first mentionings 1905–24, based on my own counting in Einstein (1905) and Einstein (1924/25)

Light quanta ("Lichtquanten")	Einstein (1905)	In Einstein (1905) 6 ×
Energy quanta ("Energiequanten")	Planck (1900)	In Einstein (1905) 17 ×
Light energy quanta ("Lichtenergiequanten")	"	Einstein (1905) 4 ×
Elementary quanta ("Elementarquanten")	"	In Einstein (1905) 2 ×
Energy projectiles ("Energieprojektile")	In Einstein (1905)	(*Berliner Tageblatt*) 3 ×
Light corpuscles ("Lichtkorpuskeln")	In Einstein (1924/25)	1 ×

The following other neologisms could be added for the same semantic field from this early period by some of Einstein's contemporaries. The Polish physicist Mieczyslaw Wolfke (1883–1947), who defended his postdoctoral thesis under Einstein at the University of Zurich in 1913, talked of "light atoms."[51] Terminologically this suggests a return to the Newtonian conception of particles, but that is not at all what Wolfke meant. The same applies to later adaptations of this compound in the Anglosaxon variant, 'atoms of light.'[52] In France, Louis de Broglie (1922) and Fred Wolfers 1926 referred to "atomes de lumière"; Gilbert N. Lewis added the variant "particle of light," and Lewis and W. Band, "light corpuscles."[53] Other formulations include the metaphorical one from 1917 by Daniel Frost Comstock (1883–1970), who proposed "bullets of energy, one might say," to describe "separate units [of radiant energy] in space."[54] Arthur Holly Compton later took over this expression as 'light bullets,'[55] and his fellow countryman Robert A. Millikan also used a similarly visual rewording in a textbook from 1935: "photon, or light-dart, theory of radiation,"[56] which the British physicist Edward Neville da Costa Andrade (1887–1971) immediately espoused.[57] Other, more exotic oddities include Arthur Lleywelyn Hughes's expression from 1914, "that light was molecular in structure, each 'molecule' or unit

---

[51]"Lichtatome." See Wolfke (1913/14) p. 1123; cf. the reply to this text by G. Krutkow (1914).

[52]E.g., Troland (1917), Ornstein and Zernike (1920), Ehrenfest and Joffé 1924, and as late as 1926 C.D. Ellis, when he delivered three lectures on "The atom of light and the atom of electricity" at the *Royal Institution* in London.

[53]See Ellis (1926), notwithstanding the changed title; furthermore Lewis (1926a) p. 22, (1926b) p. 236 and with reference to this: Band (1927), Louis de Broglie (1922) p. 422 and Wolfers (1926) p. 276; cf. here p. 30 on Wolfers and Lewis.

[54]See Comstock and Troland (1917) §10, p. 46—part I of this textbook, the origin of this quotation, was written by Comstock. In this same paragraph Comstock translated the expression "Licht Quanta" as "light quantities," which he wrongly attributed to Planck not Einstein.

[55]See Compton (1925) p. 246: "light bullets": "light as consisting of streams of little particles".

[56]See Millikan (1935) p. 259.

[57]See Andrade (1930/36) p. 128 and Andrade (1957), available on line in excerpt at http://www.uefap.com/reading/exercises/ess3/andrade.htm.

containing an amount of energy $h\nu$ which could not be subdivided."[58] He evidently overlooked that molecules are certainly *not* indivisible entities but—quite unlike 'light quanta' can be split into atoms.[59] All the same, this expression also found its followers, such as Wolfke, but he attempted to take this metaphor literally and considered real bonds between individual light quanta to explain fluctuations in the spatial concentration of radiation.[60]

Let's next take a look at Albert Einstein's usage of a sample from among these terms in a few of the statements published in his paper about 'light quanta' in *Annalen der Physik* in 1905.[61] Right at the beginning he mentions his heuristic assumption "that the energy of light is discontinuously distributed in space." He continues: "According to the assumption to be contemplated here, [... light] consists of a finite number of energy quanta [*Energiequanten*] that are localized in points in space, move without dividing, and can be absorbed or generated only as a whole." (p. 133, transl. ed., p. 87) And further down: "Monochromatic radiation [...] behaves [...] as if it consisted of mutually independent energy quanta" (p. 143, transl. ed., p. 97). These statements introducing his concept are remarkably carefully phrased. They are in conjunctive, with the emphasis placed on the hypothetical, indeed preliminary character of this 'assumption.' In substance the main weight is on discrete energy quanta (consistent with our terminological statistics). Behind his concretely applied terms there are at first a still very vague concept and another still diffuse mental model of spatially localized energy quanta or particles of light as carriers of sharply defined (hence 'quantized') energy packets.[62]

Einstein knew how bold and risky his assumption was, to quantize not only the abstract magnitude of energy but also concretely perceptible light, which the classical electrodynamics of a continuum seemed to have already described so perfectly. To his close friend Conrad Habicht[63] (1876–1958) Einstein wrote in May 1905:

> So, what are you up to, you frozen whale, you smoked, dried, canned piece of soul, or whatever else I would like to hurl at your head, filled as I am with 70% anger and 30% pity!

---

[58] Hughes (1914a, b) p. 5.

[59] It may have been in allusion to William H. Bragg's speculations at that time about γ-radiation as neutral bipolar particles. On this see here Sect. 4.6.

[60] See Wolfke (1921); cf. further Kojevnikov (2002) p. 200 for a few attempts along these lines, all of which failed.

[61] All of the following quotes are from Einstein (1905). I cite the pagination from the original paper dated 17 March 1905 as well as from the translation edition of his collected papers (CPAE vol. 2, doc. 14).

[62] On the introduction of the idea of quantization in 1900, which itself certainly did not happen suddenly either, and the initial uncertainties about this concept, see for instance Kuhn (1978), Darrigol (2000), and Gearhart (2002).

[63] Habicht was a member of the informal discussion group jokingly called their "Olympia Academy." Together with Besso, they read and discussed basic texts on mechanics and electrodynamics by Ernst Mach, Hermann von Helmholtz, Heinrich Hertz, Ludwig Boltzmann among many other leading physicist-philosophers of the day. On Einstein's contexts and contacts in Berne, see Hentschel (2005).

[...] I promise you four papers [...], the first of which I might send you soon [...] The paper deals with radiation and the energy properties of light and is very revolutionary.[64]

The title of this promised paper is searchingly cautious, by contrast: "On a Heuristic Point of View Concerning the Production and Transformation of Light." The adjective "heuristic" continues, as it ever has, to resound with connotations ranging from 'oriented toward problem-solving,' to 'in fruitful quest of novelty,' to 'tentative,' 'unprovable' and 'uncertain.'[65] The solid core of these many nuances of meaning of this expression implies a still opaque epistemic status of the given proposal. By it Einstein signaled a carefully groping search for a robust concept able to explain the photoelectric effect and a whole slew of other phenomena connected with the interaction between light and matter (see below). This explains the stress on a 'heuristic,' something not yet finished, not yet taxable, a testing trial.

2.6 The Slow Rise of the Term 'Photon'

It is typical of the early developmental phase of an as yet diffuse nascent concept for a multiplicity of terms to appear. Between 1905 and 1925 linguistically much was still in a state of flux.[66]

It is astonishing that 'photon,' the most prevalent term currently in use, is nowhere to be found in Einstein's oeuvre. This term had been introduced independently between 1917 and 1926. Within this period of ten years, the term 'photon' appeared in four different contexts: two of them in sensory physiology, one in photochemistry, and the last and most famous one in physical chemistry. Helge Kragh, who first pointed out these older instances in 2014, set them in contrast to the "Einsteinian context,"[67] making of them rather exotic but irrelevant precursors. In the following argument I shall primarily examine their hidden connections. As we shall soon see, all four of these contexts are at least indirectly related to papers by Einstein, and the four now generally forgotten protagonists were aware of his paper from 1905 and the ensuing debates about it.

---

[64] A. Einstein to C. Habicht, Berne, 18 or 25 May 1905, publ. in CPAE, vol. 5, doc. 27, pp. 31f. (transl. ed., pp. 19–20).

[65] The definition in Webster's Third New International Dictionary is: "providing aid and direction in the solution of a problem but otherwise unjustified or incapable of justification." According to Oxford English Dictionary, vol. 7 (2nd ed. 1989), p. 193: "heuristic, serving to find out or discover," with the oldest recorded instance being a letter by Samuel Taylor Coleridge (1772–1834) dated 1821.

[66] On the fluidity of concepts during their developmental phase, regarded from the aspect of programming, see Hofstadter (1995).

[67] Kragh (2014b) p. 263: "when it [the term 'photon'] was originally introduced, it was with a different meaning. It can be traced back to 1916, when it was proposed as a unit for the illumination of the retina, and ten years later the name was revived in still another non-Einsteinian context."

This term first appears in a paper by the American biophysicist and sensory physiologist Leonard Thompson Troland (1889–1932).[68] On commission of the *National Lamp Works of General Electric Company*, Troland conducted psychophysiological experiments on the way the human eye responds to visual stimuli. He proposed 'photon' as the name of a unit of photometric intensity at which certain measurable reactions by the pupil are elicited:

> The writer has expressed his intensity measures throughout in terms of a unit involving the pupillary area, and has proposed that this unit, called the *photon*, be adopted as the standard means of specifying the photometric intensity of visual stimulation conditions. [...] A *photon* is that intensity of illumination upon the retina of the eye which accompanies the direct fixation, with adequate accommodation, or a stimulus of small area, the photometric brightness of which [... is] one candle per square meter, when the area of the externally effective pupil [...] is one square millimeter. The *physiological intensity of a visual stimulus* is its intensity expressed in *photons*. The photon is a unit of illumination, and hence has an absolute value in meter-candles. The numerical value of the photon, in meter candles [...] will obviously be subject to some variation from individual to individual.[69]

The textbook he coauthored with Daniel Frost Comstock (1883–1970), a physicist at M.I.T. in Cambridge, Massachusetts, which appeared that same year under the title *The Nature of Matter and Electricity – An Outline of Modern Views*, contains an unusually early survey for America of the debates in Europe about interpretations of the quantum theories by Planck and Einstein, including the "modern doctrine of light quanta."[70] Troland, who had written this section of the textbook, hence knew about Einstein's light-quantum hypothesis, if not directly, then at least indirectly through reports, talks, and review articles in English-language periodicals.[71] Troland referred back to his earlier coinage in a survey article for the *National Research Council* in 1922. Otherwise, the term appears not to have been used by any of his professional colleagues and was forgotten thereafter.

Sensory physiology is also the context of the second form of 'photon,' and again a clear link exists to quantum theory. In 1920 the Irish physicist and geologist John Joly (1857–1933) began to develop a quantum theory for vision that attempted to correlate the visual stimulus with the input of energy in the eye, in analogy to Albert Einstein's quantization of the photoelectric effect from 1905. Joly imagined the process of vision as light quanta impinging upon the eye to trigger physiological sensory stimuli in the retina, which in turn are conducted into the brain along "visual fibres" and the visual nerves to produce visual perceptions. Joly's 'photon' was "the unit of light stimulus or sensation":[72]

---

[68]Troland studied biochemistry in addition to physics and psychology, taking his doctorate in 1915 at Harvard University with a thesis on visual adaptation, and served as president of the Optical Society of America 1922–24. See Kragh (2014b) pp. 271f.

[69]Troland (1917) pp. 1, 5 and 32 (original emphasis).

[70]Comstock and Troland (1917) pp. 182–189. After taking his doctorate 1906–07 Comstock briefly worked with J.J. Thomson in Cambridge and supported an emission theory of light.

[71]Millikan's plenary lecture before the *American Association for the Advancement of Science* in 1913 is one example.

[72]Joly (1921a, b) p. 26.

> The unit light stimulus discharged by a single visual fibre [...] must not be confused with the quantum which plays the part merely of the finger of the trigger. This minute quantity of energy discharged into the cerebral cortex evokes our unit of luminous sensation. [...] I propose to designate it a photon, using the English plural, photons. Symbolically, the letter ϕ will be assigned to it. [...] Each sensation is an accomplishment of a particular form of energy stimulus, i. e., of two, three, or of four photons simultaneously discharged.[73]

This usage of the term was still definitely situated within the field of sensory physiology but Joly saw a biunique relation between the physiological visual response (visual stimulus) and a physical input (or triggering) producing this visual stimulus—namely, incident (Einsteinian) quantized energy in the eye. Just because this energy was quantized, that is, just because it was only available in finite packets of fixed magnitude h, Planck's quantum of action, did these Joly photons become countable units of ϕ, as discharge responses triggered by this influx of energy into the eye. There is no question that this 'photon' is far away from the present usage of the term, which denotes those incident packets of energy themselves and not the physiological responses to stimuli. Irrespective of its localization within sensory physiology, we have here a clear "quantum theory of vision," which Joly tried to expand to incorporate a formulation for color vision. According to the relation $E = h\nu$, the amounts of energy E and frequency ν are strictly proportional to each other, so he set 4ϕ for blue light, with red light having the lesser value 2ϕ. An inconsistency with $E = h\nu$ resulted when instead of setting 3ϕ for the yellow light in between, he postulated $2\phi + 3\phi$, in order to be able to derive out of it the complementarity between blue and yellow in white light, which he interpreted as the sum of blue and yellow light: $(2\phi + 3\phi) + 4\phi = 9\phi$. Then one could have concluded that if blue were removed from white light, at 9ϕ, yellow light would be the exact remainder of $2\phi + 3\phi$ (and vice versa). But the strict proportionality between E and ν no longer held. This second, quite original usage of the term 'photon' remained obscure as well and—after 1921—was as quickly forgotten as Troland's.

The third proposed introduction of the term 'photon,' by the French biochemist and physiologist René Wurmser (1890–1993), moves us away from sensory physiology into photochemistry.[74] This was the field to which Einstein's subsequent papers from 1907, 1909 and 1909 referred. It was also one of the areas in which Einstein himself sought experimental evidence for his light-quantum hypothesis. René Wurmser happened to verify in 1913, together with his superior Victor Henri at the *Laboratoire de Physiologie Générale de la Faculté des Sciences de Paris*, Einstein's prediction that exactly *one* Einsteinian light quantum is minimally required to trigger a photochemical reaction.[75] Thus a clear link again exists between this third coinage context and Einstein's. During the 1920s, Wurmser was working on photosynthesis. He supposed that molecular resonance was behind the transfer of energy from incident radiation

---

[73] Joly (1921a, b) p. 29.

[74] Wurmser's doctoral thesis in 1921 bears the title: "Recherches sur l'assimilation chlorophylienne." He later presided over the *Société française de Chimie*, became a founding member of the *Société française de Biochimie et de Biologie Moléculaire*. See http://cths.fr/an/prosopo.php?id=112958 and further sources cited there.

[75] See Henri and Wurmser (1913).

on chlorophyll in leaves. In considering the process quantitatively, Wurmser argued in 1924: "This could be rigorously explained by having the activation of a molecule [...] demand the absorption of an integral but variable number of *photons*."[76] This is, in fact, a clear usage of 'photon' exactly as we understand the term today. Nevertheless, Wurmser was a singular case and did not insist on his priority later either, when he referred back to the article in 1987, pointing out: "This paper of 1925 contains some considerations which remain of interest from a modern point of view."[77] This ninety-seven-year old may not in retrospect have realized that his article was the first instance in which the term 'photon' was used in the now conventional meaning. This is quite in contrast to the next two usages of the term to be discussed here.

In 1923, the French experimental physicist Frithiof (Fred) Wolfers[78] was still a most determined opponent of the light-quantum hypothesis. He remained stolidly against it even after the publication of Compton's scattering experiments, which swayed many of his colleagues toward changing their convictions. In a paper coauthored with E. Friedel, he questioned the conclusions that Compton had drawn out of his experiment and preferred to consider another interpretation instead along the lines of the BKS theory.[79] Initially, Wolfers also explicated his own experiments according to that same theory.[80] After the BKS theory was experimentally refuted, however, destroying his own interpretational attempts in the process, we read the following passage in a paper submitted by Wolfers on 26 July 1926 to the *Comptes Rendus* of the Parisian *Académie des Sciences*:

> Assigning the name *photons* to the projectiles that supposedly transport radiant energy and also possess the character of a periodic frequency ν (atoms of light), I suggest that photons may be repelled by the atoms of matter when they pass by their immediate vicinity, at least in the case of directed atoms that form the separating surface between two media. One may imagine that the repulsion be due to a kind of resonance between the photons and the resonators that are just far enough away from the trajectory for absorption not to occur. [...] This deviation of the photons' trajectories would be of the order of a few arc minutes.[81]

The tone of the expressions 'projectiles' and 'atomes de lumière' make us immediately recognize the old Newtonian projectile theory of light (see Sect. 4.1) with its naive connotation of particulate light quanta which transport energy and can be repelled by atoms. Admittedly, it is not quite so naively conceived as a ping-pong ball bouncing off a smooth surface. It rather takes the meaning that Wurmser had

---

[76] Wurmser (1925a) p. 60 (original emphasis), submitted in Sep. 1924. Similar wording appears in Wurmser (1925b) p. 375: "cette activité particulière des radiations vertes pouvait être expliquée en admettant que l'activation d'une molécule [...] exige l'absorption d'un nombre entier, mais variable, de photons."

[77] Wurmser (1987) p. 92. How right he was!

[78] Very little biographical information is available about Wolfers, not even his years of birth and death. 1920–1940 he was employed at the University of Algers and afterwards became a professor of physics in Paris. He may have died in California, U.S.A. in 1969.

[79] See Friedel and Wolfers (1924) and Wolfers (1924) or Bohr et al. (1924) as well as here pp. 126ff. on the Compton experiment and p. 128 on the BKS theory.

[80] See Wolfers (1924, 1925) p. 366.

[81] Wolfers (1926) pp. 276–277 (original emphasis).

also presumed: molecular resonance between the incident radiation and the atoms at the light-scattering surface. Wolfers referred back to his coinage from 1926 only once, in a paper two years later. Apart from a single reference to Wolfers by Louis de Broglie, neither Henry Small nor Helge Kragh were able to identify any later quotations from Wolfers's papers.[82] His brand-new minting thus soon lost its shine as well and faded into obscurity. Another go at it was attempted just a few months later, though.

This fifth and last context in which the term 'photon' appeared was, as we shall now see, the one that unfairly attained the greatest fame and ultimately prevailed terminologically. It was unfair because the expression that the American physical chemist Gilbert N. Lewis introduced did *not* in fact bear the meaning now regularly in use—quite contrary to Wurmser's in the third context from 1925 discussed above. Lewis attached to this expression interpretative content that does not correspond to our present understanding of the 'photon.' For example, he presumed that his 'light corpuscle' conserved its number of photons, whereas according to the modern interpretation it is not a conservation quantity at all, because photons are generated in luminiferous processes and eliminated again in processes involving light absorption.[83] Consequently, G. N. Lewis is definitely not the father of the modern photon as a *concept*, even though he was one of the first to use the *term* in the physicochemical context,[84] as opposed to in the other fields of sensory physiology, photochemistry or biophysics. This demonstrates once again how important it is to distinguish between terms, concepts and mental models (cf. here the introduction and Sect. 9.5).

Gilbert Newton Lewis (1875–1946) studied chemistry at the *University of Nebraska* and worked toward his doctorate (from 1893) at *Harvard University* which he defended in 1899.[85] After conducting research under the physical chemists Wilhelm Ostwald at Leipzig and Walther Nernst at Göttingen, he returned to the United States and became a professor at the *Massachusetts Institute of Technology* (MIT) in 1905, moving on in 1913 to the *University of California* at Berkeley. While still at MIT, Lewis became the first American scientist to work intensely on Einstein's special theory of relativity. In his position as chair for physical chemistry in California, he shifted his research focus toward thermodynamics, a physical theory of chemical bonding, and atomic and molecular structure. During the early 1920s, the quantum theory by Bohr and Sommerfeld was still undergoing intense scrutiny.[86]

---

[82]See Small (1981) vol. 1, col. 1892, Kragh (2014b) p. 274 and the *Web of Science*, a consortium of scientific on-line citation and literature databases, originally set up by the *Institute for Scientific Information* (ISI), which was taken over by *Thomson Reuters* in 1992, and by *Clarivate Analytics* in 2016.

[83]See Lewis (1926c) on this supposed "conservation of photons," which he had called 'light corpuscles' half a year previously in the same journal and had related to naive quasi-Newtonian particulate notions, see Lewis (1926a).

[84]The research contexts and problems that G. N. Lewis was pursuing when he introduced this concept is treated by Roger H. Stuewer (1975b), citing the primary and secondary sources.

[85]Lewis's biography is treated in Hildebrand (1947), Stuewer (1975b) and Lewis (1998).

[86]See, e.g., Nisio (1973), Pais (1991) Chap. 10, Kragh (2012), Eckert (2014), here Sect. 3.9 and the primary sources cited there.

According to the theory, the fundamental interaction between radiation and atoms constituted the emission and absorption of energy in the form of electromagnetic waves. Bohr and Sommerfeld's idea was that the quantized packets of energy being exchanged among atoms come from energetically sharply defined permissible orbits upon which the negatively charged electrons revolve around an atom's positively charged nucleus. Quantum jumps by the electrons from one of those permissible orbits onto another create sharp (bright) spectral emission lines, when the energetic level of the initial orbit was higher than the destination orbit, and (dark) absorption lines in the opposite case.[87] The following paradox resulted when these emission and subsequent absorption processes are somewhat classically considered as sequential in time. How should the electron that is about to 'jump' off its orbit know where it is supposed to land? It has to have this information to be able to couple this jump with the emission of the energetically appropriate spectrum line. This energy difference between the upper and lower energy levels is actually only established once the process is completed, but then it is too late for the spectrum line belonging to this process to be emitted. This problem becomes even more obvious when the emitting atom and the reabsorbing atom are not close together but kilometers apart, if not light years away in the cosmos. G. N. Lewis and a number of other physicists had already noticed this paradox but had not made much of a fuss about it. Ernest Rutherford (1871–1937) in Manchester, for instance, wrote in a letter to Niels Bohr (1885–1962) on 20 March 1913: "How does an electron decide what frequency it is going to vibrate at when it passes from one stationary state to the other? It seems to me that you would have to assume that the electron knows beforehand where it is going to stop."[88] G. N. Lewis took this problem as an occasion to ponder further about this thematic complex as a whole. He believed in a basic symmetry existing between emission and absorption. They matched each other so well that he subsumed them both under transmission and started a systematic search for a temporally symmetric description of this exchange process: "it is as absurd to think of light emitted by one atom regardless of the existence of a receiving atom as it would be to think of an atom absorbing light without the existence of light being absorbed."[89] Lewis's theoretical revision of this exchange of energy between two atoms hence contained the following new mental model of this transmission process:

> We therefore say that a corpuscle of energy travels with the velocity c from atom to atom, and that there is a field (a 'retarded' field if ascribed to the emitting atom, an 'advanced' field if ascribed to the receiving atom) which determines the probability of the exchange. The invention of the corpuscle of light and these fields permit us to express the process of radiation in conformity with our ordinary spatial ideas and with the laws of conservation. [...] if the theory that I propose is correct, it should later be possible to express the probability

---

[87]See, e.g., Hentschel (2009b, c).

[88]Reproduced from Niels Bohr: *Collected Works*, vol. 2, p. 584 in Hentschel (2009b) p. 599; further Stark (1920, 1930) pp. 688f. and Sommerfeld (1920) p. 420.

[89]Lewis (1926a) p. 24. Lewis received the support of Tolman and Smith (1926); cf. Stuewer (1975b) for more details about Lewis's search for a strictly temporally symmetric description.

of an energy transfer merely as the degree of 'fit' between the emitting and absorbing atoms in their points of virtual contact.[90]

Lewis's final point thus anticipated a fundamental idea in the work by John A. Wheeler and Richard P. Feynman on the theory of absorbers within the context of QED, which was being developed just then, and the later "transactional interpretation of quantum mechanics" that John G. Cramer (∗1934) started to develop in 1986.[91] According to these notions the radiative process hence is not that an initial emission simply occurs out of the blue, so to speak, with absorption taking place in its suite at a later point in time. It is rather that the amplitude of probability of emission coming from the emitting atom meets that of the absorbing atom midway in space-time and they supplement each other, owing to the phase difference and time reversal of the latter, into probability 1 in a transition from potentiality to actuality. The merely imaginable, virtual process thus turns into a real transition with the transferral of energy and momentum.[92] Of greater import for our purposes is Lewis's decision to introduce a new designation for this 'corpuscle of energy,' this exchange particle mediating the contact between emitter and absorber and publish it in the high-profile journal *Nature*. In this article which subsequently appeared on 29 October 1926, he offered thorough justification for this choice as well:

> It would seem inappropriate to speak of one of these hypothetical entities as a particle of light, a light quantum, or a light quant, if we are to assume that it spends only a minute fraction of its existence as a carrier of radiant energy, while the rest of the time it remains as an important structural element within the atom. It would also cause confusion to call it merely a quantum, for [...] it will be necessary to distinguish between the number of these entities present in the atom and the so-called quantum number. I therefore take the liberty of proposing for this hypothetical new atom, which is not light but plays an essential part in every process of radiation, the name *photon*.[93]

Metaphorical turns of phrase always bring along their own ballast, however, and Lewis's was no exception. The reference to a hypothetical new 'atom' semi-automatically inferred the connotation 'indestructible.' And just that is what happened with Lewis. He attached the postulate of invariability to this new concept. His photons could neither be generated nor destroyed—like ping-pong balls, they could merely be volleyed back and forth between emitter and absorber without themselves being altered. This postulate marks a stark difference between Lewis's photons and the modern interpretation as massless exchange particles in QED (Sect. 3.12). That is why it is plain wrong to identify G. N. Lewis as the one to have designated the term 'photon' to Einstein's notion of the light quantum, as alleged in about 99% of all texts treating this topic. Another myth must also be debunked here: Physicists certainly did not unanimously adopt this term right after the appearance of Lewis's

---

[90]Lewis (1926b) p. 238.

[91]See Wheeler and Feynman (1945, 1949) or Kastner (2012), Cramer (2015) and the literature cited there. On Wheeler see here p. 156, on Feynman p. 88.

[92]This idea is well put in Paul (1985) pp. 70–86.

[93]Lewis (1926c) p. 875 (original emphasis).

Table 2.2 The frequency of usage of 'light quantum' and 'photon' 1926–55, after Kragh (2014b) p. 276 and the *Web of Science*. For methodical reasons, only articles originally written in English are used as a basis

	1926–1935	1936–1945	1946–1955
Light quantum / light quanta	20	0	5
Photon	19	29	243

article (1926c). Over the course of the next forty years, it was quoted altogether twice.[94]

The Danish historian of physics Helge Kragh, who took up the etymological search for traces of the term, pointed out in 2014 that 'photon' was not a predominant expression when it was chosen as part of the title of the proceedings of the fifth Solvay Conference in Brussels.[95] It was not mentioned at all, either, in the invitations or documents drafted during the preparations.[96] Interestingly enough, one of the conference participants felt the necessity to explain his preference for this neologism over its forerunner 'light quantum.' The recent Nobel laureate of physics Arthur Holly Compton (1892–1962) wrote:

> In referring to this unit of radiation I shall use the name 'photon' suggested recently by G. N. Lewis. This word avoids any implication regarding the nature of the unit, as contained for example in the name 'needle ray.' As compared with the term 'radiation quantum' or 'light quantum,' this name has the advantages of brevity and of avoiding any implied dependence upon the much more general quantum mechanics, or upon the quantum theory of atomic structure.[97]

During the period 1926–35, we find both expressions in about equal frequency in the English-speaking literature. The subsequent years including World War II are not very instructive statistically owing to various distorting effects, but the term 'photon' gained the upper hand over 'light quantum' in the Anglo-American context after 1945 by an overwhelming ratio of 243:5. It would have been misleading to take other foreign-language articles into account, including the ones in Table 2.2 or the next graph (Fig. 2.3), because the English translations by *Web of Science* of all such titles in many instances substitute the modern term 'photon' for 'light quantum.'

By 1945, the term 'photon' was firmly established in English. At the present time, it is the one practically exclusively in use by English speakers. In German,

---

[94]In 1927—albeit by the author himself, in follow-up papers! See Small (1981) vol. 1, cols. 1032-1036 and the *Web of Science*. More details about this weak reception of Lewis's work are given in Kragh (2014b) p. 276.

[95]See *Électrons et Photons/Electrons and Photons* (1927/28) and Bacciagaluppi and Valentini (2009) (including documentation and an English translation of these proceedings).

[96]Kragh (2014b) p. 276 writes: "Most likely the name entered the title of the proceedings only during the last phase of preparation, reflecting that several of the speakers and discussants used 'photons' rather than 'light quanta' in their reports. Among them were Lorentz, Louis de Broglie, Paul Dirac, Léon Brillouin, Paul Ehrenfest, and Arthur Compton."

[97]Compton (1928) p. 57, cited according to the English translation in Compton (1929b) p. 157.

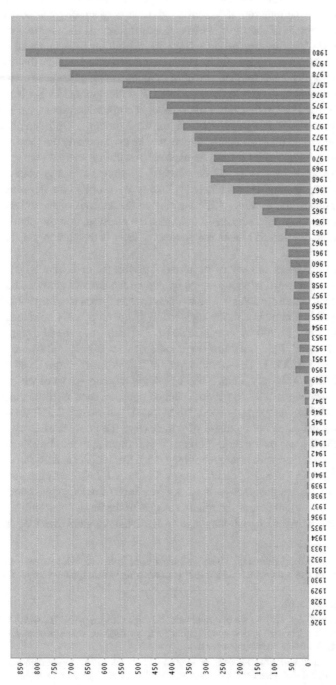

Fig. 2.3 English usage of the term 'photon' in the professional literature 1926–1980, compiled by Kragh (2014b) p. 277, based on a total of 7,325 articles in English in *Web of Science*. Reprinted by permission of IOP Publishing, © 2014, EDP Sciences and Springer Verlag Berlin Heidelberg)

reference still is occasionally made to 'Lichtquanten' but rather rarely, because the term 'photon'—in emulation of 'electron'—has become an accepted neologism.[98]

The development of the laser led to a rise in the annual number of instances in which the term 'photon' was used to above 50 in 1960. In 1964 this total surpassed 100, in 1967 it doubled again and in 1977 it exceeded 500. It seems that the person responsible for its greater frequency within the context of popular science was not so much G. N. Lewis as A. H. Compton, though. His much-noted Nobel lecture, held on 12 December 1927, referred to x-rays as "light corpuscles, quanta, or, as we may call them, photons."[99] His popular article for the February 1929 issue of *Scientific American* described the high-energy x-rays that he used to study the Compton effect as composed of photons. By setting the expression within quotation marks, Compton placed emphasis on its novelty, but the absence of a source also suggested that it had been of his own making. He considered 'photons' to be one of the three elementary building blocks of the universe: "Having carried the analysis of the universe as far as we are able, there thus remains the proton, the electron, and the photon—these three. And, one is tempted to add, the greatest of these is the photon, for it is the life of the atom."[100]

A *google* search for the key word 'photon' performed on 6 March 2016 yielded 31,800,000 hits and an analogous search by *google scholar* still generated 2,580,000 references to academic articles—which themselves had been cited over 7,000 times. The corresponding numbers for the keyword 'light quantum' only produced 102,000 *google* hits, resp., 11,900 by *google scholar*.[101] Hence, in general usage 'photon' predominates over 'light quantum' by a ratio of over 300:1; and in the scientific literature it still is 200:1. The German variant 'Lichtquant' comes to an estimated total of 20,000 *google* finds, resp. 2,690 by *google scholar*—most of which relate to older sources between 1927 and 1960. Other composites, such as, the expression that Robert A. Millikan temporarily used, "light quant," only catch 2,350 *google* items, resp. 282 by *google scholar*, many of which were the product of entry errors.[102] This boom in the use of the term 'photon' is still ongoing, including its composite derivatives connected with laser technology since the 1960s, such as, the neologism 'photonics' (with approx. 14,500,000 *google* hits, resp. 1,730,000 by *google scholar*).

Using the program *google n-gram viewer* (see https://books.google.com/ngrams), I retraced the development of the relative usages of the terms 'photon,' 'light quantum' and 'optics' in the English-language professional literature over the period 1926

---

[98] Arthur Lewis Caso argues that such borrowed neologisms from foreign languages are the second most frequent form of minting new scientific terms, following semantic extension; see Caso (1980) p. 107.

[99] Compton (1927) p. 84.

[100] Compton (1929a) p. 236. I owe this reference to Kragh (2014b) pp. 277f., which follows the career of the term 'photon' into textbook publications in English, which acquainted the rising generation of scientists with this expression.

[101] Caution: a naive search for this compound term without constraining quotation marks (which produced 2,060,000 finds) adds other disjunctive occurrences of 'light' and 'quantum.'

[102] For example, run-ons between sentences lacking terminal punctuation, which such search algorithms are unable to disregard by the insertion of quotation marks, etc.

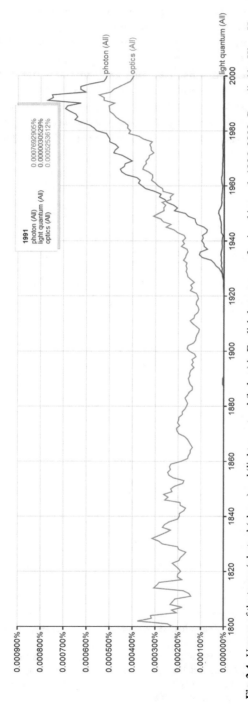

Fig. 2.4 Usages of the terms 'photon' (above) and 'light quantum' (below) in English-language professional journals 1926–2000. Compiled by Klaus Hentschel on 7 March 2016, running *google n-gram viewer*. Sources in other languages are excluded. The comparison curve (in the middle) analogously indicates usages of the word 'optics' since 1800, as a kind of calibration of the absolute frequency of texts on optics generally; the 'photon'-usage curve crosses and outdoes it after 1955

to 2000 (see Fig. 2.4). The third of these terms was plotted merely for comparison purposes, to provide a rough indicator of the overall growth or leveling off of optics texts generally. This method only yields useful results for the scientific literature in English. The relatively low number of hits for 'light quantum' limits the graph's reliability somewhat. Its maximum in 1951 is very flat and from the 1970s on returns to the negligibly low level it had gradually attained during the 1920s. The 'photon' curve (generated by a case-insensitive search) is very different. It has a disproportionately steep slope already starting in 1926 compared to the general optics curve and maintains a constant rate up to the maximum value in 1991,[103] which is almost twice as high as the very general 'optics' term for that year. By 2005 it had dropped again by a third.[104]

---

[103] Approximately 0.0007% relative to the total vocabulary of the samples of scanned texts by google entered into *n-gram viewer*.

[104] This drop down to 0.0005% could already be an artifact of *n-gram viewer's* less broad data coverage for later years. Google often has to wait many years for publishers to grant the permission to scan more recent releases. That explains *n-gram viewer's* preset condition to only count texts up to the year 2000; later ranges are unreliable.

Chapter 3
Twelve Semantic Layers of 'Light Quantum' and 'Photon'

Proceeding from my introductory propositions about semantic accretion and folds or 'convolutions' of meaning, we shall now treat the conceptual history of light quanta. This book also looks beyond the state of the so-called older quantum theory from 1900–24 to include the development of quantum mechanics from 1925 on as well as beyond to quantum electrodynamics and quantum field theory (1930ff.). Thus we shall look at the entire historical course, from the original emergence of the concept of light quanta during the first two decades of the century, to its later transformation and development. This will be based particularly upon Einstein's publications between 1905 and 1949 along with those of his few fellow specialists and experimenters in radiation theory within the scientific community (Planck, Ehrenfest, Lorentz, Lenard, Stark, Hughes, Millikan, Comstock, Compton, Debye and Raman). After carefully studying all the important texts from this period, I shall examine which semantic layers of the term 'light quantum' appear in them. Twelve layers of meaning are revealed, some of them in mutual concordance, others latently at odds:

1. Light is corpuscular, strongly localized.
2. Light propagates at a finite but very high velocity.
3. Light is emitted and absorbed by matter.
4. Light transmits momentum p, therefore, radiation pressure.
5. Light transfers energy E.
6. Energy E is correlated with its frequency ν: $E \sim \nu$.
7. Energy E is quantized $E = h \cdot \nu$.
8. The wave-particle duality applies to light.
9. Light exhibits spontaneous emission and absorption.
10. Its quanta carry angular momentum (spin).
11a. Its quanta are indistinguishable, with equal energy and spin.
11b. Bose-Einstein statistics applies to quanta.
12. Photons are virtual exchange particles of quantum electrodynamics.

© Springer International Publishing AG, part of Springer Nature 2018
K. Hentschel, *Photons*, https://doi.org/10.1007/978-3-319-95252-9_3

Each of these twelve layers will be dissected historically in order to then show how the layered stacking over time is able to depict the full historical genesis of the term.[1]

3.1 Particle Models of Light

Models conceiving light as a stream of subtle particles can already be found among the atomists of Antiquity, but Isaac Newton (1642–1727) developed the first sophisticated model of this kind. He was careful not to reveal his basic corpuscular notion of light in his early papers published in the *Philosophical Transactions of the Royal Society* from 1672 on. But there still are clear suggestions of this projectile model in his *Principia* from 1687 and *Opticks* from 1704. In his mathematical principles of natural philosophy, for example, he concludes that light refraction is caused by the stronger attraction of particles of light to a denser medium.[2] His query 29 in *Opticks* (added to the second edition in 1706) also reads:

> Are not the Rays of Light very small Bodies emitted from shining Substances? For such Bodies will pass through uniform Mediums in right Lines without bending into the Shadow, which is the Nature of the Rays of Light. They will also be capable of several Properties, and be able to conserve their Properties unchanged in passing through several Mediums, which is another condition of the Rays of Light.[3]

Albeit, please note that at this stage Newton's "projectile model" of light is formulated as a "query," not as a thesis. By assuming the existence of "minimally small bodies" he was just exploiting a mathematical analogy between the propagation of such tiniest particles of light and small material bodies, without having to make any positive statement about "whether they are bodies or not."[4] This corpuscularity was hence a cautious consideration, at best, a kind of model assumption. It was not the dogma that the projectile model rapidly became in the hands of Newton's epigones of the coming three generations of natural philosophers. His prism experiments had yielded an important insight about white light being made up of component colors, which he definitely did not want to jeopardize by also proposing a speculative model. That is why his later papers were confined to presenting a phenomenological theory of white light as composed of colored light: this was controversial enough as it was.[5]

---

[1]For earlier attempts in this direction, but only able to sample individual layers, see my papers Hentschel (2005), Hentschel (2005/07, 2015) and Hentschel and Waniek (2011) specifically on the quantum statistics (11a) of indistinguishability.

[2]Newton (1687) Sect. XIV, 141ff. See also below.

[3]Isaac Newton: *Opticks* (1704) p. 370. For a transcription of the handwritten draft of these "queries," cf. http://www.enlighteningscience.sussex.ac.uk/view/texts/normalized/NATP00055.

[4]Newton (1687) p. 626.

[5]The contemporary controversies about Newton's theory of light are discussed by Schaffer (1989) and Shapiro (1996). I discuss Newton's underlying assumptions and mental models, which he avoided mentioning in public, in Sect. 4.1.

As the historians Geoffrey Cantor, Simon Schaffer and Jean Eisenstaedt have shown, these Newtonian hypotheses about the propagation of light in a particle-like manner developed into a full-blown Newtonian corpuscular theory of light during the eighteenth century. It is also referred to as an 'emission theory of light' because its core assumption is that light is emitted in the form of very small, very rapidly moving particles. This mental model of light assumes that the emission velocity of light particles can increase and decrease with the emitter, depending on the velocity that these emitting systems have relative to the observer. According to the view of these 'emission theorists,' the velocity measured by the observer was hence the emitter's velocity $v_{emitter}$ relative to the observer plus the particle's intrinsic velocity $v_{particle}$. If the light source was moving away from or towards the observer, it was natural to assume that the measured velocity of light would decrease or increase accordingly—at least that is what the velocity-addition formula demands since Galileo's introduction of it into mechanics. There are explicit considerations and calculations in a manuscript submitted by the first professor of practical astronomy at the University of Edinburgh, Thomas Blair (1748–1828), to the *Royal Society of London* in 1786[6]:

> it appears more probable, that when light is emitted by a body in motion, the velocity of the particles projected in the direction of the motion will exceed the velocity of those, which are projected in an opposite direction, the difference being equal to twice the velocity of the moving body. And the same thing ought to take place when bodies reflect light.[7]

Blair concluded that the angles of diffraction would have to differ just a little for light coming from two light sources moving differently relative to the observer, because of the dependence of the angle of refraction on the velocity of the incident light according to the formulas for the law of refraction, formulated by the Newtonian paradigm of the particle model. Blair proposed to test this prediction experimentally using a highly dispersive instrument composed of twenty interlinked prisms on a full circle. This experiment was doomed to failure, of course, by the mere fact that such a multi-prism spectroscope would suffer a far too high total absorption of the incoming light. Consequently, Blair's proposal remained a thought experiment. Around 1800 John Robison (1739–1805), William Herschel (1738–1822) and François Arago (1786–1853) undertook to test such predictions experimentally but without being able to verify the predicted dependence of the velocity of light on the emitter.[8]

The Parisian historian of physics Jean Eisenstaedt interpreted this hypothetical influence of the intrinsic motion of the emitter of light on the velocity of light in the corpuscular theory as analogous to the Doppler effect derived today from the wave model. Obviously, it is *not* formulated in the natural reference quantity of the wave model of light: in frequency or wavelength, but in the analogous reference

---

[6]Blair (1786); communicated by Alexander Aubert, presented to *Royal Society* at the meeting on 6 Apr. 1786 and archived then under the accession label L. & P. VIII, 182. It is cited with detailed commentary in Eisenstaedt (2005, 2012) pp. 32ff.

[7]Blair (1786) p. 9, cf. Eisenstaedt (2005) p. 356.

[8]See Robison (1790). Arago (1853), based on measurements carried out 1806–10, still within the context of the emission theory. Arago later became an outstanding supporter of the undulatory theory of light: Eisenstaedt (2005) pp. 350f., 370ff.

quantity of the particulate model of light: relative velocity. If at all, only a very few scientists perceived this prediction, and it was soon returned *ad acta* after brief scrutiny. The reason probably was optical aberration, known since 1728 through Bradley, that is, the systematic apparent shift of star positions owing to the Earth's motion around the Sun, which has a constant value for all stars and for all time.[9] It already flatly contradicted a dependence of light velocity on the relative motion between a transmitter and receiver. Blair's manuscript was presumably never printed precisely because of these diametrically opposed experimental data to the prediction.

Eisenstaedt has spied other optical analogies in texts by Thomas Blair and his contemporaries as well, some of which were derived very much later in entirely different theoretical contexts (in particular, Albert Einstein's special and general theories of relativity).[10]

Blair's contemporary, John Michell (1724–1793), was another proponent of Newton's emission theory of light, which models light as a somewhat ballistic stream of tiny particles. When a cannon ball is shot horizontally, the Earth's mass deflects its path downwards. Blair, Michell and other Newtonians of the time thought that light gravitated in a similar way toward denser media. In Newton's fundamental equation between the mass m of a moving particle and its acceleration a, the gravitational force F_G is proportional to the gravity constant G, and mass m appears on both sides of the equals sign. The variable m cancels out as a consequence, and the resulting accelerations are completely independent of this unknown mass of light particles; they are proportional to the much larger mass M of the attractive object (Table 3.1):

$$m \cdot a = F_{\mathrm{G}} = G \frac{m \cdot M}{r^2} \Rightarrow a = G \frac{M}{r^2}.$$

This is why Newtonians could calculate with gravitational forces acting on particles of light without having the slightest idea about how large their mass was. John Michell took such considerations the farthest. He was a school principal in Thornhill, Yorkshire, who had published various papers since 1767 about speculative consequences of Newton's theory of gravitation. In 1772 he began to occupy himself specifically with the influence of gravitational force on light. He asked what would happen if the Earth's mass were much larger. According to the above formula, with increasing M, the force of gravity on the particles of light would increase. In order to escape such a ponderous emitter with increasing M, they would have to oppose it by an increasing counter-acceleration. The result would be a diminishing residual velocity for the particles of light. At some point they would not be able to leave the surface of such a heavy celestial body anymore. Upon emission they would be pulled back into the object by its gravitational force. Michell calculated the limit at which such a total extinction of light would happen as 497 times the radius of the Sun:

---

[9]See Bradley (1728), Melvill (1754, 1756, 1784), Wilson (1782), further Eisenstaedt (1996) pp. 144ff., (2005) pp. 348ff.

[10]See Michell (1784), Blair (1786), Eisenstaedt (1996, 2005, 2007, 2012), McCormmach (2012) and the primary or edited sources cited there.

Table 3.1 Analogies between Newton's emission theory and Einstein's theory of gravitation, compiled by K. Hentschel after Eisenstaedt (2012)

Physical effect	Newtonian emission theory	Einstein's theory of gravitation
Drag effects	Blair's thought experiment: search for velocity dependence in experiment with highly dispersive prism	Michelson experiment: search for aether wind
Effective change in velocity from relative motion	Blair-Michell effect: effective change in velocity from relative motion	Doppler-Fizeau effect: Doppler shift from relative motion
Gravitation	Reduced velocity of light from emitter's gravitational field (Michell 1784)	Gravitational red shift (Einstein 1907b)
Light attraction by masses	Newton's light deflection	Einstein's light deflection (larger by factor 2 than Michell's)
So strongly attractive ponderous matter that light cannot escape	Dark matter (Michell 1784)	Black hole (Chandrasekhar, Hawking et al.)

> If the semidiameter of a sphere of the same density with the sun were to exceed that of the sun in the proportion of 500 to 1, a body falling from an infinite height towards it, would have acquired at its surface a greater velocity than that of light, and consequently supposing light to be attracted by the same force in proportion to its *vis inertiae* with other bodies, all light emitted from such a body would be made to return towards it, by its own proper gravity.[11]

Obviously, such a star would have 122 million solar masses and a radius of 340 million km. Nowadays we no longer refer to "unobservable, dark bodies" but to a much more compact 'black hole.' Qualitatively speaking, though, predictions similar to those of Albert Einstein's general theory of relativity had already been made in this Newtonian context over a century earlier.

Although the roots of the first semantic layer defining light as corpuscular reach down to the ancient atomists, I have decided to start my discussion here with Isaac Newton (1643–1727), among other things, on the justification that Einstein was definitely familiar with Newton's texts on optics and light and appreciated and commented on them on a number of occasions.[12] Within the time frame of our examination, Einstein explicitly mentioned in at least three instances that continuities are ascertainable between his theory of light quanta and Newton's projectile model of light. In August 1913 he wrote to the astronomer Erwin Finlay Freundlich (1885–1964)—at that time an assistant at the Berlin Observatory—that the idea of a curved

---

[11]Michell (1784) p. 42; compare this Newtonian equivalent of a 'black hole' with the differing calculation by Pierre Simon de Laplace in *Exposition du Système du Monde* from 1796; see also McCormmach (1968/69), Schaffer (1979) and Eisenstaedt (2005) pp. 350f., (2005) p. 742 and their cited sources.

[12]His foreword to the the Dover reprint of Newton's *Opticks* edited by I.B. Cohen in 1952 is probably the most famous.

trajectory for light was still entirely natural within the frame of the emission theory of light.[13] In an article about Compton's scattering experiments of 1922–23 in the *Berliner Tageblatt*, Einstein also wrote in 1924: "Newton's corpuscular theory of light is coming back to life"; and in a letter to Erwin Magnus dated 22 November 1924 we read: The theory of light quanta "has points in common with the old Newtonian corpuscular theory."[14] Others were of the same opinion. In that same year, Arnold Sommerfeld concluded: "we have revived Newton's corpuscles," and in 1925 Arthur H. Compton autographed a portrait photograph of Einstein with the words: "Professor Albert Einstein. He revived the old Newtonian idea of light corpuscles in the form of quanta."[15] The close link between Newton's and Einstein's mental models is hence no figment of some historian's imagination; nor is it a symptom of the 'precursoritis' so prevalent among historians of science. A cognitive connection does evidently exist which the central actor of our story himself clearly saw. The similarities in the thinking of Newton and Einstein regarding this first semantic layer about the particulate nature of light does, in fact, leap to the eye when reading query 29 in Newton's *Opticks* quoted at the beginning of this section.

3.2 Propagation of Light at a Finite High Velocity for Any Color or Frequency

It had already been established prior to Newton that the velocity of light c was finite. Ole Rømer detected this in his observations of time delays in the transits of Jupiter's moons, although he did not give any concrete figure for it, contrary to what many accounts contend. Huygens, Cassini, Halley and other astronomers later offered the first specific estimates.[16] It certainly did not seem evident to Newton that the velocity of light is exactly the same for different colors. On the contrary, his derivation of the law of refraction described above, in which the sine of the angle of refraction was set inversely proportional to the relevant velocity of the corpuscles, seemed to indicate that the degree of refrangibility was correlated with the velocity of the light. Newton himself had shown in his *New Theory of Light and Colors* from 1672 that components of light of different colors manifest different angles of refraction. This seemed to suggest the assumption that the variously colored components of light would propagate at different velocities through the same medium. Because the

---

[13] See Einstein's letter dated mid-August 1913 in CPAE, vol. 5, doc. 468. On Freundlich and his relations with Einstein, see Hentschel (1997).

[14] Einstein (1924/25) and the letter CPAE, vol. 14, doc. 375, transl. ed., p. 364.

[15] Sommerfeld (1919c), the 4th ed. from 1924, pp. vii–viii, 57–59 and Compton (1925) p. 246.

[16] See, e.g., Huygens (1678/90b) pp. 463–466, who arrived at 232,000 km/s in modern units of measure. Wroblewski (1985) is a critical assessment of published papers about Romer's procedure.

red component of the spectrum was the least refractive, according to his theory, it actually ought to be the most rapidly moving one. Then the particles of light at the other, indigo end of the spectrum would be the slowest relative to the other colors. That is why Newton asked the Astronomer Royal, John Flamsteed (1646–1719), in 1691 about his observations of Jupiter's moons. Did the terrestrial observer first perceive the red component of the light right after their transits behind the planet and the blue component only afterwards? Flamsteed's negative reply dissuaded Newton of the hypothesis that red light must be faster than blue light.[17] He then suspected that the different color-dependent degrees of refrangibility either came from differing sizes for his light globuli or differing masses.[18] It was later confirmed that the propagation velocity c in a vacuum was constant and independent of the frequency, not only for all spectral components of light but also of other transversal waves (such as, thermal radiation, ultraviolet light, x-rays, γ-rays, and radio waves), all of which were interpreted according to Maxwell's theory as forms of electromagnetic radiation differing only in wavelength or frequency.[19]

It was only during the nineteenth century that experimental technology advanced far enough along to measure the velocity of light also in terrestrial experiments. Then sufficiently sophisticated instrumentation was developed, such as reliable rapidly rotating mirror systems, or interferometers. How the velocity of light changes when an emitter of light or the medium in which the light is propagating is itself in motion relative to the observer, is a question unrelated to these precision measurements in media at rest. The superposition principle of classical mechanics and hydrodynamic models of the spreading out of light within a fluid also led one to expect that the velocity of light c_n in a medium n would change accordingly. In his determination of the velocity of light c' in water flowing at the speed w (cf. Fig. 3.1), Hippolyte Fizeau (1819–1896) discovered in the middle of that century that a vectorial addition of the velocity of light in water at rest (c/n) with the medium's velocity w of refractive index n did not yield c', but a slight alteration of it by the factor $(1 - 1/n^2)$. He interpreted this as an only 'partial' drag of the luminiferous aether by the moving water, which corresponded precisely to what his fellow countryman Augustin Fresnel (1788–1827) had already quantified at $c' = c_0/n - w(1 - 1/n^2)$ in 1818.[20]

---

[17]See Turnbull et al. (1959–77) vol. 3, p. 202 and Shapiro (1993) p. 218, Eisenstaedt (1996, 2012) p. 30.

[18]On this episode see Bechler (1973, 1974), Hall (1993) pp. 164–166, Eisenstaedt (1996) pp. 124ff. and the primary sources mentioned there, esp. Newton's correspondence with Flamsteed and Gregory.

[19]These diverse extensions of the electromagnetic spectrum and the related ontological debates are analyzed in Hentschel (2007a) with extensive primary-source citations. For later precision measurements of c, see Roditschew and Frankfurt (1977) pp. 333ff.

[20]See Fresnel (1818), Fizeau (1851/53, 1859/60), Mascart (1889-94–94) vol. 3, pp. 91–144 and Jan Frercks (2004) as well as further primary sources quoted there on 'the relation between publication and theory and experiment in Fizeau's research program on aether drag.'

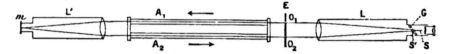

Fig. 3.1 Schematic setup of Fizeau's aether-drag experiment 1851. Beams of light emitted from the light source S at the far right are split into two. One of them follows the path: slit O_1, tube A_1, mirror m, tube A_2, slit O_2. The other follows this path in the opposite direction: O_2, A_2, m, A_1 and O_1. As soon as the water in the tubes is not at rest (relative to the light source) but in motion as indicated by the arrows, at the speed 1 m/s, interference shifts occur at S which Fizeau confirmed by this experiment. *Source* Mascart (1889-94) vol. 3 (1893) p. 101

Einstein also considered the existence of a dependence on an emitter's motion as suggested by the projectile theories of light before he arrived as his postulate of the constancy of the velocity of light in a vacuum. We know this from his later autobiographical notes as well as from comments he made about contemporary efforts by Walther Ritz (1878–1909) to develop an emission theory of just this kind.[21] Einstein's postulate of a constancy of the velocity of light in all inertial systems, which forms the basis of his special theory of relativity from 1905 as the second of two axioms besides the principle of relative motion, came as a consequence of the failure of these theoretical considerations about emission theories.[22] Here we see another subtle connection between Einstein's papers from his *annus mirabilis* 1905 (cf. Fig. 2.1).

Fresnel's drag coefficient, which classical electrodynamics could only 'explain' by making quite artificial assumptions about the luminiferous aether being only partially dragged along (see above), became in Einstein's 'electrodynamics of moving bodies' an almost trivial consequence of his relativistic superposition rule of velocities which assumed the place of the classical velocity-addition formula (cf. box 2). However, Einstein made sure to formulate the theory of relativity so that it remained compatible irrespective of whether it was viewed within the particle model or the wave model of light.

---

[21] See Ritz (1908a, b) and Ritz (1911), Ritz and Einstein (1909), Fritzius (1990), Norton (2004, 2016) p. 260, Martinez (2004) and Pont (2012).

[22] Along these lines see also Norton (2004). It is remarkable that parallels also exist in Newton's particle model to this irrefutable consequence for want of empirical evidence of some kind of dependence between the velocity of light and the velocity of the emitter. Robert Blair (1786), p. 25 concludes from experimental results: "it is at least possible for any thing we know to the contrary, that light may be emitted with the same velocity from shining bodies, whether they be at rest or in motion."

Box 2: Classical and relativistic velocity-addition formula

According to the classical conception, the velocities of a moving object and its observer are vectorially additive. If, for example, this observer sees light that is moving away from the emitter at the velocity c_0, then from his perspective the velocity of this light would amount to $c_0 \pm v$, depending on whether he is moving with the light or in the opposite direction. The motion of the aether, a hypothetical medium for the propagation of light, would likewise change the velocity of this light. Hence, in these classical emission theories of light, the speed of light is not a constant but depends on the state of motion of the observer relative to the luminiferous aether. This prediction becomes experimentally testable by measuring the velocity of light in moving media. Fizeau demonstrated in 1851 that only a slight alteration occurs, of about factor $(1 - 1/n^2)$ and interpreted this as a *partial* drag of the luminiferous aether by the moving water. The special theory of relativity provides a much simpler explanation for this Fresnel coefficient: relativistic velocity-addition, presented below in the form Max von Laue put forward in 1907 and explicated in greater detail in 1911.

If the water's flow velocity is $v_1 = w$, and the light's propagation velocity in this water (of refractive index n) is $v_2 = c/n$, by setting $v_1 = w$ and $v_2 = c/n$ in the relativistic velocity-addition formula, we have

$$v_{relat} = (v_1 + v_2)/[1 + v_1 \cdot v_2/c^2]$$
$$v_{relat} = (w + (c/n))/[1 + w \cdot (c/n)/c^2] = (w + (c/n))/[1 + w \cdot (1/n)/c]$$
$$\simeq (c/n + w) \cdot [1 - w/(nc) + 1/2 \cdot (w^2/(nc)^2)]\pm) \simeq c/n + w \cdot (1 - 1/n^2).$$

The last expression $w \cdot (1 - 1/n^2)$ contains the Fresnel drag coefficient as a factor, according to which the aether is dragged along by the medium only by the partial amount $(1 - 1/n^2)$ of its flow velocity (formulated in conformity with the classical notion that the medium partially drags the luminiferous aether along with it). In Einstein's reinterpretation of this formula, this mechanistic interpretation is dropped but the formal expression is adopted by the new theory unchanged.

I concede that I cannot produce any sketch by Einstein to illustrate a light quantum as graphically as Fig. 4.1 does for Newton. But that does not mean that Einstein's modeling was less sophisticated. It was just more abstract. But his thoughts about light as a particle were also closely associated with its propagation velocity. The 'surroundings,' Newton's aether, become for Einstein the surrounding radiation field of the other light particles (more about this later); and the constant propagation velocity of light was one of the famous axioms of his paper, submitted to the *Annalen der Physik* just three months after the one about light quanta, 'On the Electrodynamics of Moving Bodies' (received on 30 June 1905). We see here one of the hidden connections between the three famous papers from 1905 (cf. here Fig. 2.1).

3.3   Emission and Absorption of Light 'Particles' by Matter

The third semantic layer can also be traced back to Antiquity and extends further at least up to Einstein's famous papers about induced emission from 1916 and 1917 and to QED (see the subsections about layers 8 and 11 toward the end of this chapter). Here we again begin with Isaac Newton. In query 5 of the *Opticks* Newton wrote in 1704: "Do not Bodies and Light act mutually upon one another; that is to say, Bodies upon Light in emitting, reflecting, refracting and inflecting it, and Light upon Bodies for heating them, putting their parts into a vibrating motion wherein heat consists?"[23] Let me point out on the side that this corpuscular model of light had a great influence on the so-called chemistry of light during the seventeenth and eighteenth centuries and on the conceptual interpretation of the many discoveries of new kinds of rays in the nineteenth and early twentieth centuries.[24] Around 1800 when infrared and ultraviolet light were discovered, and similarly again, shortly before 1900 when α- and β - rays resp. γ- and x-rays were discovered: each time considerable insecurity arose about whether these new types of rays were waves or particles, with the prejudice weighing heavily in favor of the particle interpretation, thanks to Newtonianism.[25] The controversies were particularly hefty about interpreting roentgen rays or x-rays (preferentially named the latter by their discoverer Wilhelm Conrad Röntgen (1845– 1923)) from 1895 on, as well as the energetically even more powerful γ-rays (1900).[26]

Joseph John Thomson (1856–1940) in Cambridge, for example, interpreted x-rays as particulate because of their extremely directed and pointed action (they were also referred to as 'needle rays' for this reason). The intensity does not diminish by $1/r^2$ but remains virtually the same over larger distances r. Gas also ionized in a pointed manner, which indicated small areas of interaction, not large expanses of action zones as would have suited wave fronts described by the competing wave model for this radiation. Furthermore, it became evident that the energy of this radiation was completely independent of its intensity. For many contemporaries all of this spoke clearly against the undulatory model: "These facts seem to be completely inexplicable on any sort of a spreading wave theory [...] the emitted energy keeps together as an entity, or quantum, which may be transformed back and forth between a β-ray and an X- or γ-ray."[27] That's why in 1950 the Nobel laureate Robert A. Millikan still referred to the "semi-corpuscular or photon theory of light" in his autobiography.[28]

---

[23] Isaac Newton: *Opticks* (1704) p. 339.

[24] See Hentschel (2007a), Shapiro (2009) Chap. 9 and Principe (2008), only available on-line at http://methodos.revues.org/1223.

[25] See Hentschel (2007a) with primary-source citations.

[26] See Stuewer (1971, 1975a) Chap. 1, Wheaton (1983) with primary-source citations, furthermore Sect. 4.5 on J.J. Thomson's mental model.

[27] Millikan (1913) p. 128.

[28] Millikan (1950) Chap. 9, pp. 101–102.

It was very difficult to measure the propagation velocity of x-rays, and the experimental results remained controversial for a long time.[29] Their spreading speed remained unknown until the successful experiments by Erich Marx (1874–1956) in 1905 showing that x-rays moved as rapidly as light, at least in order of magnitude. It took a decade since their discovery for the proof of their polarizability to be forthcoming as well. Charles Glover Barkla (1877–1944) achieved this in 1906.[30] Six years later Max von Laue (1879–1960), Walter Friedrich (1883–1968) and Paul Knipping (1883–1935) in Munich as well as William Henry Bragg (1862–1942) and his son William Lawrence Bragg (1890–1971) in England demonstrated that x-rays interfere off solids, by reflecting off neighboring crystal planes.[31]

By 1913 the evidence that x-rays could polarize and interfere was reliable enough to make the current interpretation of x-rays as a high-frequency form of electromagnetic waves indisputable. "The identity in nature of X-rays and light could no longer be doubted," the American experimental physicist Arthur Holly Compton averred in historical retrospect.[32] All the same, as late as 1920 Fernand Holweck (1890–1941) was still able to submit a dissertation in Paris about experimental proof of a Wittgensteinian family resemblance between light and x-rays.[33] Compton laid the copestone with his proof that x-rays behave exactly like Einstein's postulated light quanta, therefore, like particles, also in scattering processes. He concluded: "It is clear that the X-rays thus scattered proceed in direct quanta of radiant energy; in other words, that they act as photon particles."[34] In allusion to these debates of the foregoing decade, the title of Compton's Nobel lecture in 1927 was programmatic: 'X-rays as a branch of optics' and the second paragraph of it opened with the statement: "It has not always been recognized that X-rays is a branch of optics."[35] We also see here how the actors themselves—not just we historians of science a hundred years later—were already able to decry the extended historical lines along which concepts develop.

This third layer of meaning concerning the emission and absorption of light by matter, a familiar idea latest since early modern times, merged with the seventh

---

[29] See, e.g., Blondlot (1903) and Marx (1905) for objections to the applied method.

[30] See Marx (1905), resp., Barkla (1905, 1906); further J.J. Thomson (1911) pp. 695f., Wheaton (1983) pp. 44ff.

[31] See, e.g., Tutton (1912) p. 307, who described this experiment as a "crucial test" for interpreting x-rays as electromagnetic waves. See also the Nobel lectures by Laue (1920) and Bragg (1915) on x-ray diffraction by crystals, also on-line at nobelprize.org/physics/laureates/1914, resp., /...1915. On initial failures to measure the index of refraction and wavelength of x-rays, see Compton (1924) pp. 174ff.

[32] Compton (1927a) p. 179; cf. further Wheaton (1983) pp. 199ff. and the citations there of Max von Laue et al.

[33] See the expanded published book, Holweck (1927); furthermore, Beaudouin (2005) pp. 85f., 149, 155 about the career of this experimenter, who died under torture for his political resistance to the German occupiers. His thesis supervisor was Marie Curie, no less.

[34] Quoted from a retrospective article by Compton (1961) p. 820; cf. Compton (1921), Stuewer (1975a, 1998), Silva and Freire (2011) about Compton's learning curve concerning the light quantum.

[35] Compton (1927a) p. 174 resp. Stuewer (1975a), Wheaton (1983) pp. 94ff.

layer, the quantization of this exchange of energy, a later superposition from after 1905. Thanks to Bohr's model of the atom from 1913, this third layer gained added substance. By virtue of this very successful model, the quantum-like emission and absorption of radiation, forming the basis of Einstein's light-quantum hypothesis, became known to large numbers of researchers. According to it, the transition of electrons between orbits occurs by the emission or absorption of light quanta (in so-called quantum jumps). In some special areas, however, the light-quantum hypothesis did not need this backing. It was already instructive enough about absorption processes, for instance, on its own. The light-quantum hypothesis formed the basis of the photochemical law of equivalence, by which the absorption of light can only take place in integral quanta when light triggers a chemical reaction. When Johannes Stark raised protest in 1912, claiming priority over Einstein in "applying Planck's elementary law to photochemical processes," Einstein merely remarked: "the law of photochemical equivalence is a self-evident consequence of the quantum hypothesis," which he, of course, had developed in 1905.[36] Because this hypothesis was still being heatedly disputed and had found few followers, Einstein had presented a purely "Thermodynamic Proof of the Law of Photochemical Equivalence" in 1912 and countered his critics with the argument that the quantum hypothesis was not needed to derive the equivalence law, but that this law "can be deduced from certain simple assumptions about the photochemical process, by way of *thermodynamics*."[37] Thus the law of photochemical equivalence was secured in any case and independent of the as yet uncertain modeling.

The term 'quantum yield' (*Quantenausbeute*), which Einstein never used, denotes the number of photochemically altered molecules in relation to the number of absorbed light quanta. According to Einstein's model of light absorption, this quantum yield should always be equal to 1, which John Eggert (1891–1973) and Walter Noddack (1893–1960) confirmed in the period 1921–27 applying the decomposition of silver bromide into silver and bromine. In this case, each absorbed quantum of light does in fact cause exactly one chemical reaction. But many other reactions were found to have quantum yields very much greater than 1, which led to the conclusion that other secondary reactions must follow the primary photochemical reaction, causing the action of a single light quantum to be multiplied, as in a chain reaction. The reaction mechanisms suspected by Max Bodenstein (1871–1942) in 1911 and by Walther Nernst (1864–1941) in 1918—none of which Einstein had predicted but which are nevertheless in conformity with his notions about quantum yield—were confirmed in 1922 by Carl Weigert (1845–1904) and his doctoral student Hermann Richard Kellermann (1890–?).[38]

---

[36] See the experimental paper by Stark (1908) and Einstein (1912a, b). On the priority claim, see Stark (1912a, b) and Einstein (1912c): CPAE vol. 4, doc. 6, transl. ed., p. 125; furthermore here Sect. 4.4.

[37] See Einstein (1912c) p. 888 (CPAE vol. 4, doc. 6, transl. ed., p. 125), original emphasis.

[38] The primary sources are cited, e.g., in Plotnikow (1920), Bodenstein (1942), Meidinger (1934).

Between 1923 and 1955, the quantum yields under discussion in biological photo-synthesis were between 4 and 12 absorbed photons per oxygen molecule produced.[39]

Last but not least, Einstein's modeling of the photoelectric effect also suited Millikan's experimental estimates of a very short time interval between the incidence of the electromagnetic radiation on a polished metal surface and the thereby induced emission of a photo-electron. Ernest O. Lawrence and Jesse Beams were already able to reduce this time interval to less than $3 \cdot 10^{-9}$ s in 1928, using electro-optical Kerr cells; Forrester et al. attained 10^{-10} s in 1955.[40] We know now that for radiation at the frequency $\nu = 10^{15}$ hertz, this time interval is in order of magnitude of femtoseconds (that is, 10^{-15} s and sometimes even less), which categorically excludes alternative models of this physical process in the direction of a gradual augmentation with a subsequent explosive release of energy, such as in Lenard's trigger model.[41]

3.4 Momentum Transfer (Radiation Pressure) by Light onto Matter

The fourth semantic layer, about radiation pressure, also appeared speculatively during early modernity within the context of cometary observations. Already in 1608, Johannes Kepler (1571–1630) wrote the following about the interaction between sunlight and a comet flying past the Sun:

> The Sun's rays pass through the corpus of the comet and instantly take some of its material along on its way out, away from the Sun; that is how, I think, the tail of the comet comes about, which always stretches away from the Sun.[42]

Newton rather considered that the solar rays heat up the vapors in the comet's tail, causing it to move away from the heat source, similarly to smoke up a chimney.[43]

All the early experiments performed in the eighteenth century in an attempt to verify such a radiation pressure by light particles within the framework of Newton's projectile theory of light ran up against insurmountable experimental hurdles. Many

---

[39]On the controversy between Otto H. Warburg (1883–1970) and Ralph Emerson (1903–1959), James Franck (1882–1964) and Hans Gaffron (1902–1979), see Nickelsen (2013, 2016).

[40]See Lawrence and Beams (1928) pp. 484–485: "atoms emit quanta of radiant energy practically instantly [...] non-existence of a [time] lag in the photoelectric effect," Forrester et al. (1955) p. 1691.

[41]This so brief time interval is no proof of the indivisibility of the photon, as pointed out by Roychoudhuri (2006) p. 3, (2009) p. 3.

[42]Kepler (1608c) p. 60 in his report about the 'hairstar or comet' observed in 1607.

[43]See Newton's *Principia* (1687b, c), book 3, prop. 41 (p. 528 of the Motte-Cajori edition), the continuation of 59 about comets.

of these natural philosophers failed simply because they had no way of producing a sufficient vacuum and the available suspension mechanisms were trouble-prone.[44] Nevertheless, for a while they did believe that their experiments indicated the existence of a radiation pressure just happening to follow so neatly out of the projectile model. Wilhelm Homberg (1652–1715) was one of the first. In 1708 he was able to demonstrate that it could flip an asbestos fiber placed at the focus of his Tschirnhaus burning mirror, measuring one meter in diameter, which he took as evidence "que les rayons de soleil eussent la force de presser et de pousser, même quand ils sont renis par le Miroir ardent."[45] When a beam of solar rays, concentrated by a focal lens about 30 cm in diameter, was aimed at the end of a freely hanging spring, it started to swing noticeably as if it had been struck with a rod. By the middle of the eighteenth century it was clear to experimenters that the enormous heating action at the focus of such a burning lens or mirror, in combination with the vaporization of matter and the currents it incurred, could be behind such distortion effects. The first scientific observation of a freely swinging needle in a partially evacuated glass vessel (see Fig. 3.2) was performed in 1792 by Reverend Abraham Bennet (1749–1799), but his results were sobering: "I could not perceive any motion distinguishable from the effects of heat." He immediately suspected that a basic defect existed in the projectile model of light: "Perhaps sensible heat and light may not be caused by the influx or rectilineal projections of fine particles: but by the vibrations made in the universally diffused caloric or matter of heat, or fluid of light."[46] Crookes's radiometer was long held to be a "light mill" in this context, even though Arthur Schuster's paper from 1876 already showed that the reason for the motion was actually only thermal action on the residual gases in the partially evacuated vessel.[47]

Before precision measurements were able to demonstrate, shortly after 1900, that light rays impinging on a suspended surface within a partially evacuated glass container exert "light pressure" on that surface, the Italian physicist Adolfo Bartoli (1851–1896) produced an elegant thermodynamical argument for why this light pressure must exist. His thought experiment from 1876 shows that the second law of thermodynamics, when applied to a cyclical process with thermal radiation, compellingly predicts radiation pressure (see Fig. 3.3 and Box 3).

---

[44] See the following two footnotes. Cf. further Worrall (1982) p. 141 about analogous experiments by Nicolas Hartsoeker (1696) and Mairan (1747) as well as their renderings in Pieter van Musschenbroek's *Cours de physique experimentale et mathématique* 1769.

[45] See Homberg (1708) (unfortunately only documenting Fontenelle's abstract of a presentation) and Bennet (1792) esp. experiment X, pp. 87f., as well as Principe (2008).

[46] See Bennet (1792) pp. 87f. and plate II (here Fig. 3.2) without providing the residual pressure. On Bennet and the treated problems and research contexts of these two early-modern experimenters, see Elliott (1999).

[47] This debate about Crookes's radiometer is treated by Woodruff (1966) and Dörfel and Müller (2003) on the basis of a variety of primary sources. Further publications about this episode are cited in Worrall (1982) pp. 147ff.

Fig. 3.2 Setup of the experiment by Bennet in 1792 to measure the radiation pressure on a suspended needle in a partially evacuated glass vessel. *Source* Bennet (1792), plate II

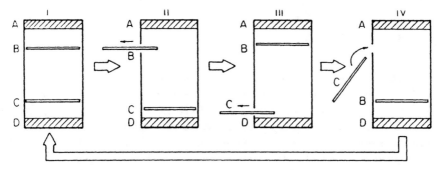

Fig. 3.3 Bartoli's thought experiment from 1876 deducing light pressure from the second law of thermodynamics: A and D are ideal black bodies at equilibrium with the cavity. Let the temperature of D (in the space CD) be higher than that of A (in space AB). By clever shifting of membrane B away from A toward D, heat could conceivably be conducted from the colder body toward the warmer one, provided there is no counterforce acting against the radiation pressure. Consequently, in order not to conflict with the second law of thermodynamics, radiation pressure is required. *Source* Carazza and Kragh (1989) p. 188. Reprinted by permission of Taylor & Francis © 1989

Box 3: Light pressure and the second law of thermodynamics

Consider an evacuated cavity (Fig. 3.3) with ideal black bodies A and D at its two ends. A membrane is positioned close to each end, that is, membrane B next to A and membrane C above D. Let the temperature of D be higher than the temperature of A. The system is in a state of equilibrium: in other words, the radiation in cavity CD is at the same temperature as D, and analogously, the radiation in cavity AB is at the same temperature as A. Through clever cyclical manipulations of membranes B and C, as illustrated in Fig. 3.3 (i.e., removing C in III, followed by a shifting of B in the direction of D in IV and reinserting C at the earlier position of B in IV and II), it would be possible to conduct heat from the colder body toward the warmer body without the exertion of work. But according to the second law of thermodynamics this is strictly forbidden. Evidently, a counterforce must be acting against this downward shift of membrane B, namely, radiation pressure; work must be exerted against it to execute the shift. Therefore, the existence of this radiation pressure is already required by the second law of thermodynamics. (See Bartoli (1884) and Carazza and Kragh (1989).)

Experimental proof of this pressure of light finally arrived at the beginning of the twentieth century. In 1901 Pyotr Nikolaevich Lebedev (1866–1912) at the *Lomonosov State University* in Moscow succeeded in producing the first laboratory proof of radiation pressure, but with a high margin of systematic error (greater than 10%). In 1903 Ernest Fox Nichols (1869–1924) and Gordon Ferrie Hull (1870–1956) at *Dartmouth College* in the U.S.A. managed to reduce this error to just 1%.[48]

3.5 Energy Transfer by Light

The theoretical comparison standards in these ballistic precision measurements at the beginning of the twentieth century were the predictions of the magnitude of this light pressure by James Clerk Maxwell (1831–1879) and John Henry Poynting (1852–1914), based on Maxwell's theory of electromagnetic radiation.[49] The energy density of the electromagnetic field results from Maxwell's equations for the electric and magnetic fields E and B proportional to: $E^2 + B^2$. The so-called Poynting vector $S = E \times B$ describes the magnitude and direction of flow of electromagnetic energy and was first computed by this English physicist in 1884. The Poynting vector, is mathematically proportional to the vector product of the electric and magnetic fields, therefore, rigorous orthogonality of the transfer of energy and momentum to both carrier fields holds: "It follows at once that the energy flows perpendicularly to the

---

[48] See Lebedew (1901), Nichols and Hull (1901, 1903a, b).

[49] See Maxwell (1873) §792–793, Poynting (1884); in addition Poincaré (1900), who also calculated the recoil of a system emitting or reflecting light.

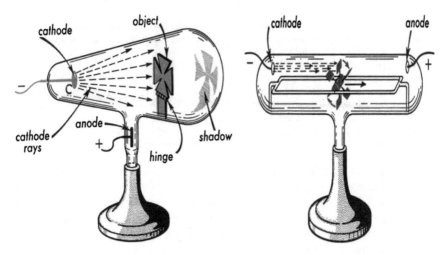

Fig. 3.4 Two demonstration experiments with Crookes tubes. Left: the shadow of a Maltese cross is cast onto the inner end of the tube opposite the cathode, which shows the rectilinearity of cathode rays. Right: a rotatable mill is driven rightwards from the left by cathode rays, which proves that they transfer energy and momentum. Since the 1870s this was regarded as indicative that cathode rays are composed of particles. Source of both images: Wikimedia, in the public domain

lines of electric force, and so along the equipotential surfaces where these exist. It also flows perpendicularly to the lines of magnetic force, and so along the magnetic equipotential surfaces where these exist. If both sets of surfaces exist their lines of intersection are the lines of flow of energy."[50]

The physical interpretation of this flow of energy, which is always associated with a flow of momentum, leading to the radiation pressure, was much less clear, however, causing intensive debate about this aspect of the theory of electromagnetism among physicists. What exactly transports the energy and momentum orthogonally to the electromagnetic field's own oscillatory direction? How should this be illustrated by a British model of the waning nineteenth century? Was this case really analogous to such a stream of particles as cathode rays have been observed in a Crookes tube leading to those impressive experiments irrefutably demonstrating the corpuscular nature of cathode rays (cf. Fig. 3.4 right).

The Dutch theoretical physicist Hendrik Antoon Lorentz (1853–1928) carefully distinguished between the two cases, that is, the transfer of energy and momentum by particles versus electromagnetic waves when he was writing his book *The Theory of Electrons* from 1909:

> The flow of energy can, in my opinion, never have quite the same distinct meaning as a flow of material particles. [...] It might even be questioned whether, in electromagnetic phenomena, the transfer of energy really takes place in the way indicated by Poynting's law, whether, for example, the heat developed in the wire of an incandescent lamp is really due to energy

---

[50]See Poynting (1884) p. 345 furthermore the Wikipedia article on Poynting's theorem and the informative website: http://www.mathpages.com/home/kmath677/kmath677.htm.

which it receives from the surrounding medium, as the theorem teaches us, and not to a flow of energy along the wire itself. In fact, all depends upon the hypotheses which we make concerning the internal forces in the system, and it may very well be that a change in these hypotheses would materially alter our ideas about the path along which the energy is carried from one part of the system to another. It must be observed however that there is no longer room for any doubt, so soon as we admit that the phenomena going on in some part of the ether are entirely determined by the electric and magnetic force existing in that part. Therefore, if all depends on the electric and magnetic force, there must also be one near the surface of a wire carrying a current, because here, as well as in a beam of light, the two forces exist at the same time and are perpendicular to each other.[51]

Einstein was well-informed about the latest experimental research. In a paper from 1909 he mentioned "the pressure of light, which has only recently been established experimentally, and which plays such an important role in the theory of radiation."[52] In the same year he also discussed the fluctuation effects on freely suspended mirrors resulting from this exchange of momentum between light and matter, which I will return to in Sect. 8 in connection with wave-particle duality.

There was another much more elementary consequence, however, which Einstein already saw in 1905 even though it was a little paradoxical. On one hand, light and other forms of electromagnetic waves signify a transfer of energy, as Poynting already knew and as more recent experiments on radiation pressure had confirmed. On the other hand, as he showed in one of his other classical papers from this *annus mirabilis* 1905, energy and mass are equivalent to each other according to the famous formula $E = mc^2$. He backed this up and emphasized it further in two follow-up papers in 1906 and 1907 as well. In 1905 he just proved a strict proportionality between an increase in energy and an increase in mass; the later extension turned it into an equivalence between energy and mass. He wrote to his friend Conrad Habicht:

> One consequence of the study on electrodynamics did cross my mind. Namely, the relativity principle, in association with Maxwell's fundamental equations, requires that the mass be a direct measure of the energy contained in a body; light carries mass with it. A noticeable reduction of mass would have to take place in the case of radium. The consideration is amusing and seductive; but for all I know, God Almighty might be laughing at the whole matter and might have been leading me around by the nose.[53]

Einstein had not been mistaken to suppose a mass-energy equivalence. Nevertheless, how exactly to interpret the way light might "transfer" mass remained unclear for a long time. Did it mean that the light quanta attributed to light have their own non-vanishing mass at rest? At first this was occasionally supposed,[54] but as it later turned out, this is *not* the case. Photons are strictly massless quanta of the electromagnetic

[51]Lorentz (1909) §18, pp. 25–26; this quote is unchanged in the second edition from 1915. We shall see later (in Sect. 4.3), that at this time Lorentz was one of the smartest critics of the concept of light quanta.

[52]Einstein (1909a) pp. 817f. (CPAE, vol. 2, doc. 60, transl. ed., p. 380).

[53]A. Einstein to C. Habicht, undated, written between 30 June and 22 Sep. 1909, CPAE, vol. 5, doc. 28, transl. ed., p. 21.

[54]E.g., Broglie (1922) p. 438, (1923) p. 508, estimating the mass of a light quantum at $< 10^{-50}$g.

field,[55] but mass can be assigned to the energy contained in this field, which in experiments like the one described above about radiation pressure, can also transfer momentum onto material particles.[56] For many of his contemporaries, it seemed self-contradictory that a 'massless' light quantum should transfer momentum like a material particle in collision processes such as the Compton effect. But from the formalism of the special theory of relativity for the squared four-vector (p, mc^2) of momentum p and the rest mass of light quanta mc^2, it follows further that for a vanishing rest mass $m = 0$:

$$E^2 = (pc)^2 + (mc^2)^2 = (pc)^2 + 0 \Rightarrow E = pc.$$

This ultra-relativistic limit inheres a strict proportionality between energy and momentum $p = E/c = h\nu/c$. We are going to encounter it again when we derive the formulas for the Compton effect, where those ultra-relativistic dynamics again prove their merit experimentally as well. As Max Planck and Louis de Broglie showed, the radiation pressure of light and other electromagnetic waves could be brought quantitatively into agreement with the increasingly precise measurements when Einstein's postulate $p = h\nu/c$ is adopted, whereas Newtonian dynamics or semiclassical electrodynamics only lead to half the measured value.[57]

3.6 Energy-Frequency Proportionality in the Photoelectric Effect

The discoverer of the waves that later carried his name, Heinrich Hertz (1857–1894) observed in 1887 in his laboratory that sparks transmitted by radio waves got smaller when he held a pane of glass in the spark gap in front of the receiver but resumed their original scope when he substituted the pane with thin lead crystal which is permeable to ultraviolet. He assigned further analysis of this physical phenomenon to Wilhelm Hallwachs (1859–1922).[58] This co-worker of Hertz connected a piece of pure zinc to an electrometer to gauge the charge. Once negatively charged, the zinc plate would discharge as soon as light was cast onto it. The more ultraviolet was mixed into the incident light, the more powerful the discharges became. When the UV component was removed by a UV-absorbing medium, the effect proceeded much more slowly; likewise if the zinc plate was initially positively charged. Thus it was Hallwachs in 1888—not Hertz in 1887—who first showed that irradiating a cathode in a special

---

[55]This is precisely the reason why they don't have a finite decay time τ owing to Heisenberg's uncertainty relation $\Delta E \cdot \tau \geq \hbar$ —different from virtual exchange particles, which do have mass.

[56]More recent research on the masslessness of photons include Okun (2008) citing precision verification experiments performed 1992–2004 down to limits below 10^{-16} eV.

[57]See Planck (1913) and Broglie (1949) p. 346.

[58]See Hertz (1887) and on the following Hallwachs and Franz (1888a, b, 1889), further Wiederkehr (2006).

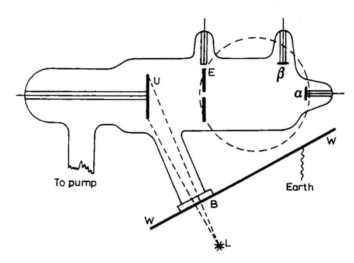

Fig. 3.5 Lenard's cathode-ray tube from 1902 with a UV-permeable aluminium window B, cathode
U, anode α and a negatively charged grid E that permits the stream of photons triggered by the UV
radiation to be regulated. *Source* The Nobel lecture by Philipp Lenard (1906b) p. 122. Reprinted
by permission of the Nobel Foundation © 1905

partially evacuated tube known as a cathode-ray tube, with ultraviolet light leads
to the emission of cathode rays. Hallwachs was still referring to 'light-electricity'
(*Lichtelektrizität*) but it soon became generally known as the 'photoelectric effect.'[59]

Hertz's other assistant at the time, the experimental physicist Philipp Lenard
(1862–1947), also examined the photoelectric effect. His experiment used a specially
designed evacuated tube (see Fig. 3.5) with an aluminium window B where the
electromagnetic radiation enters from L onto the cathode U. This tube permitted the
flow of cathode rays to be modulated by means of a negatively charged grid E.[60]

Depending on the selected preliminary tension between the cathode U and the grid
E, the electrons coming from the cathode are either permitted through to the anode
α or diverted. By varying this preliminary tension he could determine the number
of electrons as a function of their energy. The result of Lenard's measurements
was that the limiting potential Φ of the flow of photons beyond which no cathode
rays can trespass the negatively charged grid set at the tension U depends on the
"type of light" and the basic composition of the electric arc[61] but is independent
of its intensity. The photoelectric effect was therefore dependent on the kind of

---

[59] See Hallwachs (1916), cf. further Lenard (1906), Schweidler (1910, 1915), Hughes (1914a) for
surveys.

[60] For the experimental details see Lenard (1894)–(1906) and here Fig. 3.5. Einstein fully appreciated
Lenard's experimental skills at this time: see his letter to Jakob Laub, 17 May 1909, CPAE, vol. 5,
p. 187.

[61] Lenard (1902) pp. 167f. We now know that the power of an electric arc is proportional to ν, but
Lenard did not realize this at the time!

electromagnetic irradiation but not on its intensity.[62] From the perspective of classical physics, intensity is correlated with energy, making this result initially puzzling. Lenard found a way out with his trigger hypothesis: The incident UV radiation accordingly acts as a kind of trigger releasing cathode rays (i.e., electrons, as we now call them):

> There then remains the assumption of complex motion conditions for the internal parts of the body, but also the notion that seems to suggest itself for the time being, that the initial velocities of the emitted quanta do not originate from the light's energy but from present motions existing within the atoms even prior to the exposure, making the resonance motions merely play a triggering role.[63]

This is the first time that the expression 'quanta' appears within the context of the photoelectric effect, but here it still unspecifically means 'amounts of energy'; nor does Lenard provide any underlying mental model for it at this time.

Between 1900 and 1910 numerous articles about the photoelectric effect appeared, making it a focus of research for experimental physicists of the day.[64] Four months before Einstein submitted his paper on light quanta to the *Annalen der Physik* in March 1905, there appeared a survey article by Egon Ritter von Schweidler (1873–1948) in the first volume of the *Jahrbuch der Radioaktivitt und Elektronik*, edited by Johannes Stark, on photoelectric phenomena.[65] This Austrian experimental physicist was employed as an assistant to Franz Serafin Exner (1849–1926) at the second physics institute of the *University of Vienna*.[66] At the end of his review of the available publications on the subject, von Schweidler pointed out the lack of precise measurements of the relation between electron numbers resp. their initial velocity and wavelength for monochromatic light, which were indispensable for any theory of the photoelectric effect. The figures that experimenters delivered in the following years for a very narrow frequency range were incompatible, however. Was E proportional to ν^2 or to $\nu^{2/3}$, to ν or even to $\log(\nu)$? Neither Lenard nor Stark were

---

[62]See Lenard (1900, 1902) pp. 150, 163–166, (1906) p. 123 and, e.g., Niedderer (1982) pp. 41f., Katzir (2006) pp. 451ff.

[63]Lenard (1902) p. 170; cf. also Lenard (1906, 1918, 1944) p. 267, furthermore Hughes and Llewelyn (1914b) p. 48 and Stuewer (2014) p. 144. On Lenard, who later became an anti-Semitic Nazi, cf., e.g., Hentschel (1996), Schönbeck (2000) and Hagmann and Füssl (2012).

[64]By 1902 over 160 papers treating the photoelectric effect already existed. Cf. the chronological bibliography in Lenard (1906) pp. 131–134.

[65]On "lichtelektrische Erscheinungen," see von Schweidler (1904). The term 'Elektronik' in the title of this new journal did not mean what we would now associate with the term. It signified everything related to electrons, the novel discovery from 1897.

[66]On Schweidler see Karlik and Seidl (2005) and Seidl (2010); on Schweidler's interaction with Exner and his circle, see Karlik and Schmid (1982), esp. pp. 111–114. On the broader context of Austrian research on radioactivity, cf. Fengler (2014).

involved in this research.[67] Five years later, a survey then appeared on "The latest researches on the emission of negative electrons," this time by an experimental physicist who had recently taken his academic teaching degree at the *University of Breslau* (now Wroclaw, Poland), Rudolf Ladenburg (1882–1952).[68] He was also disturbed about the conflicting, inhomogeneous experimental data and closed his paper with the wish that an exact analysis be undertaken of the dependence of electron velocity on the frequency of the acting radiation, extended over the largest frequency range possible and for a variety of substances. Only Millikan and Ladenburg heeded this requirement. Many other researchers (including Stark, for instance) got caught up in secondary effects and attempted to formulate vague hypotheses far removed from any solid observational findings.[69] In Millikan's laboratory at the *University of Chicago*, the doctoral student James Remus Wright (1883–1937) even asserted that his experiments showed: "with certainty [...] the maximum photoelectron energy does *not* vary approximately linearly with the frequency," while Owen W. Richardson and his student Karl T. Compton at *Princeton University* could but reach the conclusion in 1912 that all the foregoing experiments were "very contradictory." Even after Richardson moved to *King's College* in London, the data collection continued to be ambiguous. The experimental results in Berlin by Robert Wichard Pohl and his collaborator Peter Pringsheim were indecisive about whether the dependence was linear or quadratic because the frequency range they examined turned out to be too limited.[70] This experimental situation was resolved only in the mid-1920s, particularly by Charles Drummond Ellis (1895–1980), who broadened the range up to x-ray frequencies.[71] In the midst of this general puzzlement, how pleasant it is to read Einstein's clear argumentation and appreciate his acute sense of the scope or limitations of what an individual experiment might divulge.

If Einstein had relied on the available publications about the photoelectric effect and radiation pressure in 1905, then the linear correlation between the energy and frequency of light would probably have escaped him. For shortly after 1900, Lebedev as well as Nichols and Hull set out from classical electrodynamics, whereby

---

[67] Kunz (1909, 1911) and Cornelius (1913) p. 26 pleaded for ν^2, while Karl T. Compton and Richardson (1913) criticized the inconsistency among these results and voted for $E \sim \nu$. For the references, see Schweidler (1904, 1915), Hughes (1914a), Millikan (1914).

[68] See Ladenburg (1909). Ladenburg and Einstein knew each other well after they both emigrated to the U.S.A. in 1933, but Ladenburg may possibly have already met Einstein in Bern in 1909, see CPAE, vol. 5, p. 81.

[69] Stark's mental model of light quanta is covered below in Sect. 4.4.

[70] See Richardson and Compton (1912) p. 575, resp., Richardson (1914), Richardson and Rogers (1915) as well as Pohl and Pringsheim (1913); on Richardson's failed theory cf. furthermore Katzir (2006).

[71] See, e.g., Millikan (1913) pp. 129–131 and the review articles by Hughes (1914a), Marx (1916) p. 578 supplementing Hallwachs (1916) pp. 284–299, 335–353, 500–507, 530–535, Louis de Broglie (1921/23), and Ellis (1926); further Franklin (2013), Stuewer (2014) pp. 150–153 and the primary sources cited there.

the energy U of light is proportional to the intensity I, hence: $U \sim I \sim B^2 + E^2$. Lebedev wrote explicitly: "These pressure forces of light are directly proportional to the incident amount of energy and are independent of the color of the light." Nichols and Hull thought they could confirm this result two years later (1903), because their measurements of light pressure initially also suggested a frequency-independent energy proportional to the light's intensity. It was thanks to Einstein's extraordinary flair for assessing the tenability of experimental results that he did not set out on a false track in 1905. Rather than counting on a single sector, Einstein kept the most disparate fields of experimental and theoretical research in view and combined their findings in a novel way. Instead of inching further along a single strand, he wove it together with others, quickly creating an increasingly dense web, with which he was able to secure himself against any inevitable errors or dead ends threatening any individual one of these fields. A remarkable comment that Einstein made is of relevance here. It was in response to Walter Kaufmann's finding that electrons in motion increase in mass with velocity, apparently refuting his theory of relativity in favor of competing theories by Max Abraham and Alfred Bucherer. Einstein defended his standpoint with the argument: "However, the probability that their theories are correct is rather small, in my opinion, because their basic assumptions concerning the dimensions of the moving electron are not suggested by theoretical systems that encompass larger complexes of phenomena."[72] Einstein's considerations about the light quantum span the same, almost holistic range of entire experimental complexes, not merely individual findings or even individual fields of research.

Einstein was very interested in the photoelectric effect and its interpretation and kept abreast of the debates about the latest experimental data. In 1905 he decided— after much deep thought—to reconsider the experiment using a completely different kind of model.

> our prevailing [conventional electromagnetic] conceptions cannot explain why radiation of higher frequency should produce elementary processes of greater energy than radiation of lower frequency. In brief, we neither understand the specific effect of frequency nor the lack of a specific effect of intensity.[73]

Thus a double failure of the classical theories confronted him. First, they contrafactually presumed a proportionality of 1 : 1000 for the intensity, which was definitely not experimentally establishable, not even for variations in intensity. Second, they were not able to explain this "specific effect of frequency" of the radiation either.

---

[72]Einstein (1907b) p. 439 (CPAE, vol. 2, doc. 47, transl. ed., p. 284), Holton (1984) pp. 122f.

[73]Einstein (1911/12) p. 430 (CPAE, vol. 3, doc. 26, transl. ed., p. 421) in his contribution to the Solvay conference in Brussels, published 1912, also translated and cited in Debye and Sommerfeld (1913) p. 924.

The key that made Einstein turn away from classical explanatory attempts was that his solution was able to explain not just this one experiment but a number of others at the same time.[74] Einstein regarded the photoelectric effect as an expression of energetic conversion—the incident radiation passed energy on to the particles inside the cathode. If this energy was larger than a material-dependent constant, the work of emission W_A (*Austrittsarbeit*), then some of the particles could escape from the cathode and enter the highly evacuated tube. That is how Einstein arrived at his prediction that the photoelectric effect was dependent on the frequency: "If the formula derived is correct, then [the limiting potential Φ], presented as a function of the frequency [ν] of the exciting light in Cartesian coordinates, must be a straight line whose slope is independent of the nature of the substance investigated."[75]

Lenard was not looking for a frequency dependence of this kind at all, because his trigger model led him elsewhere. He tried to establish a diffuse dependence between the limiting potential and the sort of light involved.[76] Clear experimental verification of Einstein's prediction came ten years later by Robert Millikan (1868–1953) in 1915 (published 1916). His success was due to the special precautions he took to prevent the prepared metal surfaces from oxidizing as the measurements were still on-going. The results of many other researchers had been affected by it.[77] Millikan's tube was a veritable "machine shop in vacuo." It contained a mechanism that allowed the experimenter to scrape the surface of the alkaline metal under analysis afresh from inside the evacuated tube prior to any measurement taking. Suitable filters prevented disturbing photoelectric emissions by scattered light of shorter wavelength. In addition, the wave-length range was extended fourfold, within which a linear relation between the incident radiation's frequency and the limiting potential could be checked (cf. Fig. 3.6).[78] But remember, Millikan had set out on the assumption that Einstein's prediction would be refuted, as he recalled: "I spent ten years of my life testing

---

[74]The underlying pattern is a "consilience of inductions," which distinguishes Einstein's hypothesis from Lenard's by a larger and more natural explanatory scope. This was first philosophically analyzed by William Whewell (1794–1866). Cf., e.g., Thagard (2012) pp. 88ff.

[75]Einstein (1905) p. 146, Φ here replaces Π in the original (CPAE vol. 2, doc. 14, transl. ed., p. 101). For Hughes and DuBridge (1932) p. 7 and Wright (1937) p. 65, this equation $E_{kin} = h\nu - W_A$ even became the "most important single equation in the whole quantum theory."

[76]See Lenard (1902) pp. 166–168. Nonetheless, he did notice that this limiting potential was independent of the intensity of the incident radiation, which according to the classical theory would not have been expected.

[77]For details on Millikan's instrumental technique, see Hughes (1912) p. 37, Millikan (1916b) and (1924) as well as Franklin (2013) pp. 577–587, (2013) pp. 4–21.

[78]This is indicated particularly by Franklin (2013) pp. 574–577, (2016) pp. 9–11; Ladenburg (1909) came to $57\mu\mu$, Compton's collaborator Kadesch (1914) to 170 μm; Millikan's measurements of sodium permitted analyses between 240 and 680 μm; at the same time, the margin of error for the measured limiting potentials between 265 and 577 μm were only 0.1 Volt, hence three times better than the foregoing measurements.

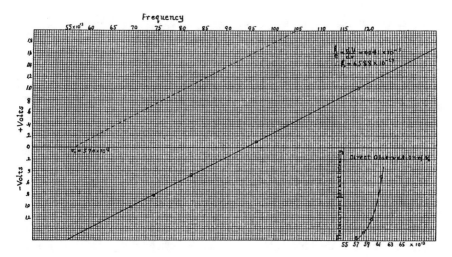

Fig. 3.6 Millikan's experimental results, 1916: The constant h, which according to Millikan's plot defines the slope of the interpolation line contained in all his measurements with a precision of $\pm$ 0.5%, came to $6.57 \cdot 10^{-27}$ erg s. Converted into modern units, it is $6.616 \cdot 10^{-34}$ Js, which is in very good agreement with the modern value for Planck's quantum of action $h = 6.62607 \cdot 10^{-34}$ Js. *Source* Millikan (1916b) p. 377. Reprinted by permission of the American Physical Society © 1916

that 1905 equation of Einstein's and, contrary to all my expectations, I was compelled in 1915 to assert its unambiguous verification in spite of its unreasonableness since it seemed to violate everything we knew about the interference of light."[79] Thus it is wrong to assume that experimenters always only confirm what they anticipate. In 1916 Millikan was forced to concede what he had originally supposed to be an untenable prediction by Einstein. His precise experiments confirmed that a rigorously rectilinear relation does in fact exist between energy E_{kin} and frequency ν in the formula $E_{kin} = h\nu - W_A$, with the work of emission W_A dependent on the material and Planck's constant h.

Anyone wondering whether it was advisable to apply new hypotheses to experiments considered this experimental result by an American experimenter, who was definitely not a friend of Einstein's, a clear indication that perhaps Einstein's heuristical considerations should to be taken more seriously after all.[80] For the Nobel prize committee, too, this experimentally direct and completely uncontestable verification of a clear prediction by Einstein was decisive in their conferral of the award for 1921 on him for his research on the theory of the light quantum from 1905— not for the theory of relativity, which was much less secure at the time or for other

---

[79]Robert Millikan (1949) p. 344, likewise also in his autobiography from 1950. On these unanticipated results, see Millikan (1916a) p. 18; cf. Holton (2000), Stuewer (1998), Franklin (2013), (2016) and Chap. 5 in the present volume on the reception.

[80]This cautious interpretation is expressed by Hughes (1914a) pp. 5f., 39–41 and Chap. III, for example, as well as by Comstock and Troland (1917) pp. 184–185 (these passages of their textbook were authored by Troland).

achievements of this multifaceted theoretician.[81] Obviously, this growing approval of Einstein's *formula* describing the photoelectric effect was not the same as approval of the underlying heuristical *model* of the concept of light quanta. Millikan complained as late as 1917: "Einstein's theory of localized light-quanta [...] is as yet woefully incomplete and hazy. Almost all we can say now is that we seem to be driven by newly discovered relations in the field of radiation to the hypothetical use of a fascinating conception which we cannot as yet reconcile at all with well-established wave-phenomena."[82] The justification for this conferral of the Nobel prize to Einstein in 1921 was very carefully phrased: "for his services to Theoretical Physics, and especially for his discovery of the law of the photoelectric effect"—Einstein's theory of light quanta was purposefully *not* mentioned.[83] When Einstein came to visit the *California Institute of Technology* in 1931, Millikan's dinner speech offered rather mixed praise: "The extraordinary penetration and boldness which Einstein showed in 1905 in accepting a new group of experimental facts and following them in what seemed to him to be their inevitable consequences, whether they were reasonable or not as gauged by the conceptions prevalent at the time, has never been more strikingly demonstrated."[84]

3.7   Strict Energy Quantization $E = h\nu$

Einstein's paper from 1905 developed the proportionality between energy and frequency already discussed above further on the basis of entirely differently situated considerations about statistical theory. The outcome was an exact equation $[E = h\nu - W_A]$ with a proportionality constant. This constant h, Planck's quantum of action, he was able to calculate theoretically in a virtuosic comparison between statistical and thermodynamic considerations.[85]

---

[81] See Pais (1982) Chap. 30 and Elzinga (2006). The awarding policy of the Nobel committee is studied by Friedman (2001).

[82] Millikan (1917) p. 260.

[83] See http://www.nobelprize.org/nobel_prizes/physics/laureates/1921/ and additionally Franklin (2013) pp. 588ff., (2016) pp. 16–19.

[84] Millikan (1931) p. 378.

[85] These subtle and highly sophisticated theoretical sections of Einstein's light-quantum paper of 1905 are discussed in Dorling (1971), John Stachel's editorial headnote "Einstein's early work on the quantum hypothesis," in CPAE, vol. 2, pp. 134–148 and Hentschel (2005).

> **Box 4: Max Planck and quantum theory.**
>
> Planck's considerations around 1900 presupposed an ideal 'black body,' that is, an object that absorbs all radiation and is at equilibrium with the surrounding radiation field. His derivation of the formula for the energy density u as a function of the frequency ν, which agreed empirically extremely well with precision measurements performed at the PTR in Berlin at the time, was intrinsically two-tiered: For the energy density u of the field as a function of frequency ν and the mean energy of a single resonator U, he used the following formula from classical electrodynamics: $u = 8\pi(\nu^2/c^3) \cdot U$. This means the oscillatory modes of the radiation field were presumed to be continuous. In the combinatorial computation of the number of "complexions" K, i.e., the number of microstates that correspond to a given macrostate of fixed energy and temperature, he first followed Boltzmann's method from 1877 by setting $K = (N + P - 1)!/N!P!$, where N is the number of resonators and P is the number of portions of energy, that is, very large integers, and $N!$ is defined as $N \cdot (N - 1) \cdot (N - 2) \cdot \ldots \cdot 3 \cdot 2 \cdot 1$. In this step, the input and output of energy by the N resonators (i.e., for instance, vibrating atoms or molecules participating in the transfer of heat from the black body to the surrounding radiation field) was therefore presumed to be discontinuous. In order to be able to compute these complexions at all, by combinatorics, the energy portions P for distribution onto the N resonators necessarily had to be finite. According to Planck this was "just a formal assumption," even though (different from the Boltzmann method) the boundary transition $h \to 0$ for the later Planck quantum of action was impossible.

For Max Planck (1858–1947), that "renitent revolutionary," energy quantization was just an emergency solution to stop the energy from being redistributed in the radiation field in infinitely small packets of energy. He himself chose for a very long time to interpret this as a secondary effect resulting out of a mysterious property of the surrounding matter to permit only defined, that is quantized oscillation modes. In other words: the walls of the black body, which Planck in emulation of Kirchhoff idealized as 'resonators' at thermal equilibrium with their enclosing radiation field, were (for as yet unknown reasons) only able to take up or release energy in finite packets.[86] Planck's notion of how this conservation of energy happens has been variously illustrated: Butter can only be bought in half-pound packets even though it could be produced and distributed in any other amount. For purely practical reasons we eat soup in small spoonfuls even though soup does not come piece-wise or have a quantized structure.[87] In both cases something appears to us (and Planck and other defenders of this semiclassical worldview) as discrete; but the reasons lie not in the structure of the object itself but in the interaction between the object and

---

[86]Cf., e.g., Darrigol (1988, 2000), Gearhart (2002), Badino (2015) pp. 58f., 99f.

[87]See Gamow (1966), cited in Weinberg (1977) p. 20 as well as here p. 23 and Paul (1985) p. 57.

the surrounding matter. This half-hearted Planckian notion of energy quantization was a large step away from true quantization of the radiation field. The insight about the reality of 'light quanta' (by Einstein (1905) but still under the cautious heading "a heuristical point of view") or of "*Lichtatome*" (by Wolfke (1913, p. 1123), of "light corpuscles" or "photons" (by Gilbert Lewis (1926a) and Band (1927) or Lewis (1926b)), developed only stepwise in a gradual process along a path that was anything but straight. In a textbook on *The Nature of Matter and Electricity*, Daniel F. Comstock, for instance, faults Max Planck in 1917 (instead of the person actually responsible: Albert Einstein!) for not providing—whether or not willfully— an appropriate picture of the emission process of such 'bullets,' as balls propelled through a gun barrel, for instance: "In Planck's theory, however, it is more as if the electron emitted bullets of radiation. Just how this occurs he does not attempt to say."[88] His coauthor L.T. Troland referred in part II of this textbook to 'light atoms' as "quantities of radiation [as] integral multiples of the units in question [...], radiated from bodies [...] in sudden outbursts."[89]

3.8 Wave-Particle Duality: First Hints and Later Deepening

After having pointed out the ionization of gas by UV-radiation (in Einstein (1905) with Stark (1908) for the empirical confirmation), and a simple explanation for the photochemical law of equivalence (in Einstein (1906a)), and a drop in specific heat at low temperatures (in Einstein (1907a)), Einstein found another experimentally accessible area in which quantum phenomena are manifest—namely, fluctuations Einstein (1909a): in (a) their spatial distribution and (b) radiation pressure. Although these fluctuations are normally far too small to be directly detectible, systems exist that augment such fluctuations by a building-up process of statistically significant freak values. The fluctuations then become macroscopically visible. In a lecture delivered at a conference in Salzburg in 1909, Einstein presented a perfect example of a system of this kind in the form of a thought experiment analyzing a mirror in a vacuum.[90]

---

[88]Comstock and Troland (1917) part I, §10, p. 47.

[89]Comstock and Troland (1917) part II, §54, p. 183. For Troland's first usage of 'photon,' see here p. 28 and Kragh (2014b, c).

[90]See also the discussions at this meeting by the Gesellschaft Deutscher Naturforscher und Ärzte about Einstein (1909a) in *Physikalische Zeitschrift* no. 10 (1909), pp. 224f., 323f., 825f.; cf. Klein (1964), Kojevnikov (2002), Irons (2004) and Duncan (2012) pp. 14–17. On box 5, see Einstein (1906a), cf. Hentschel and Grasshoff (2005) p. 25, Rynasiewicz and Renn (2006) and here Fig. 2.1 on these interrelations in Einstein's oeuvre from around 1905.

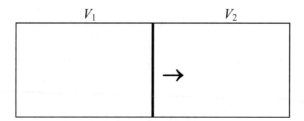

Fig. 3.7 Einstein's thought experiment of 1909 on fluctuations in a submicroscopic radiation field. When energy fluctuations occur in both partial volumes V_1, V_2, innumerable collisions of various intensities will hit the movable mirror set between the two volumes, causing it to execute trembling motions

---

Box 5: Fluctuations in energy and momentum

In the following thought experiment Einstein concluded macroscopically observable fluctuations out of statistical fluctuations of the submicroscopic radiation field. He considered a freely suspended mirror with a surface f positioned between two radiation fields V_1 and V_2. When energy fluctuations occur in the two partial volumes V_1, V_2, then the mirror at the dividing plane is bombarded from both sides with innumerable collisions of various intensities, causing it to tremble. This trembling motion of the mirror resembles Brownian molecular motion, which he had explained in his third world-famous paper from the end of 1905 (Einstein 1906a).

This consideration demonstrated again that the radiation field transmits not only the energy of electromagnetic waves onto the mirror but also momentum, which strengthened the notion of radiation pressure (semantic layer 4). One pupil of Arnold Sommerfeld, Peter Debye (1884–1966), added a supplement in 1911 with a simple thought experiment about the recoil of a hollow sphere that emits radiation only in one direction from a small opening. He also concluded: "thus we in fact cannot regard it otherwise than that the field itself be the carrier of its own momentum."[91]

Fluctuations in the radiation pressure in the two halves of the volume in Fig. 3.7 (see also box 5) cause the mirror to become displaced from its position in the boundary plane between the two partial volumes. The expectation value for the mirror's velocity at any time is equal to zero, though, because the collisions from the left and from the right are equally probable owing to the precondition of isotropic space. Nevertheless, the expectation values for the square of its velocity v^2, and therefore also the momentum $\frac{m}{2}v^2$ are not vanishing, hence are not equal to zero.

For the changes in radiation pressure Δ on the mirror surface f during the time interval τ from random fluctuations in the radiation field at energy density ρ, we have:

---

[91] See Debye (1911) p. 157.

$$\Delta^2 = \frac{f\tau \cdot d\nu}{c}[h \cdot \nu \cdot \rho + \frac{c^3\rho^2}{8\pi\nu^2}].$$

Analogously, because of independent motion or the interference "of narrowly extended complexes of energy $h \cdot \nu$" in volume V, the energy fluctuation ΔE is:

$$(\Delta E)^2 = V \cdot d\nu[h \cdot \nu \cdot \rho + \frac{c^2\rho^2}{8\pi\nu^2}].$$

Both expressions in square brackets on the right-hand side have the same form. It is a sum of two terms, the first of which could be traced back to a collection of corpuscular light quanta of energy $h\nu$, whereas the second term could be derived under the precondition of interference between waves of frequency ν. This led Einstein to a peculiar duality between wave-like and particle-like aspects of the radiation field. The second of these terms within square brackets on the right-hand side goes back to Maxwell's theory of continuous electromagnetic radiation. The other term is only comprehensible if light quanta are interpreted as mutually independent particles.[92] The correct, complete result only comes from the sum of these two terms. What did Einstein conclude out of this in 1909?

> In addition to the nonuniformities in the spatial distribution of the momentum of the radiation which arise from the wave theory, there also exist other nonuniformities in the spatial distribution of the momentum, which at low energy density of the radiation have a far greater influence than the first-mentioned nonuniformities. [...] the two structural properties (the undulatory structure and the quantum structure) simultaneously displayed by radiation according to the Planck formula should not be considered as mutually incompatible.[93]

This was the most far-sighted anticipation of what later became known as the wave-particle duality. In retrospect, it would seem quite incomprehensible to modern readers of these words by Einstein from 1909 why another decade had to elapse before physicists gained clearer insight into the dual nature of light combining aspects of waves with aspects of particles. But this thought, that the wave model and the particle model of light are not starkly irreconcilable opposites but rather two sides of the same coin—complementary aspects of reality, or Yin and Yang of a closed circle—was too radical to be recognized upon its first appearance. In the early 1920s physicists were rather annoyed and irritated by the irresolvable contradictions between the two approaches, as this sarcastic statement by William Henry Bragg shows:

> On Mondays, Wednesdays and Fridays, we use the wave theory; on Tuesdays, Thursdays and Saturdays we think in streams of flying quanta or corpuscles. That is after all a very proper attitude to take. We cannot state the whole truth since we have only partial statements, each covering a portion of the field. When we want to work in any one portion of the field or other, we must take out the right map. Some day we shall piece all the maps together.[94]

---

[92] See again Einstein (1909a b) and Klein (1964), Irons (2004).

[93] Einstein (1909ab) pp. 499–500 (CPAE, vol. 2, doc. 60, transl. ed., pp. 393–394).

[94] W.H. Bragg (1921/22) p. 11.

The next major step toward such an integrative clarification of wave-particle duality was taken by Louis de Broglie (1892–1987).[95] The de Broglie family counted
among the French nobility, yet his elder brother, the Duke Maurice de Broglie
(1875–1960), was an experimental physicist. He passed on his enthusiasm about
science to his younger brother at an early age who then also started to study mathematics and physics in Paris in 1911. The first World War interrupted these studies
and afterward Louis worked for a while in his brother's private laboratory, where he
was especially occupied with testing Einstein's equation $E = h\nu$ by experiment. He
wrote a lengthy review article about this for the Solvay conference in April 1921.[96]
In 1924 Louis de Broglie completed his studies with a dissertation on *Recherches
sur la Théorie des Quanta*. The point of departure of his reflections was Einstein's
postulate from 1906 about an equivalence between mass m and energy E. From Einstein's formulas $E = mc^2$ and $E = h\nu$ de Broglie drew the consistent conclusion
that to any mass m there must also be a corresponding frequency $\nu = mc^2/h$. Thus,
a frequency and also wavelength $\lambda = h/p$ must be assigned to each particle as well,
where $p = mv$ is the momentum of a particle of mass m and velocity v, and the
frequency ν and velocity of light c must be attached to the wavelength λ as $\lambda = c/\nu$.

In 1922, it was not yet clear to de Broglie that the rest mass of a light quantum
is exactly zero, unlike other elementary particles such as the electron. In a paper for
Journal de Physique et Le Radium, he wagered that these "atoms of light (presumed
to be of the same very small mass) seem to move at speeds varying with their energy
(frequency), but all at extremely close to c."[97] This is an astonishing parallel with
Isaac Newton's projectile theory of light 250 years before. As we have already seen
in Sect. 3.2, Newton had drawn a very similar dependence between the velocity of
light and the presumed mass of his light globuli, but rejected it again when he saw
that there was no empirical evidence of any existing speed discrepancies in light
from different regions of the optical spectrum.[98]

In 1927 Clinton J. Davisson (1881–1958) and Lester H. Germer (1896–1971)
succeeded in verifying de Broglie's bold predictions.[99] Their experiment made the
matter waves associated with electrons interfere with each other—a clear characteristic of wave-like entities!

Einstein was delighted with Louis de Broglie's publications long before this positive experimental test could be attached to it. He enthusiastically wrote to de Broglie's
doctoral advisor, the French physicist Paul Langevin (1872–1946), at the end of 1924:
"He has lifted one corner of the great veil."[100] Other laudatory words can be found in

---

[95]De Broglie's important research (1924/25) and his mental path to wave-particle duality are discussed by Kubli (1971), Milonni (1984) and Darrigol (1986, 1993).

[96]See Louis de Broglie (1921/23).

[97]Broglie (1922) p. 438 and (1923) p. 508, where he estimated the mass of a light quantum to
be $< 10^{-50}$g. On this episode and on vain efforts to determine the dimensions of light quanta
cf. Stuewer (1975a) pp. 306–312.

[98]See here the sources quoted in notes 18f. on p. 45 of Chap. 3.

[99]See Davisson and Germer (1927), Davisson (1937), Russo (1981), Darrigol (1986).

[100]A. Einstein to P. Langevin, 16 Dec. 1924, reprinted in CPAE, vol. 14 (2015) transl. ed., doc. 398,
p. 393.

Einstein's publications of the period.[101] This achievement was appreciated by others as well. De Broglie received awards in 1926 and 1927 from the *Institut de France*, and two years later the highly regarded *Medaille Henri Poincaré* conferred by the Parisian *Académie des Sciences* as well as the Nobel prize in physics.[102]

In a speech to an association of mathematicians and physicists at the University of Berlin on 23 February 1927, Albert Einstein described the complicated, if not deadlocked situation that theoretical and experimental researchers found themselves in by then. The constant vacillation between undulatory and corpuscular properties of light, and now also of matter, hopelessly overtaxed the "intellectual powers of physicists"—including his own:

> The problem that we presently have, which is of a principal nature in the area of luminous phenomena, comes down to showing either that the corpuscular theory grasps the true essence of light, or that the undulatory theory is right and the quantum-like aspects are merely apparent, or, finally, that both interpretations correspond to the true nature of light and that light has characteristics that are both quantum-like and undulatory.[103]

Interference phenomena of light led to the conclusion that the transmission process had larger spatial and temporal extension; however, needle radiation, the photoelectric effect and Compton scattering pointed to "something abrupt, projectile-like." But how could these two diametrically opposed models possibly be combined conceptually? "What Nature is demanding of us is not quantum theory or wave theory; Nature is rather demanding of us a synthesis of both these interpretations, albeit exceeding the intellectual powers of physicists thus far."[104]

The very year in which these rather resigned lines were written, Werner Heisenberg and his mentor Niels Bohr entered the arena in the public competition over the predominant interpretation of the new quantum mechanics, putting forward hypotheses that shed new light on the wave-particle duality. Heisenberg published his uncertainty relation explaining why it is only possible for us to conduct precise measurements on one of the two measurable quantities in pairs of variables, such as location and momentum or energy and time, but not both simultaneously. Bohr presented his complementarity postulate that this uncertainty relation is merely an expression of a much deeper irreconcilability between two approaches in nature that might be referred to as analytic spatial description versus causal energetic explanation.[105]

Later, in 1941, when the debate over these two central recipes in interpreting quantum mechanics had abated somewhat, Pascual Jordan (1902–1980) answered

---

[101] Such as Einstein (1924/25b, 1927).

[102] De Broglie's later career and his successful search for pilot wave theories are discussed in Bohm and Hiley (1982).

[103] Einstein (1927) p. 546. This is a report about Einstein's talk before the *Mathematisch-Physikalische Arbeitsgemeinschaft an der Universität Berlin*, with passages occasionally quoted almost verbatim.

[104] Einstein (1927) p. 546. Cf. Sect. 4.2, on p. 96 about Einstein's mental model of the light quantum.

[105] This contest over authority in what later came to be known as the Copenhagen interpretation of quantum mechanics is treated by Bohr (1933), Heisenberg (1927, 1930), Halpern and Thirring (1928/29), Weizsäcker (1931, 1941), Born (1969), Jammer (1974), Wheeler and Zurek (1983), Beller (1999), Howard (2004), Zeh (2012), Friebe et al. (2015) Chap. 2 with further citations there.

the question of how light is constituted very differently. In a popular essay titled "Physics and the Secret of Organic Life" he wrote:

> What *is* light? A wave process or a particle beam? But gradually we arrived at the insight that the word 'is' just simply doesn't fit here—that one must rather ask what radiation can *become*, depending on the kind of observational analysis of this object. Yet it can *become* *both*. [...] The truth is that light is neither a wave nor a corpuscle, but a 'third' alternative that escapes mental reconstruction as a graphic notion and is only describable in abstract mathematical terms.[106]

The former either/or dilemma between wave *versus* particle, is replaced by a new, not-yet-fixed category of an object-in-the-making that depending on the experimental arrangement is capable of becoming a wave just as easily as it is capable of becoming a particle. In later essays, Werner Heisenberg attempted to provide this with a deeper foundation by reanimating the old Aristotelian opposition between actuality and potentiality.[107] This wave-particle duality became an official part of the standard interpretation of quantum mechanics latest since the mid-1920s, acknowledged by the leading physicists in Copenhagen, Göttingen and Munich. We have thus learned to live with this paradox but few have really taken to this weakening of the strict dichotomy. The categories of particle versus wave are too deeply entrenched in our intuition as contrasting classes which we use to distinguish between them categorically and verbally. Einstein himself always considered this Copenhagen interpretation of quantum mechanics and the wave-particle duality as unsatisfactory. "This interpretation, which is looked upon as essentially final by almost all contemporary physicists, appears to me as only a temporary way out."[108]

3.9 Spontaneous and Induced Emission: 1916–17

In 1916 Einstein strode forward again in the direction of "quantum-theoretical contemplations on the interaction between matter and radiation." Following the style of the Berlin theoreticians Kirchhoff and Planck, Einstein wanted to avoid incorporating concrete models of matter and keep his reflections as general as possible, "without specialized suppositions about the interaction between radiation and those structures." In a first step Einstein modeled the material in the walls of the black body in the Planckian manner: as 'monochromatic resonators' that can absorb and release energy from the electromagnetic field. In a second step he modeled it quantum-theoretically as 'molecules' moving statistically back and forth between a finite number of quantized states.[109]

---

[106] Jordan (1941) pp. 29 and 38.

[107] See, e.g., Heisenberg (1959) Chap. X.

[108] Einstein (1949) p. 51. There are similar statements in Einstein and Infeld (1938a) p. 278.

[109] See Einstein (1916a) received on 17 July 1916 (CPAE, vol. 6, doc. 34, esp. transl. ed., pp. 212 and 213).

Let us first briefly consider the first electrodynamic modeling. Energy E of a structure oscillating at exactly one frequency can change during the time τ, which is significantly longer than the oscillation period of those resonators, in two ways: On one hand, this energy can decrease by emitting energy in the form of radiation. For this spontaneous emission of radiation Einstein set, in conscious analogy to Rutherford's formula for radioactive decay:[110]

$$\Delta_1 E = -AE\tau.$$

If only this spontaneous emission were possible for the resonators, then their energy would decrease exponentially as a function of time, because their energy loss would always be exactly proportional to the remaining residual energy. The counteracting process of energy absorption from the surrounding field also exists though. "This second change increases with the radiation density [ρ] and has a 'chance'-dependent value and a 'chance'-dependent sign."[111] Hence, it is at times positive and at times negative. By combining electromagnetic and statistical considerations, one obtains as the mean value:

$$\Delta_2 E = B\rho\tau,$$

with ρ denoting the radiation density. The mean energy $\bar{E}$ of the Planck resonators at equilibrium must be independent of the duration τ; therefore after the period τ the mean value must be equal to the original mean value: $\bar{E} + \Delta_1 E + \Delta_2 E = \bar{E}$. From this, entirely along the lines of Planck, it also follows that

$$\bar{E} = \frac{B}{A} \cdot \rho.$$

In a second, quantum-theoretical derivation of Planck's energy density for cavity radiation, Einstein introduced coefficients for the transition probabilities of the atoms of gas which must be at thermal equilibrium with the surrounding radiation. They can make the transition (cf. Fig. 3.8) from the state Z_n at the energy ϵ_n into the state Z_m at the energy ϵ_m by transmitting radiation at the frequency $\nu_{nm} = (\epsilon_m - \epsilon_n)/h$, whereby depending on this term's sign, either emission takes place or absorption. Given the case of equilibrium, statistically considered, just as many molecules must perform the transition from the Z_n state into the Z_m state as vice versa. Einstein set the case of spontaneous emission of energy at the frequency ν_{nm} as proportional to the number N_m of molecules in the initial state Z_m and described the statistical (non-stochastic) probability for this decay by the coefficient A_m^n, which yields the number of transitions $m \to n$ per unit time for $A_m^n N_m$. According to Einstein, the probability of absorption at the frequency ν_{nm} and hence the transition stimulated

---

[110]Ernest Rutherford had postulated in 1900 that the rate of radioactive decay $\mathrm{d}N/\mathrm{d}t = -\lambda N$ was strictly proportional to the number of not yet decayed radioactive atoms, which leads to an exponential decrease.

[111]Einstein (1916a) p. 319 (CPAE, vol. 6, doc. 34, transl. ed., p. 213). See below for the interpretation of this term 'chance' in this context.

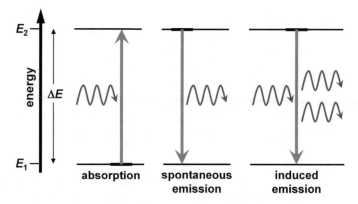

Fig. 3.8 Term diagrams of absorption and spontaneous and induced emission. The latter increases proportionally with the density of the radiation field, which is why once set off, such emission processes continue to augment. Figure modified from http://www.seos-project.eu/modules/laser-rs/images/two-level-system-en.png (22 Jan. 2017). Reprinted with kind permission of Dr. Rainer Reuter (Oldenburg)

by it from state n into the higher-energy state m, also implies the possibility of a transition from a higher-energy state into one with lower energy. There likewise, results $B_m^n N_m \rho$. Different from spontaneous emission, this latter case corresponds to "induced emission," stimulated by the surrounding radiation field.[112] For all the transitions taken together, Einstein obtained the total at thermal equilibrium:

$$A_m^n N_m + B_m^n N_m \rho = B_n^m N_n \rho.$$

In order to be able to reproduce Planck's formula for the mean energy density of the radiation field by this proposition in conjunction with Boltzmann's formula for the probability of state Z_n as a function of its energy ϵ_n, Einstein was compelled to introduce two other coefficients for the induced emission besides the coefficient A_m^n for spontaneous emission. Unlike this latter these two coefficients are proportional to the density of the radiation field. But in the thinking of the day this was counter-intuitive. Assuming Einstein's formula, though, numerous cross-links could be tied to existing elements of knowledge:

- Continuity with Planck's considerations in the context of classical mechanics and electrodynamics seemed secured.
- The proposition for spontaneous emission was rigorously analogous to the law of radioactive decay already well established by Rutherford's experiments, $dN/dt = -\lambda N$.
- The photochemical law of equivalence also followed easily from it.

---

[112]This term was coined by Vleck and Hasbrouck, in 1924. Einstein and Ehrenfest (1923) referred to 'negative irradiation', Fabrikant in 1939 to 'negative absorption': see Lukishova (2010).

- Bohr's frequency condition $\epsilon_m - \epsilon_n = h\nu$ for the transitions between the energy levels ϵ_m and ϵ_n was also automatically satisfied, which guaranteed a link between radiation theory and Bohr's successful atomic model.

This tonal harmony between many independent strains of research (another instance of 'consilience of inductions' here—cf. p. 20) motivated Einstein in 1916 to conclude: "the simplicity of the hypotheses, the generality with which the analysis can be carried out so effortlessly, and the natural connection to Planck's linear oscillator [...] seem to make it highly probable that these are basic traits of a future theoretical representation."[113]

Einstein was to be proven right—perhaps more than he later would have liked, though. His modeling of quantum theoretical connections based on transition probabilities between a finite number of quantized initial and final states was used by the young Heisenberg as orientation in conceiving matrix coefficients for his matrix mechanics.[114] Paul A. M. Dirac (1902–1984) then showed in 1927 that spontaneous and induced emission have a place in quantum field theories.[115] Heisenberg's uncertainty relation was also translated into relativistic quantum mechanics in 1931 and later into quantum electrodynamics as well.[116] Einstein's steadfast opposition to Heisenberg's matrix mechanics, Dirac's operator algebra and other variants of later quantum mechanics with its associated stochastic interpretations is well known.[117] But aren't there a number of allusions to "chance" in Einstein's own paper from 1916? It is remarkable that Einstein enclosed the two instances of this word there in scare quotes (see p. 72 above). He wanted to imply that he did not think that completely undetermined chance events were involved, in other words, stochastic processes like what radioactive decay appeared to be, which he took as his mathematical model. They were rather merely statistical effects in the sense employed by Boltzmann and other nineteenth-century physicists in classical statistical mechanics, where the governing conditions are ultimately assumed to be unknown to the observer. Nonetheless, Einstein was sure that he was on the right track because it "let the establishment of a quantum-like theory of radiation appear as almost unavoidable," whereas its greatest remaining weakness was "that it does not bring us closer to a link-up with the undulation theory."[118] Einstein's paper "On the Quantum Theory of Radiation," republished in *Physikalische Zeitschrift*, came in second among the most-cited physical papers between 1920 and 1929.[119] Laser development led to another overwhelm-

---

[113] Einstein (1916a) p. 322 (CPAE, vol. 6, doc. 34, transl. ed., pp. 215–216).

[114] See Heisenberg (1925) as well as Mehra and Rechenberg (1982) vol. 3, Darrigol (1992) part B.

[115] See Dirac (1927a, b, 1930) Chap. X, pp. 232–239 and, e.g., Heitler (1936b), Duncan (2012) p. 35.

[116] Siehe Landau and Peierls (1931), Bohr and Rosenfeld (1933) resp. Widom and Clark (1982).

[117] See Pais (1982) Chap. 25, Wheeler and Zurek (1983), Home and Whitaker (2007) pp. 28ff., 83ff.

[118] Einstein (1917) pp. 127f. This article is essentially a reprinting of his contribution to the Physical Society in Zurich, Einstein (1916a) esp. p. 61: CPAE, vol. 6, doc. 38, transl. ed., p. 232.

[119] See Small (1986) pp. 144–145: between 1920 and 1929, Einstein (1917) was cited altogether 76 times in the 20 leading international physics journals (only Compton (1923a, b, c) exceeded this with a total of 78 citations). The earlier study by Small (1981) of *Physics Citation Index 1920–29* (which

ing renaissance for Einstein's papers from 1916 and 1917 on the theory of induced emission. My electronic search in *google scholar* for Einstein's papers (1916a and 1916b) on 18 March 2016 produced 364 resp. 118 documentable current citations plus a total of 1,815 for Einstein (1917).[120]

Modern laser technology is based entirely on this 'induced' or stimulated emission of additional light quanta (photons) by the radiation available inside the laser resonator (cf. Fig. 3.9).[121] The mechanism for million-fold amplification of this effect is reflection. A mirror is positioned at each end of the resonance area to reflect light quanta back into the area and stimulate more emissions in the laser's active medium each time they traverse it. A cascade-like amplification of this coherent radiation results that is then either sent through a minimally semipermeable mirror or by means of 'Q switching' are released pulse-wise out of the resonance area. They exit as a high-energy, coherently oscillating and virtually monochromatic laser beam. The original idea for such resonance amplification of light in standing waves came from a Russian by the name of Valentin Aleksandrovich Fabrikant (1907–1991), who wrote his dissertation on the topic in 1939 in Moscow and applied for a patent (no. 123209) there on 18 June 1951.[122] The laser became technically feasible in 1960 within the context of the intensified research on radar and microwaves during World War II.[123]

Einstein explicitly pointed out in another paper that appeared in the same year as the original paper, that in addition to the energy $h\nu$, momentum $h\nu/c$ is also assignable to light quanta and that atomic emission and absorption of light quanta are directed processes. Semantic layers 1 through 8 had become established in Einstein's thinking by 1916 even though at this time he was still almost the only one to take the existence of light quanta really seriously. Einstein was fully aware of the consequences of this statistical approach, which ultimately led to stochastic quantum mechanics which he would not support. "God doesn't play dice" remained Einstein's famous maxim to the end of his days.[124] Nevertheless, Einstein was unable to find a theoretical alternative that would permit one to predict the time and direction of elementary atomic processes. Neither was anyone else able to do so. It is part of

---

closes the gap left by *Web of Science* with its research limited to citations as of 1945)—conducted around 1980—compiled citations only as of 1955.

[120] https://scholar.google.de/scholar?cites=6861658168745659959&as_sdt=2005&sciodt=0,5& hl=de and https://scholar.google.de/scholar?cites=12548366261550285506&as_sdt=2005& sciodt=0,5&hl=de and https://scholar.google.de/scholar?cites=8985365509834917851& as_sdt=2005&sciodt=0,5&hl=de.

[121] Gordon Gould (1920–2005) introduced this acronym 'laser' for 'light amplification through stimulated emission of light' at the end of the 1950s. The history of the laser starts with experiments on the development of masers in 1954, which function similarly but with microwaves. Cf., e.g., Bromberg (1991), Lemmerich (1987), Bertolotti (1999), Hecht (2005).

[122] See, e.g., Lukishova (2010) and further Russian primary sources cited there in Engl. transl.

[123] See especially Bromberg (1991, 2006).

[124] From among the copious literature about Einstein's critique of the Copenhagen interpretation, I recommend Einstein's correspondence with Max Born (1969) pp. 118ff. (esp. the letters 1924–26), as well as Wheeler and Zurek (1983) with reprints of all the important papers about the Bohr–Einstein debate.

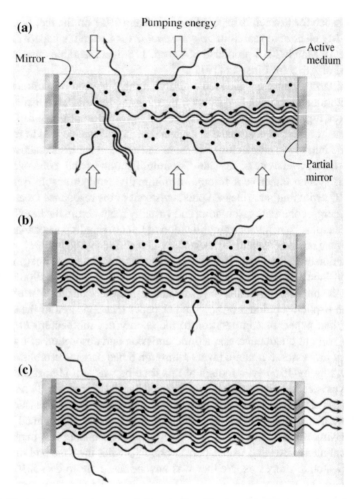

Fig. 3.9 Illustration of how lasers work. Rays volleyed back and forth between two mirrors cause an amplifying cascade of induced emissions in the laser's gain medium, which form a coherent beam of light upon exiting out of the resonance area (through a partial mirror on the right), resulting in the emission of coherent and approximately monochromatic radiation. This diagram from http://abyss.uoregon.edu/~js/images/laser_pump.gif (17 Mar. 2016) utilizes the depiction of light as a wave and the analogy to standing waves

the personal tragedy of this thinker to have made such a crucial contribution to a development that he felt so compelled to reject.

3.10 Light Quanta Carry Internal Angular Momentum (Spin)

Spin is the tenth layer in the complex semantic stratification of the light-quantum concept. The history of this notion is specially convoluted: From the modern point of view, it is actually genuinely quantum mechanical, even though it took shape within the context of the old, semiclassical quantum theory.[125]

Spin avant la lettre 1921–24 The first person to presume the existence of an intrinsic but quantized rotational motion of negatively charged electrons was Arthur Holly Compton in 1921 at *Washington University* in St. Louis. His motivation was to find an explanation for the magnetic moment of electrons and their spiral-shaped trajectories on his cloud-chamber exposures.[126] An experimental physicist in his laboratory, Frank W. Bubb (1892–1961), sought to explain the scattering of electrons by polarized x-rays. His experiments defined the exit angle of the electrons as a function of the incident x-rays and their direction of polarization. His result was exactly what the classical theory suggested: The plane of oscillation of the electromagnetic waves must always be the one on which the charged particles of matter are set into oscillatory motion by the radiation. Bubb's finding was incompatible with the naive interpretation of x-rays as "bullets of energy," which propel particles out of the material in their own direction of motion. For this reason he interpreted his experiments as counter-indicative of the quantum theory. "The results are in accord with the classical theory. To explain them on the quantum theory we must assume that the quantum is a vector bundle of energy, for it explodes, so to speak, at right angles to its direction of motion."[127] Seen from a later perspective, these are obviously the first indications of what was subsequently to be called spin: The statistically preferred direction of emission of electrons was exactly perpendicular to the polarization plane.

Spin as the fourth quantum number of the Bohr-Sommerfeld atomic model While these scattering experiments were being performed, the old quantum theory was already in a state of serious crisis. The Danish theoretician Niels Bohr (1885–1962) in Copenhagen and the mathematical physicist in Munich, Arnold Sommerfeld (1868–1951), had already developed their model of the atom dating back to 1913.[128] This atomic model was based on the assumption that negatively charged electrons revolve around a positively charged nucleus along definite orbits, resembling the planetary orbits around the Sun except that the stabilizing centripetal force countering the centrifugal force is electric not gravitational. In such a case of electrons in rotation, hence in accelerated motion, classical electrodynamics predicted a release

---

[125]On the history of spin, see Goudsmit (1965, 1971), Dirac (1974/75), Jammer (1966), Tomonaga (1974/97), Milner (2013) and Hentschel (2009a, b, c, d) with primary citations.

[126]See Compton (1921) p. 155: "the electron itself, spinning like a tiny gyroscope, is probably the ultimate magnetic particle"; see also p. 126 below about his research on Compton scattering.

[127]Bubb (1924) p. 127; Simonsohn (1981) p. 262 provides a polar diagram of the intensity and direction of the electric field during photoionization.

[128]On the history of the Bohr-Sommerfeld model of the atom, see Jammer (1966), Nisio (1973), Kragh (2012), Eckert (2014), as well as p. 31 above, with citations of the primary sources.

of energy, which would have made the electron orbits unstable. The then still youthful Bohr simply made the bold postulate that this prediction of classical electrodynamics no longer applied to the interior of the atom. He was able to deduce from spectroscopic regularities, such as Balmer's formula for the spectral series of hydrogen, that energy levels are not continuous but rigorously quantized. To first approximation these quantized energy levels came to $E_n = E_0/n^2$, where n is the so-called principal quantum number, an integer greater than or equal to 1. Spectrum lines of atoms were interpreted as the result of transitions by electrons between these energy levels. Emission occurs if the initial energy is higher than the final energy of these quantum jumps by electrons, and absorption occurs in the opposite case.[129]

The Bohr-Sommerfeld atomic model also yielded elliptic orbits in addition to circular ones. Electrons moving along such elliptic orbits travel at higher velocities when they are closer to the nucleus. Unlike in the classical theory of gravitation, relativistic corrections to the mass, which is dependent on velocity, are not negligible. So an asymuthal quantum number l had to be introduced to describe the degree of eccentricity of the electron's orbit and it yielded relativistically calculable adjustment terms for the energy levels. These were also detectable in spectra, which had already been very precisely measured by then. Another accomplishment came when Bohr and Sommerfeld were able to show in 1915–16 that the splitting of these spectrum lines in an electric or magnetic field, known as the Zeeman and Stark effects, could be incorporated into their model if one assumed that these circular or elliptic orbits were tilted at specific angles relative to those fields. Thus in addition to energetic quantization there was also spatial quantization, because certainly not every tilt was physically 'permissible' in their model, only those described by a third 'magnetic' quantum number m. Like n and l, this number also could only take on integer values, albeit all integers (including zero) between $-l$ and $+l$, yielding Zeeman multiplets of $(2l + 1)$ spectrum lines. This agreed very well with the finding of the 'normal' Zeeman effect, according to which most spectrum lines split up into three, five, seven or nine components. That was why these three quantum numbers and a few rules about permissible and forbidden transitions between those energy levels, were able to explain the majority of spectroscopic findings of atomic physics between 1916 and 1921, sometimes in surprisingly good agreement between experiment and theory.[130]

However, at the end of the 1920s Miguel A. Catalán (1894–1957) and other experimenters began to find more and more instances of the so-called anomalous Zeeman effect. The spectral splitting produced even numbers of components instead of odd numbers, often with peculiarly different intensity distributions between these observed components. Doublet splits with only two component lines were particularly frequent. According to the multiplicity formula $2m + 1$, therefore, m should not be integral here but half-integral. Alfred Landé (1888–1976) had, in fact, been experimenting since the early 1920s on the assumption of half-integer quantum numbers to explain the doublet structures of alkaline metals along with other spectral

---

[129]On the problematic interpretation of these quantum jumps, see p. 31 above.

[130]For details see, e.g., Hentschel (2008, 2009e) citing primary sources.

subtleties of this "term zoology" and "Zeeman botany."[131] Additional experiments by Otto Stern and Walther Gerlach in Frankfurt in 1922 on atomized beams of highly heated silver sent through an inhomogeneous magnetic field yielded spectrum lines split into two components. A certain Werner Heisenberg (1901–1976), at that time still a mere student of Sommerfeld in Munich without a degree, speculated in 1922 that this half-integer quantum number was a kind of average between two actually integral quantum numbers, one half of which could be associated with the atomic core and the other half to the electronic shell. Another young student of Sommerfeld, Wolfgang Pauli (1900–1958), arrived at the even more outlandish interpretation of a half-integer quantum number: a "mechanically indescribable ambiguity" characteristic of the outermost optically active electron (*Leuchtelektron*). In January 1925 Pauli introduced a new quantum number into a mathematical description of this ambiguity, initially denoted by the letter μ, with $\mu = \pm 1/2$ and also postulated that a new selection criterion $\Delta\mu = 0$ or ± 1 must apply for physically permissible transitions. Each electron was hence described by a set of four quantum numbers n, l, m and μ (sometimes also denoted as n, l, j and s). Pauli considered the build-up of electron shells surrounding an atom as basically the energy levels all being occupied one after another, starting with the energetically most efficient one (at $n = 1$). The number of electrons per shell and many spectroscopic details, some of them very subtle, could be represented extremely well when another requirement was imposed: None of the electrons of a given atom may have the same four quantum numbers.[132] This Pauli exclusion principle yielded in a natural way the electron occupation numbers 2, 8, 18, 32 which correspond to the periods of the system of the chemical elements so familiar in atomic physics. This was a huge accomplishment for those two young upstarts in the physics profession. It was bought at the expense of another radical, if not audacious abrogation in classical physics, though. Pauli considered classical physics fundamentally incapable of covering that "mechanically indescribable ambiguity." That was why such quantum numbers were not allowed to be interpreted mechanically. At this point Heisenberg and Pauli were standing on the very threshold between the old quantum theory and quantum mechanics, which they would both begin to develop a few months later. But they had not yet taken that step beyond. Nevertheless, they already saw that Bohr's and Sommerfeld's semiclassical assumptions, such as the notion of defined electron orbits, had to be retracted in the new physics. Wolfgang Pauli, being a godchild of the philosopher Ernst Mach (1838–1916) in Vienna and brought up in the spirit of positivism, had no trouble banishing from physics this model-based concept of "metaphysical" electron orbits.[133]

---

[131] See in this respect esp. the 2nd ed. of Sommerfeld's textbook (1919b), furthermore Forman (1968), Tomonaga (1974/97) lecture 1, and the primary sources cited there. Sommerfeld joked about "number mysteries" (*Zahlenmystik*) and his colleague Carl Runge no less ironically about "witches times-tables of quantum physics" (*Hexen-Einmaleins der Quantenphysik*).

[132] See Pauli (1925), Meyenn (1979/85), vol. 1, (1980/81) as well as Waerden (1960) and Tomonaga (1974/97) lecture 2, Serwer (1977), Heilbron (1983).

[133] This intellectual influence of Mach's phenomenalism on Pauli is discussed by Popper (1935) pp. 168f.

In this mood of crisis at the turn of the year 1924/25, Ralph de Kronig (1904–1995) wrote a letter to Wolfgang Pauli to suggest that the new fourth quantum number be interpreted as the expression of an electron's fourth degree of freedom—namely, a rotation around its own axis. Pauli wrote back immediately to smartly reject this suggestion on the strength of three arguments:

1. A factor 2 was missing in de Kronig's calculation, without which a discrepancy remained for the observed doublet splitting of alkaline spectrum lines.
2. The magnetic moment of the atomic core following out of de Kronig's calculations was too small.
3. The intrinsic angular momentum assigned to this quantum number s and hence the velocity of the electron rotating around its axis was too large, making it contradict Einstein's special theory of relativity, according to which no material object may move faster than the speed light.[134]

Completely unaware of this exchange, two young postdocs under Paul Ehrenfest (1880–1933) at Leyden, George Eugene Uhlenbeck (1900–1988) and Samuel Abraham Goudsmit (1902–1978) considered a very similar interpretation of the doublet splits in the anomalous Zeeman effect. They assumed that each individual electron bears its own magnetic moment M that results from its intrinsic rotation with angular momentum (very soon afterward referred to as spin s): $M = s \cdot \frac{2}{2mc} S$, where the new quantum number s could only be ± 1. Considered quantitatively, this magnetic moment would be twice as large as would be expected in a naive semiclassical model of a point charge e, Bohr's magneton $e\hbar/2mc$. This assumption, however, might explain why the magnetic moments from the rest of the electron shell J, as well as the anomalous momentum M, all of which are coupled with the spin, could lead to the anomalous Zeeman effects.

The first documentable reaction to these as yet unpublished reflections was a letter by Hendrik A. Lorentz (1853–1928) from 19 October 1925 pointing out objections to de Kronig's arguments that were very similar to Pauli's: the rotational velocity v of such an electron, because of $\mu = \frac{e}{m} \frac{v}{r}$, would have to be roughly ten times the velocity of light, going against Einstein's theory of relativity. But Paul Ehrenfest had already encouraged his two young coworkers to submit these ideas two days prior to Lorentz's letter went out, to the less choosy semipopular journal *Die Naturwissenschaften*, where it subsequently appeared on 20 Nov. 1925.[135] The problem with the missing factor 2 was solved soon afterwards. At the beginning of 1926 Lewellyn Hilleth Thomas (1903–1992) was able to show that this factor 2 followed out of a Lorentz transformation from the electron's frame of reference into the observer's frame, which

---

[134]On the foregoing, see Pauli's letter to de Kronig, 9 Oct. 1925 in Meyenn (1979/85) pp. 242–249. Friedrich Hund (1896–1997) recalled the saying making the rounds among contemporary physicists: "De Kronig would've discovered spin, if Pauli hadn't made him shrink" ("De Kronig hätt' den Spin entdeckt, hätt Pauli ihn nicht abgeschreckt"), Friedrich Hund's interview with Klaus Hentschel on 15 Dec. 1994; see Hentschel and Tobies (1996).

[135]Uhlenbeck and Goudsmit (1926), Goudsmit and Uhlenbeck (1976) as well as Goudsmit (1971) report that Ehrenfest consoled them with the words: "You are both young enough to afford a stupidity like that."

was lacking in Goudsmit's and Uhlenbeck's calculation.[136] By then the 'old' quantum theory and the Bohr-Sommerfeld atomic model had already been replaced by the 'new' quantum mechanics. In this later context statements could be made about the spin of elementary particles and connections drawn between spin and statistics from more profound assumptions about symmetries. Quantum mechanics could adopt this tried-and-true spectroscopic phenomenology almost without alteration; and many of the formulas for spectral splits and shifts were also reproducible within the new framework virtually unchanged. Thus the concept of spin was more deeply founded through the efforts of Heisenberg, Schrödinger, Fermi and Dirac but continued to hold its own in the new conceptual frameworks of quantum mechanics as well as in quantum field theory built upon it, particularly quantum electrodynamics.[137]

Spectroscopic evidence provided good reasons for the assumption that apart from electrons and other material elementary particles, light quanta or photons also have an intrinsic spin. Direct experimental confirmation that the spin of a photon is closely connected with its angular momentum had to wait until 1936, however. Richard A. Beth (1906–1999) in the *Palmer Physical Laboratory* at *Princeton University* measured the angular motion of a leaf of quartz suspended from a thread 25 cm in length also of quartz upon interaction with circularly polarized light. The thickness of this double refracting leaf was chosen to create a difference of exactly one wavelength in the path between the regular and irregular beam so that each photon transfers angular momentum $\hbar$ onto the leaf. The sum of many such changes in angular momentum could then be demonstrated by measuring the torsion.[138]

3.11 Indistinguishability of Light Quanta – Origin of Bose-Einstein Statistics

In principle, it is possible to assign a number to each classical particle or to give it some other identifying marker because (theoretically at least) it is distinguishable from all the others. In the twentieth century when statistical mechanics was linked to the early quantum theory, it became evident that this no longer applies to the world of quanta. The indistinguishability of quantum particles stymies any attempt to identify or earmark them.[139] A Polish physicist in Cracow, Ladislas Natanson (1864–1937), was the first to realize this.[140]

---

[136] A clear derivation of this factor is provided particularly by Tomonaga (1974/97) Chaps. 2 and 11.

[137] See Landau and Lifschitz (1947/73) pp. 191–240, Dirac (1974/75), Tomonaga (1974/97) Chaps. 3–12 and Milner (2013); on the spin statistics theorem and QED, see Sect. 3.11.

[138] See Beth (1936) esp. note 1, p. 115 and note 11, p. 121 for indications that Einstein personally participated in the theoretical computations and practical performance of this experiment.

[139] See, e.g., Fraser (2008), French (2015) and Holger Lyre in Friebe et al. (2015) pp. 89–112.

[140] On Natanson see, e.g., Weyssenhoff (1937), Średniawa (1997, 2007) and Hentschel and Waniek (2011) as well as further sources cited there.

In 1911 Natanson pointed out in an article on a statistical theory of radiation that Planck's energy distribution of black-body radiation only results "if during the process of calculating the probability we presume that it is possible to treat the receptacles of energy as identifiable and that it is not possible to treat the units of energy, which are the same in every respect, as identifiable."[141]

These "receptacles of energy" (for instance, the atoms constituting the black body) would accordingly still be identifiable—in other words, enumerable—whereas the "units of energy" (i.e., quanta) would be indistinguishably similar and therefore not individually identifiable. With this insight Natanson himself, or one of his more attentive readers could just as well have developed then in 1912, the statistics of those indistinguishable particles later to be named after Bose and Einstein.[142] Planck's radiation formula was derivable *only if* this complete indistinguishability among the energy quanta be presumed. Otherwise Boltzmann's distribution results. Natanson proved this by posing the combinatorial problem: How many ways P can energy quanta ε be distributed among N "receptacles of energy," where these receptacles of energy—similar to Planck's "resonators"—are material systems (not necessarily particles) that can absorb or emit these energy quanta? Natanson adopted Planck's presupposition from 1900: "the energy is not infinitely divisible but is composed of an aggregate of discrete elements or units."[143] However, Natanson recognized the centrality of this indistinguishability among energy quanta more clearly than Planck, Einstein and other physicists had up to then: "We regard the elements or units of energy as undifferentiably the same. If we were capable of perceiving each one of them individually, the conditions of this case would change fundamentally. This must be pointed out first and foremost."[144] Other physicists then began to look at this statistical theory of radiation, which Natanson right at the beginning of his paper from 1911 had clairvoyantly predicted "would count among the profoundest discoveries in the field of molecular physics.[145] The Russian Yuri Krutkov (1890–1952) and the Pole Mieczyslaw Wolfke (1883–1947) got embroiled in a hefty dispute in 1914–15 about how much Einstein's assumption about light quanta presupposed their independence. (Wolfke called them *Lichtatome*.) This hardly suitable concept of independence meant different things to each party, though: to Krutkov it meant identity; to Wolfke, spatial separability.[146] A short paper by Paul Ehrenfest (1880–1933) and the discoverer of superconductivity Heike Kamerlingh Onnes (1853–1926) that was published in the *Annalen der Physik* in 1915 provided some clarification. It offered an elegant and comprehensible "derivation of the combinatorial formula grounding

---

[141]Natanson (1911) quote on p. 662. Cf. Darrigol (1988) esp. pp. 243ff., Kastler (1983) esp. pp. 616ff.; critically: Stachel (2002) pp. 438f., furthermore Bergia (1987).

[142]Cf. Delbrück (1980), Kastler (1983), Monaldi (2009), the papers by Borelli , Saunders, Huggett and Imbo in Greenberger et al. (eds. 2009) pp. 299ff., 311ff., 611ff. and the lit. cited there.

[143]Natanson (1911) p. 660.

[144]Ibid., p. 660.

[145]Ibid., p. 659.

[146]Comp., e.g., Mehra and Rechenberg (1982) vol. 1, pp. 559f., Navarro and Pérez (2004) pp. 130ff.

Fig. 3.10 Distributions of
three energy quanta for two
resonators

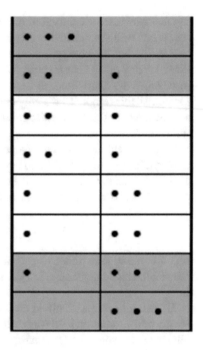

Planck's theory of radiation."[147] These physicists, at that time both employed at
Leyden, represented the P energy quanta by P (indistinguishable) dots and concrete
energy distributions of those P quanta over N Natanson receptacles or carriers of
energy (resp. Planck resonators) as a row of symbols that contained in addition to
these P energy quanta also $N - 1$ separation bars between the energy carriers. The
row of symbols ● ● | | ● ● ● ●| ● | ● ● ●● signifies, for example, a distribution in which
a total of 10 energy quanta are distributed over 5 energy carriers, such that one resonator absorbs 2 energy quanta, the next one 0 energy quanta, the next 4 and 1, and
the last 3. Planck's energy distribution problem is in this way turned into a purely
combinatorial task, to calculate the probability of the occurrence of particular com-
binations of the two types of symbols: the dot and the bar, $(P \times ●)$ and $(N - 1) \times |$.
The solution to this simple combinatorial problem is $(P + N - 1)!$ permutations of
these $(P + N - 1)$-symbols, where duplications of indistinguishable combinations
have to be corrected for, in that this expression be divided by the permutations $P!$ of
all dots and $(N - 1)!$ of all bars |. A simple computation can illustrate Natanson's
considerations (see Fig. 3.10).

Given P energy quanta on N distinguishable resonators (i.e., elementary oscillating systems). How many *a priori* equally probable possibilities are there (which
would thus also reveal the average resonator energy)? The solution depends on
whether the energy quanta are distinguishable from one another or not. The quanta
are represented by dots, the resonators by rectangles. If the energy quanta are indis-

---

[147] Ehrenfest and Kamerlingh Onnes (1915) p. 1021; on Ehrenfest's role cf. Pérez and Sauer (2010).

tinguishable, there are only four distinguishable states (shaded fields). If the energy quanta can be distinguished from one another, then eight different possibilities exist for distributing the energy over the two resonators. The average energy would be either 1.5 or 2 quanta per resonator; hence the average energy also differs depending on whether the distributed portions—this is the root of the term 'quantum'!—are distinguishable or not. That's why quantum statistics differs basically from classical statistics, where this distinguishability is assumed. It breaks down within the range of quanta.

Ehrenfest and Kamerlingh Onnes established: "Each of $(N - 1)!P!$ arrangements yields the same distribution symbol and lets each resonator stay on its energy level."[148] Thus the number W of indistinguishable cases of one combination of P energy elements and N energy carriers is:

$$W = (N + P - 1)!/[P!(N - 1)!]$$

In a few more steps this expression yields Planck's energy distribution—as Max Planck himself demonstrated in 1900 and as can be found in any textbook on quantum theory.

This simplified and distilled derivation of Planck's formula for the mean energy density of the radiation field was more than just a clever teaching trick, even though it still appears in some modern textbooks on quantum mechanics.[149] Natanson, Ehrenfest and Kamerlingh Onnes very acutely recognized the basic difference between Planck's and Einstein's derivations: "For Einstein they really are P equivalent, mutually sundered quanta [...]; for Planck, on the other hand, they aren't really P mutually sundered quanta; their introduction must be taken *cum grano salis*; as equally formal a trick as our permutation of the symbols [...]. The true object of the counting remains the number of all mutually different distributions of N resonators over the energy levels 0, ε, 2ε, 3ε, ... for a predetermined total energy $P\varepsilon$."[150] Planck had focused on the classical, idealized resonators in the material walls of the black body, which are consequently not problematic. He had only hit upon energy quantization because he had been prevented from making the boundary transition $\varepsilon \to 0$ that Ludwig Boltzmann (1844–1906) had performed 1877 according to his classical statistics.[151] Natanson pursued the thought further that Einstein had already prompted in 1905 of examining the radiation field itself and found the real core assumption of indistinguishability for a statistical derivation, which led him to the solution of treating his energy receptacles as identifiable and the energy units as not identifiable. He concluded: "As our procedure is ultimately based on this assumption, it seems natural to refer to it directly as the foundation of the theory. Apparently, not enough importance has been attached to the circumstance that we do not, in fact, have another alternative for proving the

---

[148]Ehrenfest and Kamerlingh Onnes (1915) p. 1022.

[149]See, e.g., Landau and Lifschitz (1947/73) vol. III, Hund (1984) pp. 30f.

[150]Ehrenfest and Kamerlingh Onnes (1915) p. 1023, original emphasis.

[151]Cf. Darrigol (1988/90, 1992, 2000/01) and Gearhart (2002).

Fig. 3.11 Contrasting Boltzmann's and Planck's statistics. Boltzmann statistics follows out of the basic distinguishability of its 'classical' particles; Planck statistics follows out of the basic indistinguishability of quanta

distinguishability	**vs.**	**indistinguishability**
Boltzmann statistics		Planck statistics

The number N_j of particles that occupy the state j is:
$$N_j = N_0 \cdot e^{-\beta \cdot E_j}$$
with the number of particles N_0 in the 0th state.

The occupation number $n(E)$ is equal to:
$$1/(e^{(\beta \cdot (E-\mu))} - 1).$$

legitimacy of Planck's method to calculate the probability."[152] Natanson's prophetic words again came too soon. Einstein's provocative considerations of 1905 were the subject of much controversy at least until the Compton effect was experimentally confirmed in 1921–22.[153] In the middle of that decade, it finally dawned on Albert Einstein that Planck's energy distribution is actually based on an entirely new kind of statistics. The inspiration for this insight was not Natanson's article, however, even though they were exchanging letters, but the manuscript of a Bengali physicist whom he had never heard of before: Satyendra Nath Bose (1894–1974). It subsequently appeared in the proceedings of the *Prussian Academy of Sciences* in 1924 together with Einstein's own comment.[154] Bose divided up the phase space of each light quantum into cells of the volume h^3 and calculated the entropy and radiation density from the number of possible distributions of the light quanta among these cells, similar to how Natanson had proceeded. Seven decades later, in 1995, it was finally possible to examine experimentally the "strange" characteristics of "degenerate quantum gases," which Einstein had very briefly alluded to and which came to be known as Bose-Einstein condensates.[155] Coming to think of it, the quantum statistics named after Bose and Einstein should more accurately have been called 'Planck-Natanson-Bose-Einstein statistics.' Be this as it may: we find confirmed yet again the so-called 'first law of the history of science'—namely, that (almost) no scientific result is named after its true discoverer.

As the small computation example in Fig. 3.11 demonstrates, Boltzmann statistics leads to different results than does Planck-Natanson-Bose-Einstein statistics: The distribution of distinguishable particles over N cells is $(N-1)!$ times more probable than their concentration inside exactly one cell. In the case of indistinguishable particles, it is N times more probable than their being evenly distributed over all the cells. In short: distinguishable particles tend toward a homogeneous distribution of the energy; indistinguishable ones tend toward a clustering or 'condensation' in the state of lowest energy. This Bose-Einstein condensation has become observable by

---

[152] Natanson (1911) p. 662.

[153] See, e.g., Stuewer (1975a) Chaps. 6–7 and Brush (2007).

[154] See Bose (1924), and, e.g., Einstein (1924/25) pp. 4f.

[155] See, e.g., Klaers et al. (2007) about massive bosons, Anglin (2010) and Klaers et al. (2010) on photons, furthermore Greenberger et al. (2009) pp. 299ff. as well as the literature cited there.

modern experimental means in a photon gas or other elementary particles subject to Bose-Einstein statistics.[156] However, unlike bosons which have mass, photons are particularly hard to observe, because the photon number is not a conservation quantity and at low temperatures photons tend to disappear inside the cavity walls instead of occupying the energetic ground state. This behavior deviates radically from our physical intuition about how other classical particles act, so it is not surprising that some twenty years had to pass before the full bearings of this new kind of statistics was recognized. The penny having once dropped, in 1925 Einstein aptly spoke about a "paradox," about a "degeneracy" of this 'quantum gas' and about particles of a "very puzzling kind" when discussing the strange property of light quanta, or other indistinguishable particles subject to the same statistics, to cluster or bunch.[157] A radically new "quantum theory for ideal gas" seemed to him "legitimate when one starts out from the conviction that a light quantum (disregarding its polarization property) differs from a monatomic molecule essentially only in that the quantum's mass at rest is vanishingly small." This cast a new bridge between gas theory and quantum theory. Bose's statistics of light quanta could later be transposed onto all elementary particles that have integer spins. It was later revealed that particles with half-integer spins (such as, electrons) satisfy Fermi-Dirac statistics. These statistics, discovered in 1926, basically differ from Bose-Einstein statistics in that the Pauli exclusion principle applies to fermions. Accordingly, two identical fermions never assume exactly the same quantum state, which means all the quantum numbers of that fermion must never perfectly match those of another fermion nearby. This exclusion principle prevents fermions from 'condensing' the way bosons do at the lowest quantum state and has the effect, for example, that the Z electrons of an atom distribute themselves in the Z lowest electron states.[158]

3.12 Photons as Virtual Exchange Particle of Quantum Electrodynamics

Quantum electrodynamics (QED) formed out of the interconnection between relativity theory and quantum mechanics.[159] It is an exact theory of point charges, such as electrons or positrons, and their interactions with photons and has been confirmed by experiment to very high precision and thus describes microphysical processes

---

[156] See, e.g., Ketterle (1997), Anglin (2010), Klaers (2010) and the web site recommendations at the end of this book.

[157] Einstein (1924/25a) pp. 266, 267 and (1924/25b) pp. 3, 18 (CPAE, vol. 14, transl. ed., docs. 283 & 385, and doc. 427, p. 418).

[158] The further history of quantum statistics up to the general proof of the spin-statistics theorem by Wolfgang Pauli and his assistant Markus Fierz in 1939–40 is recounted in Meyenn (1958) vol. II and doc. 30 in Schwinger (1958), further Landau and Lifschitz (1947/73) pp. 218–240, Tomonaga (1974/97) lecture 8, Miller (1994), Blum (2014) and the primary sources cited in each of these.

[159] For the early history of QED up to around 1953, see Schweber (1994) and the primary sources cited there; many of those papers are compiled in Schwinger (1958) or Miller (1994).

in distances of the order of 10^{-18}m. For example, the QED prediction for the electron's magnetic moment agrees with the value determined by experiment up to the eleventh digit after the decimal point, making it one of the very best verified theories in physics.[160] In QED, the electron is interpreted as a spinor field with the charge $-e$, and the photon is interpreted as the associated gauge field. The most important characteristics and limiting conditions of their interaction are derived from requirements such as gauge invariance and invariance in Lorentz transformations.[161]

This development began in 1926 when Max Born, Werner Heisenberg and Pascual Jordan attempted to apply the brand-new quantum mechanical methods of matrix mechanics not only to material atoms and their energy levels but also to the radiation field enveloping them. Their purpose was to reproduce results from the old quantum theory, such as Einstein's derivation from 1909 of fluctuations in the electromagnetic field leading to the trembling of a freely suspended mirror.[162] Paul Dirac (1902–1984) formulated in 1927–30 a first closed theory of the emission and absorption of electromagnetic radiation by electrically charged particles. It led to the conclusion that besides the negatively charged electron another equally ponderable but positively charged positron must also exist. Soon afterwards this could be experimentally confirmed by cloud-chamber exposures.[163] Unlike the older semiclassical theories, QED built upon this pioneering research and from the late 1940s began to develop apace. A quantization of energy occurs not only for matter (classical electromagnetic wave, but quantized matter, the so-called "first quantization"), but also for the electromagnetic field itself (thence the expression "second quantization").[164]

The photon was thus re-interpreted as the exchange particle of electromagnetic interactions. This may be imagined as a ping-pong ball (the photon) that is volleyed back and forth between two players (the charged particles), whose actions are related to each other only by these volleys. Ping-pong balls and other material exchange particles have a non-vanishing mass of their own. This is not the case with photons, which are known to have no rest mass. That's precisely why electromagnetic interactions have such an infinite range of force. The weak nuclear force in the interior of the atomic nucleus, which is responsible for beta decay, for example, is relayed by ponderous exchange particles. That's why its active range is so short, not going beyond the radius of an atom. According to QED, the exchange of photons of fre-

---

[160]See Schwinger (1958), docs. 10–12, Feynman (1985) Chap. 3, pp. 115ff., Duncan (2012) Chap. 10 and Han (2014).

[161]For systematic derivations of results from QED, see Heisenberg and Euler (1936), Feynman (1961) and Jauch and Rohrlich (1955), which only proceed from abstract premises. Feynman (1985) provides a more visualizable introduction at an elementary level.

[162]See Born et al. (1926) and Schweber (1994) pp. 10f., Duncan (2012) pp. 19ff.

[163]Dirac's research papers between 1927 and 1934 are at the beginning of the anthology Schwinger (1958). On Dirac's life and works, cf. Kragh (1990), Schweber (1994) pp. 11–32, 70ff. with further sources cited there.

[164]The different conceptions by Dirac and Jordan of this and the oscillation between particle-based and wave-based quantization in the decades that followed are discussed in Schweber (1994) pp. xii–xxvii, 25ff., 33ff., Scully and Zubairy (1997) pp. 27ff., Brown (2002), Duncan (2012) p. 42, Han (2014).

Fig. 3.12 Three possibilities
for an elementary interaction
between an individual
electron and electromagnetic
radiation. *Source* Feynman
(1985) p. 97. Reprinted by
permission of Princeton
University Press © 1985

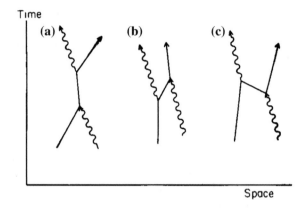

quency ν is limited in many cases to a virtual exchange for a very short interval
of time $\Delta t \leq h/\nu$. Richard Feynman (1918–1988) hit upon an ingenious way to
visualize those extremely lengthy calculations schematically in 1948. They soon
became known as 'Feynman diagrams.'[165] Straight lines represent elementary particles, particularly electrons, and wavy lines represent photons (see Fig. 3.12). The
time axis always runs from below upwards. Hence in such diagrams real particles are
always recorded as almost vertical figures and virtual particles relaying interactions
are visualized as approximately horizontal figures. In the case of an electron scattering light in the lowest order of perturbation theory, the following three diagrams in
Fig. 3.12 depict the three physically conceivable possibilities for the interaction of
an individual electron with electromagnetic radiation:

In case (a) the electron initially absorbs a photon and emits another photon later on.
In case (b) the opposite occurs: the electron emits a photon and then absorbs another
one later on, thereby increasing its energy a little. Case (c) is the most baffling (c in Fig.
3.12). At a particular point in time, the electron emits a photon, then appears to travel
back in time and at a second point in time—before that emission has occurred—
absorbs another photon and resumes its advance into the future. This last case is
physically conceivable only if the time interval Δt during which the electron is going
back in time is shorter than the time $\Delta E \cdot \Delta t \simeq h/2\pi$ in Heisenberg's uncertainty
relation, with ΔE the uncertainty of the electron's energy. Whereas the instances (a)
and (b) of incoming and outgoing electrons or photons describe real particles, the
third case (c) describes a virtual electron in-between two interactions.[166]

The Feynman diagram on the left-hand side of Fig. 3.13 shows a normal Coulomb
interaction between two electrons transmitted by a virtual photon that is almost
horizontal. Therefore it is propagating within the very short time interval Δt between
points 5 and 6, while the electrons are moving from 1–5–3 and 2–6–4, respectively.

---

[165]On Feynman's life and work see, e.g., Mehra (1994) and Schweber (1994) Chap. 8. The history
of Feynman diagrams is recounted by Kaiser (2005) and Wüthrich (2011).

[166]In 1948–49 Feynman interpreted the particles moving backwards in time as their antiparticles
(with opposite electric charge): see, e.g., Schweber (1994) pp. 428ff.

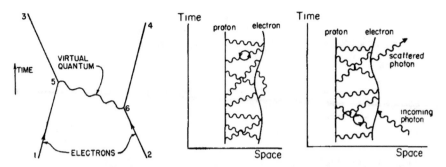

Fig. 3.13 Left: Coulomb interaction between two electrons via a virtual photon. *Source* Feynman (1949a) p. 772, reprinted by permission of the American Physical Society © 1949. Right: Diagrams illustrating the electromagnetically conveyed bond between an electron and a proton in the nucleus as well as the scattering of light off an atom. *Source* Feynman (1985) p. 100. Reprinted by permission of CCC Publications © 1985

The bends at the interaction points 5 and 6 symbolize the exchange of energy and momentum transmitted by the virtual light quantum. It would be mistaken, however, to regard these lines as depictions of trajectories or 'particle traces' in real space-time.

The diagram on the left has just one virtual quantum. The other two diagrams have many virtual exchange particles (eight) between the two solid lines signifying the two charged elementary particles. These exchange particles create the stable bond between the electron and the nucleus (represented here by the positively charged proton). The diagram in the middle shows a real photon returning to the electron (on the right side), and the two diagrams on the right depict vacuum polarization (drawn as an arrowed circle) when a virtual electron-positron pair very briefly forms. This latter process leads, among other things, to the QED corrections to the masses of elementary particles.

As just these last diagrams demonstrate, QED calls for a reinterpretation of the photon as a massless exchange 'particle' of electromagnetic interactions. In some cases it has a *real* effect on a charged particle in the form of the emissive or absorptive processes. In many other cases, however, one must account for a *virtual* process in order to be able to make an accurate calculation of such establishable observables as mass, charge and interaction probabilities. Within the limits of Heisenberg's uncertainty relation $\Delta E \cdot \Delta t \simeq \hbar$, which lead to typical nuclear energy transfers at time intervals of $\Delta t \leq 10^{-24}$ s, ultrashort infractions of the conservation of energy and momentum are 'permissible,' so long as they cancel out again within this time interval. Consequently, these virtual processes or virtual particles are modeling tools introduced in order to be able to interpret physically every term in perturbation theory or every Feynman diagram. But they are often unduly interpreted or misunderstood ontologically. The philosopher Hans Vaihinger (1852–1933) would have called them fictions,[167] and other philosophers of science criticized the introduction of a basically unobservable, virtual particle: "we should either give it up or abstain from

---

[167] On Vaihinger's fictionalism in physics, see Hentschel (2014a) and the literature cited there.

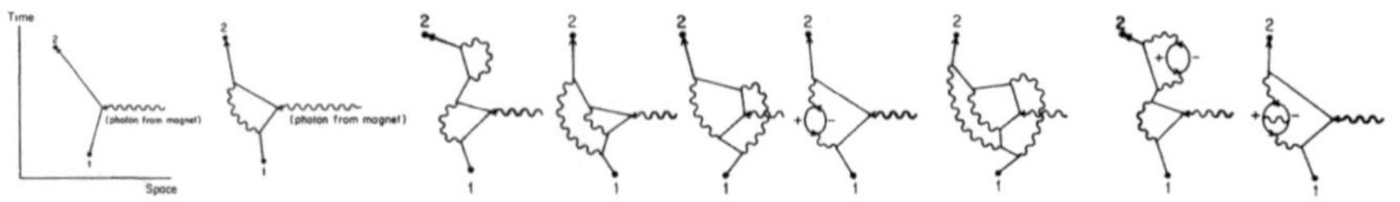

Fig. 3.14 Some Feynman diagrams to calculate an electron's magnetic moment up to sixth-order perturbation theory. *Source* Feynman (1985) pp. 115–118. Reprinted by permission of Princeton University Press © 1985

assigning it a physical meaning: we should regard it instead, at best, as a computational intermediary."[168] They have become a routine tool of modern quantum field theories nonetheless.

The Feynman diagrams in Figs. 3.12 and 3.13 (left) only show the physical processes in first-order perturbation theory. All other possibilities introduce an even higher number of vertices (the bends between the straight and wavy lines), and according to Feynman each vertex adds a factor into the estimate of the probability of that process: the fine-structure constant $\alpha \simeq 1/137$. Consequently, processes of second order or more have only slight impact. For highly precise calculations it is indeed possible to depict all these higher-order possibilities (and it is much simpler than the brute-force calculations by Schwinger and Tomonaga). In the end, Feynman diagrams are 'simple' summands: each diagram corresponds to one term in the perturbation computation practiced by Dyson, Tomonaga and Schwinger. So each diagram depicts one of the possible physical processes which, in turn, contributes to the real event, even if the higher-order contributions (i.e., with more and more vertices) become increasingly less probable. Figure 3.14 illustrates some terms for calculating an electron's anomalous magnetic moment to higher orders of perturbation theory—up to the sixth order. To get an idea of the work involved in these calculations, which occupied physicists for many years: there are a total of some 10,000 Feynman diagrams with up to six vertices. The figure for the electron's magnetic moment resulting out of them (in 1983) was: 1.001,159,652,46, with an uncertainty of 20 in the last two digits. The figure obtained by experiment at that time was 1.001,159,652,21 with an uncertainty of 4 in the last digit. As Feynman explained in one of his lectures: Scaled against the distance between New York and Los Angeles, this would correspond to an experimental margin of error of a hair's breadth.[169] The CODATA value from 2014 comes to 1.001,159,652,180,91(26), which shows how very precise this measurment and calculation has become.

These very straightforward Feynman diagrams greatly simplify the extraordinarily precise calculations. It explains the great success of an otherwise so

---

[168]Bunge (1970) p. 508 and analogously Shrader-Frechette (1977) pp. 415, 419f.; cf. further Weinberg (1977) p. 24, Hendrick and Murphy (1981) pp. 458ff., Weingard (1982, 1988), Esfeld (2012), Passon (2014), Blum and Joas (2016) along with their cited texts exploring the meaning of the 'existence' of virtual particles.

[169]See Feynman (1985) p. 118 and Ibid. pp. 143f. on similar calculations for a myon.

abstract and highly formalized theory as QED.[170] Not only the magnetic moment of elementary particles was empirically confirmed to great accuracy but also what was called the Lamb shift in the years 1947–51 in the hydrogen spectrum between the energy levels $2S_{1/2}$- and $2P_{1/2}$. According to Dirac's theory these energy levels should be degenerate but according to QED they should manifest very slightly differing energy levels from vacuum fluctuations.[171] Pursuant to the correspondence principle, QED contains classical electrodynamics as a limiting case for high quantum numbers, thus when the measurement data can be regarded as continuous. But obvious deviations from classical electrodynamics appear, given weak fields, as well as when calculating the self-energies of the electron and photon.[172] The divergences that occurred led the involved researchers astray for a long time and they only succeeded in eliminating them by calculational argument in the years 1947–51 within the context of renormalization theory. The least harmful of these mathematical problems was what was referred to as 'infrared divergences,' which Felix Bloch was the first to treat in 1936: With increasing wavelength the probability also increased for the emission of a 'soft' photon, that is, an infrared photon, which led to a first divergence.[173] However, the corrections for the radiation also diverged at the same time, when an integration was performed over the energies of all the virtual photons. Josef Maria Jauch (1914–1974) and Fritz Rohrlich (*1921) were able to show in 1954–55, though, that these two infinite expressions cancel each other out exactly in all orders of perturbation theory.[174] Other divergences involving electromagnetic self-interaction and vacuum polarization were less easily eliminated. In other words, divergences involving the formation of an electron-positron pair out of a high-energy γ-quantum. (For examples see the closed loops in Figs. 3.12 and 3.14 —cf. further Fig. 8.11.) These entirely new kinds of effects of QED led to substantial corrections to the electron's mass and charge, or partly even to strongly divergent expressions. The only recourse they found was to interpret the observable masses, charges etc., as the sum of calculable raw masses or charges with their corresponding correction terms.[175] This juggling with terms, each of which diverges on its own, but summed together would yield finite values again, initially encountered powerful resistance among many physicists.[176] The circumstance that the calculated final results could then be confirmed so accurately by experiment finally led to high acceptance of the

---

[170]The application of Feynman diagrams in teaching and training are discussed by Kaiser (2005), Wüthrich (2011), both citing further sources on the reception and assertion of QED.

[171]See, e.g., Kragh (1985), Darrigol (1988) pp. 23–26, Schweber (1994) Chap. 5 and Duncan (2012) pp. 54–55.

[172]For the history of the early debates about such divergences before the rise of renormalization theories, see Pais (1948), Weinberg (1977) pp. 24ff. and the primary sources cited there.

[173]See especially Darrigol (1988) pp. 11–13 and further primary sources quoted there.

[174]See Jauch and Rohrlich (1955/76b) Chap. 16, pp. 390ff., Feynman (1961) pp. 128ff.

[175]See Jauch and Rohrlich (1955) Chaps. 9–10 and suppl. 2, resp., Feynman (1985) Chaps. 3–4, Schweber (1994) pp. 434ff. (on Feynman), 564ff. (on Dyson), 595–605 (on Schwinger and generally).

[176]Pioneers of QED among them, such as, Heisenberg and Dirac (1974/75) p. 9: "quite dissatisfied with it."

theory. QED is now the most precisely confirmed theory in physics generally and is a bearing pillar of the standard model of particle physics. Students were often given the advice to keep out of debates about the fundamentals and interpretational problems and to just perform the calculations and measurements.[177] This also explains why we are still burdened with the problem of what exactly photons are, or whether they really exist (more on this in Chap. 9).

Is QED a synthesis of all theories and mental models about the light quantum so far, as is occasionally alleged?[178] We find traces of the semantic layers 1, 2, 5, 6, and 9; but layer 1 (the particulate nature), for instance, has been decisively modified in this new semantic layer of the photon as a virtual exchange particle. The problems of inner consistency, in connection with the necessity for renormalization procedures, and numerous attacks on QED, along with the relentless attempts to replace it with semiclassical alternatives (cf. here Sect. 9.5 and Chap. 10) all show that the mixture of mental models existing in QED (wave and particle depictions) and approaches by mathematical physics of disparate provenance (Schwinger's formalism or Feynman's diagrams) represents an unstable conglomerate that but few believe will endure as such unchanged. This is despite it being the empirically most precisely tested theory of physics and, owing to its excellent experimental confirmation, despite it surely remaining a kind of 'low-energy limit' of a yet-to-be-found 'theory of everything' (TOE).[179] As Niels Bohr has already pointed out, physical models and interpretations are always based on human speech, the mathematics notwithstanding. Language contains fundamental mental models (particles and waves, for instance), which have been accompanying us since the beginnings of humankind and to which our intuition, our thinking, and our conceptuality have adapted. But they don't fit the world of quanta. In the following chapter we shall look at the most important mental models that some of the main protagonists of our history of the light quantum built during its first decades.

---

[177]For instance, in the autobiography of Hanbury (1991) pp. 121–123: "there is no satisfactory mental picture of light [...] and the only way of getting the right answer was to do mathematics," and Tegmark (2007): "shut up and calculate." For a protest against this "just compute," see Roychoudhuri (2015) p. 169.

[178]See, e.g., Jauch and Rohrlich (1955) p. v or Kidd et al. (1989) p. 33.

[179]On QED as a methodological model for the later quantum field theories as well as for quantum chromodynamics (QCD) see, e.g., Han (2014) and the literature cited there.

Chapter 4
Early Mental Models

Having just gone through the twelve semantic layers of the complex concept of the 'light quantum,' I would now like to move to another question. What lay behind all these considerations by central thinkers such as Newton and Einstein and skilled experimenters such as Johannes Stark or Robert Millikan, who were involved in the debate about the status and nature of light quanta? How exactly did these actors imagine light quanta? To infer such mental models from textual traces is a non-trivial endeavor, comparable to reconstructing magnetic fields from the positions of iron filings or of hydrodynamic currents from the transport of small particles. But it is possible, provided there are enough textual traces generated on the basis of these mental models and that the historical actors were explicit enough in speaking about their assumptions and models. Such a cognitive analysis certainly can be done in the cases of the twelve scientists picked for this chapter.[1]

4.1 Newton's "Globuli of Light"

We have already become acquainted with the principal features of the Newtonian projectile theory of light, at least insofar as Newton was willing to reveal it publicly. Newton modeled light as a stream of very fine, minuscule 'corpuscles' or particles that under normal conditions propagate in space strictly rectilinearly and at very high speed. According to Newton's ideas, when these particles hit a boundary surface between two media of different densities, reflection occurs (similar to the way a tennis ball bounces off a smooth surface) or else there is either refraction or diffraction of the light, both of which effects he interpreted with recourse to his theory of gravitation, quantifying the attractive force of unevenly distributed masses in space exerted upon the mass borne by each light corpuscle. It is more difficult than one might expect

---

[1]For further examples and for a methodological discussion, cf. Gentner and Stevens (1983).

© Springer International Publishing AG, part of Springer Nature 2018
K. Hentschel, *Photons*, https://doi.org/10.1007/978-3-319-95252-9_4

to infer from these phenomenological assertions or laws of nature what Newton was really thinking or imagining. Newton concealed such thoughts from his most obstinate critics, including Robert Hooke (1635–1703) and other members of the *Royal Society*, who were constantly measuring him up against the empirical model that he himself had erected as a standard.

Some wanted to pin him down to promoting a projectile model of light. That was Robert Hooke's objection, for example, in his first critique of Newton's "new theory of light and colours" in 1672. Newton replied that he drew a distinction between secure facts and hypotheses:

> that light is a body [...], it seems, is taken for my Hypothesis. 'Tis true, that from my Theory I argue the Corporeity of Light; but I do it without any absolute positiveness, as the word perhaps intimates; and make it at most but a very plausible consequence of the Doctrine, and not a fundamental Supposition, nor so much as any part of it.

The reason for this caution was clear: Newton knew very well that he could not prove this model of light on the basis of observations alone; other interpretations were still possible.

> But I knew, that the Properties, which I declar'd of Light, were in some measure capable of being explicated not only by that, but by many other Mechanical Hypotheses. And therefore I chose to decline them all, and to speak of Light in general terms, considering it abstractedly, as something or other propagated every way in straight lines from luminous bodies, without determining, what that Thing is.

Newton did not want to invent hypotheses out of the blue, so to speak, in the Cartesian manner. But that did not prevent him from often using such hypotheses on a trial basis. Clearly, someone who has declared "hypotheses non fingo" so publicly would have qualms about admitting that heuristically he certainly did work with such 'hypotheses.' He took great pains, though, to distinguish what he 'really' believed from what he was just toying with as a mental exercise.[2]

Notwithstanding this public reticence, personal notes found among Newton's papers provide deeper insight. Particularly in his notebook *Questiones quædam Philosophiæ*, which could be dated to 1664–65, we find a sketch of a very small "globulus of light" (hence a particle, see Fig. 4.1) that is surrounded by a cone of "subtile matter" (a kind of luminiferous aether), "which carrys before it the better to cut the ether." Such unpublished manuscripts show that—contrary to his motto "hypotheses non fingo"—Newton definitely did develop elaborate models, which are intimately connected with his concepts and his style of thinking.[3] I would like to subsume them under the label 'mental models.'

The drawing shows this spherical corpuscle of light enveloped in "subtile matter." The motion of the globulus of light from the left rightwards makes the luminiferous

---

[2]Newton's complex methodology is discussed by Harper (2011) and Achinstein (2013) with further literature cited there.

[3]MS Add. 3996, Cambridge University Library, Cambridge, UK, fol. 104v, on-line at http://www.enlighteningscience.sussex.ac.uk/view/texts/normalized/THEM00092 ; cf. ibid., fol. 98r and Herivel (1965) p. 122 for an exact hydrodynamical analogue to the above sketch: a sphere with water flowing around it with a head wave in front (right) and eddies behind (left).

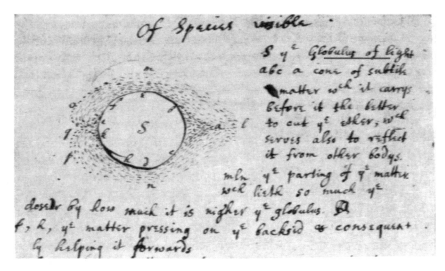

Fig. 4.1 Newton's "globulus of light" 1664–65 in *Questiones quaedam Philosophiae*, fol. 104v, MS Add. 3996, Cambridge University Library, Cambridge, UK. Reprinted by permission

aether flow around it. A compression zone depicted on the right-hand side poses due resistance. Newton presumed that the turbulent zone on the left-hand side generated a kind of push ahead, "by pressing on the back side ... consequently helping it forward."[4]

A later stage in the development of this particulate theory of light is apparent in a paper Newton published in 1675 as well as in his *Principia* from 1687. In book I, section XIV he offered a physical explanation for refraction based on the assumption that light is a stream of material particles. The refraction of light as it passes from one medium into another of different optical density or index of refraction n, had long been known. Ptolemy and Ibn al-Haytham had already examined it by experiment. Newton interpreted it as the attraction of corpuscles of light to the denser (hence more massive) medium. Although Newton had no way of knowing the scale or velocity of these hypothetical corpuscles of light, he could still draw conclusions about the curvature of their trajectory toward the medium of greater density.[5]

The law of refraction that Ptolemy and Kepler had looked for in vain was discovered by Thomas Harriot (1560–1621) and Willebrord van Roijen Snel (latinized as Snellius 1580–1626) and first published by René Descartes (1596–1650).[6] Newton was able to derive the law by assuming that the component of the light corpuscle's propagation velocity parallel to the refracting surface v_{par} remains unchanged whereas the transversal component at the transition into the optically

---

[4]The above quotations are taken from Newton's *Questiones*, dated 1664–65 in the critical edition of this notebook ed. by J.E. McGuire and Martin Tamny, Cambridge Univ. Press, 1983, pp. 384–385.

[5]For details see Newton (1675a) pp. 256ff., resp., (1675b) pp. 186ff.; furthermore Hall (1993), Sepper (1994), Shapiro (2009).

[6]See Hentschel (2001) on Snel's discovery of the law of refraction.

denser medium v_t increases. The ratio of the sine of the angle of incidence θ_{in} to the sine of the angle of reflection θ_{ref} equals the ratio v_{par}/v_{medium}. Upon dividing the two equations, the unknown velocity v_{par} cancels out:

$$\sin(\theta_{in})/\sin(\theta_{ref}) = [v_{par}/v_{air}]/[v_{par}/v_{water}] = [v_{water}/v_{air}].$$

As you can see, Newton's derivation always presumed that the propagation velocity of light corpuscles *increases* inside the denser medium because of their gravitational attraction to the more numerous corpuscles of that denser medium. This presumption clashed with the requirement of the wave model of light that the velocity of light should *decrease* in the optically denser medium. It only became possible to decide between these two fundamentally different conceptions on experimental grounds in the nineteenth century. That is why during early modern times both models could remain standing even though, owing to his unrivaled authority during the eighteenth century, Newton's model was far more influential than Huygens's competing wave model.[7]

4.2   Einstein's Mental Model of Light Quanta Around 1909

A similar discrepancy between the public and private Newton that we have just seen can also be ascertained with Albert Einstein, although it was perhaps not quite as drastic, because Einstein did not fabricate such a distorted image of his own methodology. Most of the quotations of Einstein that we have seen in earlier sections of this book were intended for the public. As we saw, Einstein thought very carefully, if not strategically about what to express in his publications and what to withhold. When referring to and conducting thought experiments about light quanta, what exact notions did he have? What do we know about Einstein's private mental models of light and light quanta that were not intended for public consumption? When one considers the flood of publications about every aspect of Albert Einstein's life and works, we know astonishingly little. One reason is the scarcity of sources from the early period of his professional activity, before he had secretaries and other assiduous souls to archive almost everything he wrote or laid his hands on. We have only a few unpublished notes and draft documents from those early years before 1916 compared to the later period and much fewer letters, although some of those that have survived are very instructive.[8] Existing correspondence with trusted friends including Besso, Solovine and Zangger, and influential colleagues including Lorentz, Planck or Max von Laue, allow a glimpse behind the scenes. For reasons of space I shall have to limit these glimpses to a few examples. One of the most telling

---

[7]See Cantor (1983), Eisenstaedt (2007) and standard histories of optics such as Park (1997) or Darrigol (2012) as well as here Sect. 3.1.

[8]See his collected papers (CPAE). Initially a comprehensive publication of all of Einstein's surviving letters, the later volumes are unfortunately becoming increasingly selective for reasons of economy.

passages in Einstein's oeuvre is a rare incidence in which he speaks not only about phenomenological conclusions but also about the causal entities behind them, in our context: light quanta. The following quotation is from his lecture before the eighty-first convention of the Society of German Scientists and Physicians on 21 September 1909 in Salzburg.

> Still, for the time being the most natural interpretation seems to me to be that the occurrence of electromagnetic fields of light is associated with singular points just like the occurrence of electrostatic fields according to the electron theory. It is not out of the question that in such a theory the entire energy of the electromagnetic field might be viewed as localized in these singularities, exactly like in the old theory of action at a distance. I more or less imagine each such singular point as being surrounded by a field of force which has essentially the character of a plane wave and whose amplitude decreases with the distance from the singular point.[9]

What epistemic status does this mental model of light quanta have for Einstein in 1909? The conjunctive mood and the many restrictive utterances ("Still, for the time being the most natural interpretation seems to me to be ..."; "It is not out of the question ...") show that Einstein was arguing extremely carefully, as if he himself were not yet entirely convinced—as if he were expressing conjectures ("I imagine it this way"), totally oblivious of his audience. There happens to be an eye-witness account of Einstein's idiosyncratic style of lecturing on this occasion. The spectroscopist and experimental physicist Heinrich Kayser (1853–1940) was in attendance in Salzburg: "I also saw and listened to Einstein here. He delivered a long lecture as he roamed back and forth on the podium filling the whole narrow side of the hall, ever staring at the floor in front of him; when he turned around, this happened in a way that his back was turned to the audience. He definitely left the impression of a dreamer completely unaware of his surroundings."[10] This peculiar mannerism is only partly explained by the fact that this had been Einstein's very first solo presentation at a major scientific convention, after having spent the last few years working on his own at the Bernese patent office with hardly any academic or public contact. After brief intermezzos in Zurich and Prague, Einstein went to Berlin in 1914 as director of the brand new *Kaiser-Wilhelm-Institut für Physikalische Forschung* which had been specially cut to size for him (which special treatment was repeated again in 1933 with his permanent fellowship at the *Institute for Advanced Study* in Princeton). That meant he was not required to teach regularly and the public lectures he did hold were generally for fundraising purposes. It is a great privilege for us to be able to experience this spinning out of a mental model before our eyes, as it were, thanks to this reminiscence and the verbatim proceedings of such talks and discussions at this convention of the *Gesellschaft Deutscher Naturforscher und Ärtze*.

---

[9]Einstein (1909a) quote on p. 499 (CPAE, vol. 2, doc. 60, transl. ed., p. 394). And similarly, in the subsequent discussion to Einstein (1909a) p. 826 (CPAE, vol. 2, doc. 61). In a letter to Sommerfeld on 29 Sep. 1909 (CPAE, vol. 5, transl. ed., p. 135), Einstein writes analogously not about particles but about an "arrangement of light energy around discrete points moving with the speed of light."

[10]Heinrich Kayser's memoirs (1936), ed. by Matthias Dörries and K. Hentschel in 1996, pp. 228f., or pp. 250f. of the original typescript.

Among his close friends, for instance during meetings of their discussion group jocularly named "Olympia Academy," Einstein often did enjoy spinning out such models, as Besso and Solovine recalled. In his papers and talks this occurred much more rarely, however, which rather took on Kirchhoff's style of maximal generality and independence from concrete models of matter. Continuing the just-quoted passage of 1909, Einstein developed this mental model even a little further.

> If many such singularities are present at separations that are small compared with the dimensions of the field of force of a singular point, then such fields of force will superpose, and their totality will yield an undulatory field of force that may differ only slightly from an undulatory field as defined by the current electromagnetic theory of light. I am sure it need not be particularly emphasized that no importance should be attached to such a picture as long as it has not led to an exact theory.[11]

As if he felt caught in the act of thinking aloud, Einstein cut this train of thought short and—as far as I can see—never came back to it again. The epistemological constraint in the final sentence draws another parallel to Newton's cautious way of expressing himself (e.g., in his controversy with Hooke mentioned above). Cautious—indeed, scrupulous—though Einstein was, even in private correspondence with his mentors and close friends, the more naturally at ease was this technical expert from the patent office in public confrontation with professional scientists.

Einstein's talk was published in *Physikalischen Zeitschrift* together with the subsequent, very frank discussion, in which Planck, Stark, Rubens and Einstein himself also participated. Max Planck began with a longer comment about whether it really was necessary—as Einstein was demanding— "to conceive the free radiation in vacuum, and thus the light waves themselves, as atomistically constituted, and hence to give up Maxwell's equations," or—as Planck was just trying to construe in his second quantum theory (see above)—to regard this quantization as merely an epiphenomenon of the interaction between radiation and matter:

> Perhaps we may be allowed to assume that an oscillating resonator does not have a continuously variable energy, but that its energy is a simple multiple of an elementary quantum instead. I believe that by using this theory one can arrive at a satisfactory theory of radiation.[12]

There then followed something like Planck's swan song for the ideal of classical physics pursued particularly fervently by British physicists like Maxwell and his followers—namely, the ideal of the possibility of building a mechanically graphic model of phenomena describable by theory and experiment. Such intuitive visualizability had come increasingly under fire after 1900, and by 1909 Planck was ready to concede that perhaps one must finally give up this ideal:

> The question is, then: How does one visualize something like that? That is to say, one asks for a mechanical or electrodynamic model of such a resonator. But mechanics and current

---

[11] Einstein (1909ab) pp. 499–500 (CPAE, vol. 2, transl. ed., p. 394). Kojevnikov (2002) pp. 188ff., interprets these passages as a mental extension of the nonmechanical, noncorpuscular conception of electrons by Lorentz. I rather view them as anticipating quantized field theories. 'Undulatory' generally means wave-shaped with the meaning here of "described by Maxwell's equations."

[12] Max Planck discussing Einstein (1909aa) p. 825 (CPAE, vol. 2, doc. 61, transl. ed., quotes on pp. 395 and 396).

electrodynamics do not provide for discrete elements of action, and hence we cannot produce a mechanical or electrodynamic model. Thus, mechanically this seems impossible, and we will have to get used to that. After all, our attempts to mechanically represent the luminiferous ether also have failed completely. [...] In any case, I think that first of all one should attempt to transfer the whole problem of the quantum theory to the area of *interaction* between matter and radiation energy; the processes in pure vacuum could then temporarily be explained with the aid of the Maxwell equations.[13]

Johannes Stark objected to this attempt by Planck to limit the damage and theoretically immunize it with reference to the characteristic curve of angular distribution for hard x-rays (see Sect. 4.4). Planck swung the next club with the argument that the phenomenon of interference works even "at the enormous phase differences of hundreds of thousands of wavelengths." If a quantum interfered with itself, it would have to have an extension of hundreds of thousands of wavelengths. "This is also a certain difficulty."[14]

4.3 Einstein's Own Doubts About the Light-Quantum Concept 1910–15

In 1911 Einstein confessed to the Dutch theoretician Hendrik Antoon Lorentz (1853–1928), "I am not the orthodox light-quantizer for whom you take me."[15] Lorentz's interference arguments were the background to this remark. He had noted that the ability of light to interfere over very long distances proves that, given appropriate experimental conditions, light 'quanta' were occasionally capable of having considerable spatial extension. This went against any naive assumption about their being point-shaped. In his letters and publications, Lorentz cited experiments by Lummer and Gehrcke at the PTR in Charlottenburg, Berlin, indicating that interferences could occur with path differences of over 80 cm.[16] That was why Lorentz found it

hard to subscribe to the view that the light quanta retain a certain individuality even during their propagation, as if one were dealing with "punctiform" energy quantities or at least energy quantities concentrated in very small volumes. It seems to me that it can easily be shown that a light quantum can have a considerable extension in the direction of propagation as well as perpendicularly to it, and that under certain circumstances only a *part* of a light quantum reaches the retina and brings about the perception of light.[17]

---

[13]Einstein (1909a) pp. 825, 826 original emphasis (CPAE, vol. 2, doc. 61, transl. ed., p. 396).

[14]Einstein (1909a) p. 826 (CPAE, vol. 2, doc. 61, transl. ed., p. 397). For Stark's reaction to this, see Sect. 4.4 below.

[15]A. Einstein to H.A. Lorentz, 27 Jan. 1911, CPAE, vol. 5, doc. 250, p. 276 (transl. ed., p. 175).

[16]See, e.g., Lorentz to Einstein, 6 May 1909, CPAE, vol. 5, doc. 153, Kox (2008) as well as Lorentz (1910a) pp. 354–355 and Lorentz (1910b) p. 1249. Such coherence lengths of trains of interfering waves can currently be made to measure some meters with modern lamps; ones with lasers can be kilometers long. That's a far cry from 'point shaped.'

[17]Lorentz to Einstein, 6 May 1909, CPAE, vol. 5, doc. 153, p. 174 (transl. ed., p. 110), original emphasis. He was to be proven right with the former but not with the latter.

In putting forward these weighty arguments, Lorentz was certainly not being polemical or triumphant. He was merely skeptical because he thoroughly understood what had motivated Einstein, Johannes Stark and a few others to consider such a radical interpretation. Lorentz himself preferred to adopt a diluted fictionalistic 'as-if' reading, with Einstein's light quantum being no more than a fiction:

> It is a real pity that the light quantum hypothesis encounters such serious difficulties, because otherwise the hypothesis is very pretty, and many of the applications that you and Stark have made of it are very enticing. But the doubts that have been raised carry so much weight with me that I want to confine myself to the statement: "If we have a ponderable body in a space enclosed by reflecting walls and filled with ether, then the distribution of the energy between the body and the ether proceeds *as if* each degree of freedom of the ether could take up or give off energy only in portions of the magnitude $h\nu$."[18]

These counterarguments, which the persistent Dutchman expressed in confidential letters, later also in his publications during this time,[19] did impress Einstein. Lorentz had been one of Einstein's most important mentors when he was developing his special theory of relativity, even though he did not always follow his advice. When the two made each other's personal acquaintance, a kind of friendship developed. Owing to the strong asymmetry in their personalities and the generational gap, it acquired somewhat the nature of a grandfather toward his grandchild.[20]

In 1910 Einstein began to have serious doubts about his light quantum. He wrote to Jakob Laub: "At the moment I am very hopeful that I will solve the radiation problem, and that I will do so *without light quanta*. I am awfully curious how the thing will turn out."[21] It soon turned out, though, that this alternative was far too costly— it came at the expense of energy conservation. One year later Einstein confessed to Michele Besso that he had retreated into applying the light-quantum hypothesis purely instrumentalistically. He was implementing it in the sense of his so very cautiously presented "heuristic point of view" of 1905, merely as a mathematical tool; he himself no longer believed in its existence:

> I don't ask myself anymore whether these quanta really exist. I'm not trying to construe its structure either anymore, because now I know that my brain can't get anywhere that way.

---

[18]Ibid., doc. 153, p. 176 (transl. ed., p. 112), original emphasis. The draft of this letter (Einstein archives, EA 16-417) states in Dutch: "Deze bezwaren jammer want theorie lichtquanta wel mooi." On Vaihinger's fictionalism and its application in physics, see Hentschel (1990) sec. 4.4, Hentschel (2014a) and the primary sources cited there.

[19]See, e.g., Lorentz (1910a) p. 354: "The aforesaid ought to suffice to show that there can be no question of light quanta which while propagating remain concentrated within small spaces and always undivided."

[20]Prior to this, Einstein described Lorentz to others in glowing terms, such as, in his letter to Jakob Laub on 17 May 1909, CPAE, vol. 5, doc. 160, p. 187 (transl. ed., p. 119): "H.A. Lor[entz] and Planck. The former is an amazingly profound and at the same time lovable man."

[21]A. Einstein to J. Laub, 4 Nov. 1910, CPAE, vol. 5, doc. 231, pp. 260–261 (transl. ed., p. 166), original emphasis. Cf. Einstein to Laub, 27 Aug. 1910, ibid., doc. 224, p. 254 (transl. ed., p. 162): "I have not made any progress regarding the question of the constitution of light. There is something very fundamental at the bottom of it."

But I am combing through the consequences as thoroughly as possible, in order to become informed about this notion's area of applicability.[22]

During the discussion at the Solvay conference in 1911, Einstein spoke of an "unsolved puzzle" and described his own light-quantum hypothesis dismissively as a "provisional attempt," as an "auxiliary idea, which does not seem compatible with the experimentally verified conclusions of the wave theory."[23] The scientific community soon caught wind of these serious objections by Lorentz, Planck and other authorities in theoretical physics and Einstein's own qualms were not a secret either.

In a plenary talk at the eighty-third meeting of the Society of German Scientists and Physicians in 1911, Arnold Sommerfeld contrasted Einstein's already trusted (special) theory of relativity of 1905 against the much more speculative theory of light quanta of the same vintage:

> The theory of energy quanta is an entirely different and problematic current issue [...] The fundamental concepts here are still in a state of flux and the problems are countless. [...] Einstein drew the farthest-reaching consequences of Planck's discovery [...] without, as I believe, maintaining his bold standpoint of the time anymore now.[24]

Robert A. Millikan, in late 1912 still among the most convinced critics of Einstein's light-quantum hypothesis, put it more directly in his plenary lecture before the *American Association for the Advancement of Science* in Cleveland, Ohio in 1913.

> Lorentz will have nothing to do with any ether-string theory, or spotted wave-front theory, or electro-magnetic corpuscle theory. Planck has unqualifiedly declared against it, and Einstein gave it up, I believe, some two years ago. [...] In conclusion then we have at present no quantum theory which has thus far been shown to be self-consistent, or consistent with even the most important of the facts at hand.[25]

However, Einstein eventually overcame these doubts, latest with his paper about induced emission from 1916.[26] His new confidence was primarily grounded in "the simplicity of the hypotheses, the generality with which the analysis can be carried out so effortlessly, and the natural connection to Planck's linear oscillator (as a limiting case of classical electrodynamics and mechanics)."[27] In 1917 he wrote to his friend Michele Besso (1873–1955) with reference to the same paper: "The quantum paper I sent out has led me back to the view of the spatially quantumlike nature of radiation energy. But I have the feeling that the actual crux of the problem posed

[22]Einstein to Besso, 13 Mai 1911, in Speziali (1972) pp. 19–20, resp. CPAE 5, doc. 267, p. 295, our translation.

[23]Einstein (1911/12) p. 347 (CPAE, vol. 3, doc. 26, transl. ed., p. 419) and the discussion, p. 359 (doc. 27, transl. ed., p. 431). The German translation of these French proceedings appeared late, in 1914.

[24]Sommerfeld (1911a) p. 31.

[25]Millikan (1913) pp. 132–133; similarly in Millikan (1916a) p. 384: "Einstein himself, I believe, no longer holds to it."

[26]Einstein (1916a, b) is discussed further in Sect. 3.9.

[27]Einstein (1916a) p. 322 (CPAE, vol. 6, doc. 34, transl. ed., pp. 215–216).

to us by the eternal enigma-giver is not yet understood absolutely. Shall we live to
see the redeeming idea?"[28] In 1921 he told his friend Paul Ehrenfest (1880–1933)
in desperation: "Pondering about light quanta is driving me crazy." And toward the
end of his life Einstein, who by then was conducting his research largely in isolation
with a few assistants at the *Institute for Advanced Study* in Princeton, still declared:
"All those 50 years of careful pondering have not brought me closer to the answer to
the question: 'What are light quanta?' Today any old scamp believes he knows, but
he's deluding himself."[29]

4.4 Johannes Stark's Mental Model of Light Quanta

The next actor whose mental model we shall be taking a closer look at is a particularly ambiguous figure in the history of physics. Johannes Stark (1874–1957)
published important experimental research in the fields of gaseous discharge physics
and spectroscopy. He also discovered the Doppler effect in canal rays as well as the
splitting of spectrum lines in an electric field (known since as the Stark effect) and
in 1919 was awarded the Nobel prize in physics for these exceptional achievements.
He was the editor of his own journal, *Jahrbuch für Physik und Elektronik*, and was
soon appointed an associate professorship in physics, in 1906—first at Göttingen,
later at the polytechnic in Hannover. In 1908 he was called to fill a chair, first at the
polytechnic in Aachen, then from 1917 at Greifswald, and 1920–22 at Würzburg.
This frequent change of location points to one of his faults: Stark had an unpleasantly
abrasive manner and quickly got embroiled in extremely aggressive polemics wherever he went, making himself many enemies along the way. After a hefty dispute
about his favored student and coworker, Ludwig Glaser, whose application to qual-
ify for academic teaching had been declined, Stark left the University of Würzburg
in protest and founded a private laboratory in his hometown, Ullersricht near Wei-
den in the Upper Palatinate, financed with his Nobel prize money, and worked on
commercial research as a manufacturer of porcelain and tiling. His hopes of soon
receiving another call back into academia or an appointment to a directorship at the
bureau of standards (PTR) in Charlottenburg, Berlin, were in vain. From 1920 on his
already latent anti-Semitism became increasingly vicious, which only led to further
professional isolation. Stark joined the Nazi Party early on, in 1930, and publicly
applauded the demise of the Weimar Republic in 1933. In May of that same year,
he was appointed president of the PTR as well as president of the German scientific research foundation, *Deutsche Forschungsgemeinschaft*. But he lost the latter
position two years later to an SS-officer, and the former one three years later. He did
not succeed in fulfilling the coveted role of head science policy-maker during the

---

[28] Albert Einstein to Michele Besso, 9 Mar. 1917, in Speziali (1972) p. 103 (CPAE, vol. 8, doc. 306,
transl. ed., p. 293).

[29] Albert Einstein in a letter to Michele Besso, 12 Dec. 1951, in: Speziali (1972) p. 453 (transl. in
Hentschel and Grasshoff (2005) p. 60).

Nazi period otherwise either. When the "thousand-year" *Reich* collapsed prematurely in 1945, Stark had to appear before a denazification tribunal and was condemned in July 1947 as a 'major offender.' His appeal in 1949 reduced his charge to 'fellow traveler' and he was let off with a fine.[30]

Paradoxically, this rabid Nazi had, as a young academic, once been a most passionate backer of Einstein's new light-quantum hypothesis. In 1909 Stark sided with Einstein in his debate with Planck at the scientists' convention in Salzburg (see Sect. 4.2).[31] So this experimental physicist counted among the earliest adherents of Einstein's conception that the radiation field itself—not just its interaction with matter—was quantized. Stark's best indicator for this thesis was the phenomenon called 'needle rays.' According to Stark's own experimental evidence, this hard (or high-energy) x-radiation, which due to Einstein's equation $E = h \cdot \nu$ is also high-frequency, "can still achieve concentrated action on a single electron" when discharged from an x-ray tube into the surrounding space over distances of 10 meters and more. "I believe that this phenomenon does represent a reason for considering the question of whether the energy of electromagnetic radiation should not be considered as concentrated even where it occurs detached from matter."[32] To Planck's counterargument that the ability of electromagnetic radiation to interfere over many multiples of a wavelength should lead one to assume that those purportedly atomic-sized light quanta must have enormous extension (cf. Sect. 4.2 above), Stark responded by pointing out that all the experiments to date had used very high radiation densities, with the consequence that "a very large number of quanta of the same frequency were concentrated in the beam of light; this must be taken into account when discussing those interference phenomena. With radiation of very low density, the interference phenomena would most likely be different." Therefore, it was not yet proven that individual light quanta had to be so extended.[33] In a subsequent article in that same year on "Roentgen Rays and the Atomistic Constitution of Radiation" Stark elaborated his own mental model of light quanta further:

> From the assumption, therefore, that the conversion of energy at the individual resonators follows the law of light quanta, it can be concluded on the basis of experimental results that the individual energy quantum of electromagnetic radiation at frequency ν, if it is moving at the velocity $c = 3 \cdot 10^{10}$cm sec^{-1} in a vacuum, remains concentrated in a volume whose linear extension is of the order of c/ν, hence of the wavelength $\lambda = c/\nu$. Accumulation of oscillatory electromagnetic energy at frequency ν within a volume larger than λ^3 means a spatial aggregation of individual radiation quanta of the magnitude $h\nu$; oscillatory electromagnetic energy at frequency ν can be subdivided in experiment only up to the amount of the elementary quantum $h\nu$.[34]

---

[30] On Stark's career and complicity in the Nazi regime, see Hoffmann (1982), Kleinert (1983), Hentschel (1996) and Eckert in Hoffmann and Walker (2007).

[31] This early contact between Stark and Einstein is covered by Hermann (1969/71), see also CPAE, vols. 1–2.

[32] Johannes Stark in the discussion following Einstein (1909a) on p. 826, resp. CPAE, vol. 2, doc. 61, p. 586 (transl. ed., p. 397).

[33] Ibid. Thereupon Einstein developed his hypotheses about singularities surrounded by vector fields already described above in Sect. 4.2.

[34] Stark (1909a) p. 583. For consistency I have substituted ν for Stark's n for the frequency.

This shows that Stark was among the first to regard Einstein's light quantum as more than an incidentally "effective heuristic auxiliary tool." He elevated it to the level of a veritable mental model that he continued to hone in the coming years. His model found beneficial applications of Einstein's light quantum to the following research areas, in which this experimental physicist was actively employed:

- the aforementioned atomistic structure of x-rays including its characteristic curve of angular distribution and its energy distribution,[35]
- photochemical processes of light absorption,[36]
- the photoelectric effect,
- spectroscopy in the optical spectrum, the UV, x-ray and gamma-ray ranges,[37]
- the Doppler effect with canal rays in which Stark found "certain new phenomena" that he thought "could be interpreted as a direct experimental verification of the light-quantum hypothesis."[38]
- the vaporizing effect of light where, different from the photoelectric effect, it is not just electrons that are dislodged by the incident electromagnetic radiation but entire molecular surfaces that are dissociated from the material and thus "vaporized" ("*zerstäubt*"),[39] and finally,
- modeling of electromagnetic processes of emission and absorption.[40]

Einstein himself had already pointed out the usefulness of his light-quantum hypothesis in providing qualitative and quantitative explanations for the first three of these points. As a theoretical physicist, Einstein tended to base his thermodynamical and statistical proof on deductions from considerations about energy distribution in radiation coming from overarching models of matter. Stark, by contrast, had his own approach to all these topics coming from experimental induction, often coupled with models of matter that were very detailed, indeed pedantically specific. The fact that despite these completely different ways of thinking they often both ultimately arrived at very similar results and conclusions was actually a very good sign and could have led to a productive resonance between theoretician and experimenter. Stark let this chance at such a fruitful relationship slip by very early on, though. Passages like the following one from a paper from 1909 are indicative. He was actually trying to bolster the light-quantum hypothesis against the criticism raining in on Einstein from almost every quarter. But the tone chosen by Stark in 1909 simply could not sit well with Einstein, who had himself been drawing attention to various experimental situations in which the light-quantum hypothesis might well prove its merit since 1905. In Stark's eyes, nobody but he personally as experimental expert was qualified to pronounce the empirical validity of Einstein's speculative heuristical hypothesis and perhaps even someday to warrant its "reality":

---

[35] Stark (1909a, b, 1910a, b).

[36] Stark (1908b, 1912a, b).

[37] Stark (1909b c, 1912c).

[38] Stark (1908b) p. 889 and Stark (1908a).

[39] On the latter, see Stark (1908c).

[40] Stark (1927, 1930).

> The hypothesis of the atomistic constitution of electromagnetic radiation energy initially seems too alien, seems far too contradictory to the long-accepted wave theory of light, making one possibly be inclined to reject it from the outset without closer inspection. Even a more discerning mind would receive it with great hesitation because of its far-reaching theoretical significance.

> As long as we bear in mind the hypothetical character of the light-quantum hypothesis, however, and as we trace its consequences never forget the hypothetical point of departure, it cannot do any harm or cause confusion. Its great theoretical importance and its in many instances already proven systematic and heuristical power, on the other hand, compel us to continue to investigate it; albeit, in doing so it would be advisable to shift the contest between the light-quantum hypothesis and the opposing older views as early as possible away from the sphere of speculation and theoretical debate onto the firm ground of experiment.[41]

Stark eventually did issue this warrant of "reality" for "the corpuscle of luminous energy, which can detach itself from an electron's energy field, can continue to move of its own accord without a medium, and can deposit itself onto an approaching free electron with an increase in its kinetic energy" as late as 1950, in his last book, in which he attempted to record the current state of research "after fifty years of experimental wrestling for knowledge about physical reality."[42]

In 1912 when Albert Einstein published in *Annalen der Physik* the connection between the Planck constant h and the chemical disintegration of molecules upon absorption of light, Stark reacted promptly with a short paper "On the Application of Planck's Elementary Law on Photochemical Processes," in which he lined up Einstein's findings against his own. Both of them had independently concluded: "a gas molecule that decomposes under the absorption of radiation of frequency ν_0 absorbs (on average) the radiation energy $h\nu_0$ in the course of its decomposition."[43] On one hand, this was a veiled priority claim, as Stark's own publications on this topic dated back to 1908.[44] On the other hand, Stark wanted to point out the complementarity between Einstein's approach to the laws governing the chemical absorption of light and his own approach: "The connection between the work of dissociation (V) for

---

[41] Stark (1909a) p. 584; cf. also the contrast set between a "pragmatic" and "dogmatic way of working" in Stark (1922) Chap. I and Stark (1950) Chap. VII, where the methodical opposition alluded to here is brought to a polemical extreme.

[42] Stark (1950) p. 22 in sec. II.5 headed: "Die Existenz von Lichtkörperchen" or in the motto on p. 5. Stark chose a more careful formulation on p. 50 (ibid.), however: as "merely a suggestion that might stimulate further observations" ("lediglich ein Vorschlag, der zu weiteren Beobachtungen anregen mag"). But this passage comes from a draft paper that Stark had subsequently incorporated into his final book. It had originally attempted to persuade his "dogmatic opponents"; hence it is rather strategic in character. Nevertheless, it shows how tactical Stark was to the very last—as far as the ontological status of "light vortices" was concerned (see below).

[43] Einstein (1912a) pp. 837–838 (CPAE, vol. 4, doc. 2, transl. ed., p. 94), directly quoted by Stark (1912a) p. 468.

[44] See Stark (1908a) p. 889: "It is in keeping with the fundamental importance of such a general principle as the light quantum hypothesis that it allow the prediction of new phenomena as well as allow one to recognize the importance of processes that had hitherto received little attention. [...] in the second part of the present communication an attempt is made to apply this [light-quantum] hypothesis to photochemistry for the first time, which application yields three fundamental photochemical laws."

the individual absorbing valency location and the absorbing light quantum is the basic precondition for me; for Mr. Einstein it is a consequence of his theory of photochemical equilibrium founded on the assumption of the validity of Wien's distribution law."[45] He may have been right about this but his hope that Einstein would agree with him without qualification was dashed. Stark had forgotten that Einstein had pointed out those very photochemical laws as a consequence of his light-quantum hypothesis as early as 1905 and in greater detail even in a survey article in 1907. Einstein's response to Stark's brief article was terse:

> J. Stark has written a comment on a recently published paper of mine for the purpose of defending his intellectual property. I will not go into the question of priority that he has raised, because this would hardly interest anyone, all the more so because the law of photochemical equivalence is a self-evident consequence of the quantum hypothesis. But I see from Stark's remark that I did not bring out the purpose of my paper clearly enough. The paper was supposed to show that the derivation of the law of photochemical equivalence does not require the quantum hypothesis, but that it can be deduced from certain simple assumptions about the photochemical process by way of *thermodynamics*.[46]

Instead of relenting and dropping the issue, Stark had to present yet another proof of his hot temper, which earned him the nickname: "Giovanni Fortissimo" among physicist circles. In an "Answer to Einstein" he reinterpreted his foregoing notice as simply establishing the different paths by which the two had arrived at similar results. This was directly followed by a spiteful protest that Einstein's claim to a purely thermodynamic derivation was based on two assumptions which, Stark argued, "are not thermodynamic and together with all the associated explanations are not simpler than the application of Planck's elementary law provided by me, either."[47] That was the end of any chance at alliance between experimenter and theoretician, even though at that time precisely these two physicists had the best purview over the far-reaching implications of the light-quantum hypothesis. Stark's difficult character, quick temper and impetuousness brought him many enemies in the scientific community. Whereas from 1910 on Albert Einstein tended to view this personal fault of Stark's with humor, a kind of archenmity developed between Stark and Sommerfeld that lasted for life.[48] But even neutral observers, such as Hantaro Nagaoka from Japan, who visited Johannes Stark at the end of 1910 in Aachen, had nothing more encouraging to say about him. To Ernest Rutherford in Manchester Nagaoka reported: "Stark is propounding his Lichtquantentheorie; there is some doubt whether he will succeed in explaining the interference phenomena, or not. The Germans say that he is full of phantasies, which may be partly true".[49]

Unfortunately Stark's tendency to reject current trends in the physics of his day and to construct wrong-headed models of matter with narcissistic conviction got worse

---

[45] Stark (1912a) p. 468; analogously also Stark (1909a) p. 583: "On the grounds of applications of the light-quantum hypothesis to experiments on x-rays, I arrive along a different route from Einstein's, which is perhaps shorter and simpler, at the same conclusion as Einstein's."

[46] Einstein (1912c) p. 888 (CPAE 4, doc. 6, transl. ed., p. 125); further details in Stark (1908b).

[47] Stark (1912a) p. 496.

[48] See Hermann (1968).

[49] H. Nagaoka to E. Rutherford, 22 Feb. 1911, quoted after Stuewer (2014) p. 147.

with age. In 1922 Stark's book on "The Current Crisis in German Physics" appeared, in which he vituperated about the growing imbalance between theoretical and experimental physics, about Einsteinian propaganda and public sensationalism.[50] In 1927 Stark published his monographic rebuttal to the quantum theory of Bohr and Sommerfeld. He postulated an "axiality of light emission and atomic structure": not only material "quantum vortices" (*Quantenwirbel*)—in other words, electrons and atomic nuclei—but also light quanta—renamed by Stark "light vortices" (*Lichtwirbel*")— are axial in structure. Light vortices accordingly have a rotational axis around which its electromagnetic field is arranged in the form of a rotator." The angle between the inner rotational axis and the direction of motion would then also yield the state of polarization of such a light vortex.[51] The processes of emission and absorption of radiation in matter also got a new, more graphic reinterpretation by Stark, wholly in the style of what he was promoting as "Aryan physics" (*Deutsche Physik*)—that is, devoid of foreign words, intuitive in content, and purely descriptive without arcane mathematics:

> The light vortex is formed by the detachment of a part of the electromagnetic energy of a quantum vortex (electron), and in the reversed process a light vortex can fuse together again with a quantum vortex into a uniform electromagnetic body.[52]

Stark regarded the aether wave of classical electrodynamics as a mere fiction but "the corpuscular discreteness of luminous energy as reality."[53] Every process of light emission was directed and the energy of light was concentrated in volume elements of the edge length of a wave $\lambda = c/\nu$ until it was absorbed. Hence it stayed "together in the form of a corpuscle."[54] Because these corpuscles of light do not transmit charge, all the electric lines of force had to lead back into the particle. This compelled the conception that they be "particles of light shaped like a vortex."[55] The position of this vortex plane relative to the direction of propagation then determined the polarization of the light. Light vortices ought to have led to a slight deviation from the straight path of propagating light, which toward the end of his life

---

[50] See Stark (1922) and the scathing book review by Max von Laue (1923a) p. 30 (transl. in Hentschel (1996) p. 7): "All in all, we would have wished that this book had remained unwritten, that is, in the interest of science in general, of German science in particular, and not least of all in the interest of the author himself."

[51] See Stark (1927) p. 29 and Kleinert (2002) for further sources and related analyses. In a subsequent article in *Annalen der Physik*, Stark (1930) p. 687, Stark revealed that this concept "had been developed [...] out of the Newtonian idea of the light corpuscle."

[52] Stark (1927) p. 33. These light and quantum vortices could be viewed as Stark's graphic reinterpretation of photon and electron spin, which is conserved in emissive and absorptive processes. It is indicative that even Stark's coworkers Robert Döpel and Rudolf von Hirsch refused to associate themselves with this manuscript and declined being named as coauthors: see Hirsch and Döpel (1928) and Kleinert (2002) p. 217.

[53] Stark (1950) pp. 61ff.

[54] Stark (1950) p. 22; analogously also ibid., p. 31 and pp. 62f.

[55] Ibid., p. 40. These passages remind one of Michael Faraday's visual thinking.

Johannes Stark believed he had succeeded in observing. Any independent corroboration of this predicted effect of Stark's never appeared, though.[56]

Stark's outright promotion of the National Socialist cause from 1930 on decimated any remaining sympathy among physicists for this extraordinarily talented experimenter. The "Aryan physics" movement that Stark and his mindmate Philipp Lenard initiated remained a small group of two-to-three dozen dogmatic opponents of modern physics (particularly relativity theory and quantum mechanics), even during the Nazi period. After a brief rush of success during the first few years of that regime, it lost its influence from 1935 on just as quickly, and when the "Third Reich" came to an end, the physics community completely shunned its members as active party members and Nazi propagandists.[57]

4.5   J.J. Thomson's Mental Model of Hard X-Rays

When Einstein went public with his "heuristical point of view" in the *Annalen der Physik* in 1905, Joseph John Thomson (1856–1940) had already begun to draft a kind of atomistic theory of radiation. The earliest form of this hypothesis, which was to occupy J.J. Thomson for many years to come, appeared in the *Silliman Lectures* on "Electricity and Matter" which he delivered as a guest lecturer at *Yale University* in 1903. It became available in German translation just a year later. An American experimental physicist pointed out this historical priority: Robert A. Millikan, who in 1916 considered Einstein's light-quantum hypothesis merely a "very particular form of the ether-string theory."[58] J.J. Thomson applied in this theory the notion of localized corpuscular quanta of light in order to explain conspicuous aberrations in the propagation of x-rays (discovered toward the end of 1895):

(i) Their effect was extremely directed and point-like. The great penetration force of such hard 'needle' rays was still punctiform over a distance of 50–100 m from its source.

---

[56]Ibid. pp. 41–50, which reproduces Stark's eight-page manuscript, "Experimentelle Untersuchungen über die Natur des Lichtes," for which he had been unable to find a publisher. At this time his improvised laboratory on his son's farm in Eppenstatt had been requisitioned as housing for refugees. Stark had been expelled from his own property in Traunstein by Military Governor Thom in 1947 (ibid., p. 61).

[57]On this 'Deutsche Physik' group and its weak influence see, e.g., Hentschel (1996) pp. lxx–lxxvii, Hentschel (2005) pp. 90–95, Eckert in Hoffmann and Walker (2007) and Schneider (2015) along with the cited secondary sources. Lenard and Stark also served as scapegoats for the physics community's bid of being otherwise uncensurable.

[58]See Millikan (1913) p. 130, Millikan (1916b) esp. p. 383.

(ii) Their intensity did not decrease as $1/r^2$ but remained almost the same over larger distances r, disregarding the incidental ionization of a directly hit gas molecule.[59]

J.J. Thomson imagined electromagnetic energy spreading through the aether along Faraday lines of force, which signified more to him than mere mathematical constructs. He visualized the electromagnetic aether as having a stringy structure. J.J. Thomson traced all electromagnetic phenomena back to changes in the positions or shapes of 'Faraday tubes,' whose beginning and end points were electric charges but were indestructible even when stretched or compressed:

> From our point of view, this method of looking at electrical phenomena may be regarded as forming a kind of molecular theory of Electricity, the Faraday tubes taking the place of the molecules in the Kinetic Theory of Gases: the object of the method being to explain the phenomena of the electric field as due to the motion of these tubes, just as it is the object of the Kinetic Theory of Gases to explain the properties of a gas as due to the motion of its molecules. The tubes also resemble the molecules of a gas in another respect, as we regard them as incapable of destruction [...] This view of the Electromagnetic Theory of Light has some of the characteristics of Newtonian Emission Theory; it is not, however, open to the objections to which that theory was liable, as the things emitted are Faraday tubes, having definite positions at right angles to the direction of propagation of the light. With such a structure the light can be polarized, while this could not happen if the things emitted were small symmetrical particles as in the Newtonian Theory.[60]

Few were willing to accept this polished but therefore hyperspecific model of J.J. Thomson's, which was easily falsified for precisely that reason. Vehement critics of Einstein's light-quantum hypothesis remained skeptical. Robert A. Millikan, for example, dismissed J.J. Thomson's mental model point blank in his survey talk on 'Atomic Theories of Radiation' although he did not condemn more complex modelings of the aether per se:

> It may be difficult, not to say repugnant, to some of us to attempt to visualize the universe as an infinite cobweb spun by a spider-like creator out of threads that never become tangled or broken, however swiftly electrical charges may be flying about or however violently we enmeshed human flies may buzz, but such is the hypothesis [...]. That we shall ever return to a corpuscular theory of radiation I hold to be quite unthinkable. [...] But I see no a priori reason for denying the possibility of assigning such a structure to the ether as will permit of a localization of radiant energy in space, or of its emission in exact multiples of something, if necessary, without violating the laws of interference.[61]

From Millikan's perspective, Einstein's light-quantum hypothesis represented a modeling along the same lines. (But he was grievously underestimating the differences

---

[59] For these experimental findings see, e.g., Thomson (1903, 1908a), Barkla (1906, 1907, 1908a, b, 1910), W.H. Bragg (1907, 1908a, b, 1912/13), Sommerfeld (1911a), Millikan (1913) p. 128 and further primary citations there.

[60] Thomson (1893) pp. 4 and 43. On J.J. Thomson's model of light and aether at this time, see McCormmach (1967), Navarro (2005), Bordoni (2009, 1890) and further primary sources cited there.

[61] Millikan (1913) pp. 130 and 133; analogously also Millikan (1916b) p. 383: "we must abandon the Thomson-Einstein hypothesis of localized energy [...] which seems at present to be wholly untenable."

between Einstein's and J.J. Thomson's mental models!) That was why he continued for years to refer to the "Thomson-Planck-Einstein conception of localized radiant energy (i.e., the corpuscular or photon conception of light)."[62] Consequently, speculations about certain forms of electromagnetic radiation being like particles are older than Einstein's 'heuristical point of view' from 1905, and these pre-Einsteinian models continued to be upheld far into the 1920s.

4.6 W.H. Bragg's Neutral Pair Model of γ-Rays

Hard x-radiation was not the only bone of contention. In 1899 the discovery of γ-radiation by Paul Villard (1860–1934) raised new questions. Because they do not undergo electromagnetical deflection, many experimental physicists in 1904 agreed with Madame Curie that: "γ-rays are penetrating rays that are not influenced by the magnetic field and are comparable to x-rays."[63] This similarity between roentgen rays and γ-rays was uncontrovertible, nevertheless what those x-rays were still wasn't clear. This uncertainty, too, was translated onto the interpretation of γ-rays. At the beginning of the 1920s Charles Glover Barkla and William Henry Bragg (1862–1942) engaged in a hefty dispute about it.[64] Barkla advocated the thesis that x-rays and γ-rays were electromagnetic waves of very high frequency ν (therefore, according to Einstein's $E = h \cdot \nu$, also of very high energy), whereas Bragg contended that both cases involved streams of very small neutral particles which he called 'neutral pairs'.[65] Bragg did not deny that very good reasons existed for assuming that x-rays and γ-rays shared the same nature as light.[66] His opponent Barkla had just presented indicative proof that they were polarizable. All attempts to detect diffraction or interference had failed, however. They also resembled particles much more in other respects as well: "The absence of reflection and refraction give the [X and γ] rays more resemblance to the α and the β rays than to the radiation of light, and there is a still further similarity to the material rays in the way in which the X rays are irregularly scattered by every substance through which they pass."[67] Electromagnetic

---

[62]See, e.g., Millikan (1916b) p. 383, (1917) p. 237, (1924) in his Nobel prize lecture, p. 61 and the quote on p. 64, furthermore in his autobiography from 1950.

[63]Curie (1904) p. 41.

[64]On W.H. Bragg's background as an Australian trained in England who later immigrated permanently in 1909 after having returned home to Australia, and on his work, see Wheaton (1983) pp. 81ff., Jenkin (2004, 2007). See above in Sect. 3.3 about his opponent Barkla, whose experiments confirmed the polarizability of x-rays in 1905.

[65]For the first time in Bragg (1907) pp. 440f. These pairs of particles that Bragg occasionally also called 'neutrons' are *not* identical with what are currently being referred to by that name: those chargeless nuclear components also known as hadrons!

[66]Bragg (1912/13) p. vi: "the three forms of radiation ["α, β and X or γ rays"] differ in degree rather than in kind." Cf. Bragg (1912/13) pp. IV–V, where he argued that from this aspect, no difference apart from wavelength existed between x-rays, γ-rays and light.

[67]Bragg (1912) p. 106 the want of proof of diffraction is also discussed there.

waves dispersed over large volumes of space in circular wave fronts away from the center of excitation. Hard x-radiation and γ-radiation propagated very differently. They remained virtually point-shaped and carried the entire energy from the emissive focal point to the point at which they were absorbed by other matter in an inelastic process of collision. For Bragg, "In a very real sense X radiation is a "corpuscular" radiation consisting of entities or quanta, each of which moves uniformly in a straight line without change until in some encounter with an atom the X ray energy disappears and β ray energy takes its place."[68] He even thought his hypothesized neutral pairs could explain Barkla's indications that x-rays can polarize. He added the assumption that those externally neutrally charged pairs rapidly rotated around their shared center and had a preference for interacting with atoms rotating in parallel to their own rotation.[69]

Obviously these idiosyncratic considerations would not be left uncontested. Barkla pointed to serious problems with Bragg's hypothesis as regarded his interpretation of the results of Barkla's polarization experiments. Barkla was positive that they yielded "quite conclusive evidence in favour of the ether pulse theory."[70] Bragg countered this objection with his own experimental results. They demonstrated among other things that the secondary cathode rays released upon moderation of γ-rays by the absorbing matter, had a clear preference for the orientation parallel to the incident radiation, which flatly contradicted Huyghen's principle of spherically symmetrical waves.[71] It was not an easy task for defenders of the wave picture of roentgen and γ-radiation to model such directed 'needle rays' mathematically within the framework of Maxwell's theory of electromagnetism, although in this form Bragg's argument was false, because the spherical wave must refer to the emissive body's coordinate system at rest.[72] Robert Wichard Pohl (1884–1976) refused to treat Bragg's corpuscular theory of x-rays at all in his textbook "On the Physics of Roentgen Rays" from 1912, "because I see no possibility of interpreting even just the most important properties of those rays with its assistance." And these 'most important' properties were, of course, precisely the ones that proved the wave-like character of x-rays, as "aether radiation" (*Ätherstrahlung*), that is, for example, not to experience "the least deflection" in a magnetic field.[73]

If both of these rivaling conceptual models were excluded, the remaining possibility was to interpret γ-rays as a new kind of radiation with its own essential characteristics being a mixture of corpuscular and undulatory properties. Some went so far as to chop up γ-radiation into other types based on their powers of penetration,

---

[68] Bragg (1912) p. 138. Further analyses and references to Bragg's corpuscular theory of x-rays and γ-rays are provided by Stuewer (1971, 1975a) pp. 6–23 and Wheaton (1983) pp. 81ff.

[69] See, e.g., Bragg (1907, 1908a) p. 270.

[70] Barkla (1907) p. 662.

[71] See, e.g., Bragg (1908a) p. 270: "the kathode [sic] radiations from a given stratum of matter traversed by γ rays possess momentum in the original direction of motion of the rays, and this shows that the rays are material."

[72] The controversy between Sommerfeld and Stark on this point is discussed by Hermann (1968), and Wheaton (1983) pp. 116ff., 135ff.; see also the cited primary sources and correspondence there.

[73] See Pohl (1912) pp. VI, 18.

while others rather saw continuous variabilities within this unique class of γ-rays.[74] At the beginning of the 1920s the classification of γ-rays was still anything but unproblematic. The experimental physicist Erich Regener (1881–1955) used a relatively complicated argument by analogy to justify his idea of integrating γ-radiation within roentgen radiation, as a more thoroughly examined range:

> If we now look at the γ-rays of radioactive bodies, we must draw a parallel between them and roentgen rays. Just as is the case with roentgen rays, their nature is not as certainly known as are the natures of α- and β-rays. One is nonetheless generally inclined to presume them to be irregular wave motions of the luminiferous aether, so-called aether pulses. However, whereas the emission of roentgen rays occurs when cathode rays are *absorbed*, γ-rays are primarily known as a concomitant phenomenon upon the emission of β-rays by radioactive bodies. As with roentgen rays, a parallelism also exists between the penetrating power of γ-ray 'hardness' and the velocity of β-rays, for which they occur as a concomitant phenomenon.[75]

In Regener's taxonomy of, x-rays and γ-rays are by their nature 'related' in kind (*wesensverwandt*), this analogy (or as he put it: this "parallelism") outweighed the disanalogy between the penetration capacities of the two types of rays. It was much higher for γ-rays than for x-rays. In Regener's view, this difference legitimated the introduction of a separate class of rays. W.H. Bragg, on the contrary, always presented x-rays and γ-rays together, stressed their "strong family likeness" and even decided to regard γ-rays as part of the notion of x-rays.[76] In an attempt to motivate his mental model of a material x-ray or γ-ray, Bragg presented an interesting thought experiment. It is a form of transmutation or conversion argument:

> If we try to construct a theory which shall make the explanation of the interchangeability its principal feature, we are first led to conceive of a more material X ray. The electron of the β ray may be imagined as capable of attaching to itself enough positive electricity to neutralise its own charge and of doing this without appreciable addition to its mass. This is the transformation from electron to X ray: the reversed transformation occurs when the electron puts down its positive again.[77]

This thought experiment explains the charge neutrality of hard x-rays or γ-rays as well as the strongly asymmetrical angular distribution of the secondary radiation triggered by these rays.[78] But it does not explain its polarizability, not to mention the propagation of both at the speed of light (which experience had not yet finally confirmed). At the end of his examinations of radioactivity, Bragg pleaded that his corpuscular model of x-rays and γ-rays be taken as seriously as Stokes's aether pulse model, despite the still open problems, for: "A hypothesis is not to be set aside because

---

[74]On the former: Kleeman (1907) pp. 638, 662; the latter developed into γ-spectroscopy.

[75]E. Regener, c. 1912, p. 103: 'Die Strahlen der radioaktiven Substanzen,' separate offprint of a chapter from an unidentifiable book, among Regener's papers at the Archive of the University of Stuttgart. Similarly also in Regener (1915) p. 8: "In essence, so-called 'γ-rays are similar to roentgen rays." Egon von Schweidler (1910) unsuccessfully tried to reach an experimental decision.

[76]Bragg (1912) pp. vi, 105f., Chap. XII; see similarly Bragg (1907) p. 442, Bragg and Marsden (1908a) pp. 938 and 670: "The x-rays resemble the γ rays so closely that it is practically inconceivable that the two radiations should be essentially different."

[77]Bragg (1912/13) p. 191.

[78]On this point see esp. Bragg and Marsden (1908a) p. 670, (1908b).

it does not supply an immediate explanation of every fact." The pulse theory may have had some limited success in providing qualitative explanations for x-radiation early on in its history, but very little progress had been made since then and it stayed far behind general advancements. He continued,

> [...] we ought to search for a possible scheme of greater comprehensiveness, under which the light wave and the corpuscular X ray may appear as the extreme presentments of some general effect. To do this, the extreme views should be applied to all the phenomena of both light and X rays in order to find out how far each can be made effective.[79]

In 1914 Rutherford and a collaborator finally succeeded in making a first indirect determination of the wavelength of soft γ-radiation from radium B and C from interference off a crystal plate.[80] Just a year later it had already become established in the scientific community that x-rays and γ-rays were particularly energy-rich forms of electromagnetic radiation. But Bragg's mental model flared up again from time to time in later proposals about the structure of the light quantum: as a "wavelet" (Arthur Stanley Eddington in 1928), as a neutrino pair (Pascual Jordan and Ralph de Kronig in 1935–37), or even—physically very implausibly—as a matter-antimatter binding state (Bruce Wayne in Wayne (2009)). But all of these incredible constructions had obvious problems with indispensable limiting conditions from relativity theory and quantum field theory; and they never could take hold.[81]

4.7   Energy Packet Model by Planck, Debye and Sommerfeld

As already described in Sect. 2.3, Max Planck took fright at his own boldness after his interpolation of two approximation formulas for the radiation density of black-body radiation had worked so well. In retrospect, his paper from December 1900 seemed to him to be an "act of desperation" that he, "by nature, peaceable and not inclined to dubious adventures," had only undertaken because "a theoretical interpretation had to be found at any price, no matter how high." [82] Whereas Planck himself was intent on finding the best possible fit for his work with the methods and findings of classical physics, Einstein's papers from 1905 onward went straight to its sore spot to expose the inconsistency between the light-quantum hypothesis and classical Maxwellian electrodynamics. That was why afterwards Planck tried so hard to find a somewhat mellowed reinterpretation of his own conclusions of 1900—in other words, to 'salvage the phenomena' by changing the interpretative frame. How did Planck restyle in retrospect what his ominous quantization actually meant physically? His

---

[79]Bragg (1912/13) pp. 192–193; cf. Bragg (1912/13) pp. 237f.

[80]See Rutherford and Andrade (1914a) and supplementarily Meitner (1922) p. 382.

[81]Bruce Wayne, an American plant physiologist at *Cornell University* who also developed some alternative models of his own, offers a good survey of the literature in Wayne (2009) pp. 23ff.

[82]Planck to Robert Williams Wood, 7 Oct. 1931, quoted by Hermann (1969/71b) p. 23. Cf. p. 14 above for the context of Planck's research around 1900.

letter to the American physicist Robert Williams Wood provides interesting hints also in this regard. Planck's interpolation formula from 1900 prevented a divergence of the radiation density into infinity by virtue of its construction to fit the long-wave and short-wave limits of the radiation curve. The mathematical formalism thus halted what Paul Ehrenfest described as the "ultraviolet catastrophe" and an analogous "infrared catastrophe" at the other end of the spectrum.[83] But what is the physical significance? Planck's interpretation was as follows: The prescription that oscillators only emit radiation in finite packets of a minimum size given by Planck's quantum of action h prevented the radiation released by atoms or other oscillating systems from being subdivided into indefinitely many infinitesimal packets. Such a divergence in the numbers of packets of radiation, with individual magnitudes approaching zero, would have constituted the infrared catastrophe:

> Because, according to it [classical physics], over time the energy in matter must convert entirely into radiation. In order for it not to do so, we need a new constant [Planck's quantum of action h] that assures that the energy not disintegrate. [...] one finds that this dissipation of energy as radiation can be prevented by the assumption that energy be compelled from the outset to stay together in specific quanta. That was a purely formal assumption and I did not really consider it much, just that I must, under all conditions, cost what it may, force a positive result.[84]

According to Planck, the radiation field was not the underlying reason for the limitation to tiniest possible packets of energy but the oscillation properties of the resonators. This basic idea that the quantizing prescription actually referred not to the radiation field itself but only to the systems emitting that radiation, Planck developed further during the period 1906–13 into what came to be known as the 'second quantum theory.'[85]

Not just Planck, but a number of other physicists also saw the attraction of such a reinterpretation of quantization. It differed "from the light-quantum theory in its relation to electrodynamics insofar as it comes nowhere into conflict with electrodynamics, rather extending it as regards the course of processes that electrodynamics on its own knows nothing about."[86] This formulation by Planck's fellow theoreticians in Munich, Peter Debye and Arnold Sommerfeld, already implies the different kind of mental model that all supporters of this 'second quantum theory' conceived of such a "course" of emissive processes of electromagnetic energy at the atomic level. Debye and Sommerfeld articulated this in a lengthy paper on the "Theory of the photoelectric effect from the point of view of the quantum of action" as follows:

> An atom accumulates incident vibrational energy in the motions of its electrons until the action quantity $\int (T - U)dt$ has attained the value $h/2\pi$. [...] Once this action quantity $h/2\pi$

[83] See Ehrenfest (1911) p. 92.

[84] Planck to Robert Williams Wood, 7 Oct. 1931; cf. p. 14 above for a text-critical evaluation of the full quotation.

[85] On Planck's second quantum theory, see Planck (1910b)–(1913), as well as Needell (1980), Whitaker (1985), Gearhart in Hoffmann (2010) p. 116 and Kragh (2014c).

[86] Debye and Sommerfeld (1913) p. 875.

has been reached, the electron is released from the atomic bond with the just-attained kinetic energy T.[87]

Planck's quantum of action h was thus made into the critical fundamental constant of quantum physics.[88] This minimum volume of phase space had to be completely full before a vibrating atom could emit this minimum packet h in the form of radiation of the energy $E = h \cdot \nu$. As long as this minimum energy-packet size was not reached, the vibrational energy had to gather (or to use Debye's naively mechanistic wording: to "accumulate" (*aufgehäuft*)) inside the atom. Debye and Sommerfeld imagined this atom as a "self-contained system,"

> which can store energy from the surrounding fields by means of a resonating electron. The constant h determines when this self-containment collapses and energy is released out of the interior of the atom. The physical significance of h would accordingly be that h determines when a quasi-elastic bond breaks up or a valve of the atom opens. Disregarding possible radiation damping, owing to this valve effect the release of energy happens discontinuously and in a quantum-like fashion.[89]

Consequently, Planck, Debye and Sommerfeld considered the quantization of energy in a radiation field as a mere epiphenomenon caused by the strange "valve effect" of material vibrating systems. Unlike the early convert Stark (Sect. 4.4), Planck, Debye and Sommerfeld did not support Einstein's interpretation of light quanta at all. This also explains why Arnold Sommerfeld would write in one of his replies to Stark from 1909 that it was his intention "to strengthen our confidence again in the validity of the electromagnetic theory also for elementary processes in the electric field, which certainly does seem to have been somewhat shaken by the most recent speculations about light quanta."[90] The above passages by these three German theoreticians only described the minimum output of energy. It was known, though, that this energy could also be released in larger natural-number multiples of that minimum packet size: $E = n \cdot h \cdot \nu$, where $n = 1, 2, 3$, etc. We find one model that also includes such larger energy packets spelled out in Robert A. Millikan's address on 'Atomic Theories of Radiation' at the annual convention of the *American Association for the Advancement of Science* in Cleveland in December 1912:

> Planck now assumes that emission alone takes place discontinuously, while the absorption process is continuous. At the instant at which a quantity of energy $h\nu$ has been absorbed, an oscillator has a chance of emitting the whole of its unit, a chance which, however, it does not necessarily take. If it in this way misses fire, it has no other chance until the absorbed energy has arisen to $2h\nu$, when it has again the chance of throwing out its 2 whole units, but nothing less. If again it misses fire, its energy rises to $3h\nu$, $4h\nu$ etc. The ratio between the chance of not emitting when crossing a multiple of $h\nu$, and the chance of emitting, is assumed to be proportional to the intensity of the radiation which is falling upon the oscillator. This, then, is at present the most fundamental and the least revolutionary form of quantum theory,

---

[87]Debye and Sommerfeld (1913) p. 874 (original emphasis omitted).

[88]In this sense, see also Planck (1907), Sommerfeld (1911a).

[89]Debye and Sommerfeld (1913) p. 923; cf. further Niedderer (1982) p. 43.

[90]Sommerfeld (1909) p. 976.

since it modifies classical theory only in the assumption of discontinuities in *time*, but not in *space*, in the emission (not in the absorption) of radiant energy.[91]

As we saw (in Sect. 3.9), Einstein came back to the idea suggested in this quote in his famous paper of 1916: To distinguish between spontaneous and induced emission and make it dependent on the density of the surrounding radiation field. Millikan's comment about discontinuities in space and time meant that Planck's second quantum theory simply postulated a kind of temporary delay in the process of emission but no spatial quantization of the energy in the form of concentrated particulate or point-shaped singularities. To that extent, Einstein's mental model could hardly be more different from Planck's second quantum theory. There was one point, however, on which all these German-speaking theoretical physicists agreed at that time: chance played a leading role. For Planck it was the number of complexions, which was only statistically calculable; for Einstein it was the emissive and absorptive processes supplied with probability coefficients; for Sommerfeld, Debye and Planck it was the processes of emission conceptualized as random processes. Debye and Sommerfeld explicated this in their paper from 1913:

> A factor of chance [...] certainly must be added in order not to run up against a contradiction: If the electrons in all the atoms were equally disposed to the photoelectric effect, then the process of accumulation would begin at the same time for all; and at the instant corresponding to the maximum [...] a photoelectric catastrophe, so to speak, would occur in which a disproportionately large flow of photons would appear. That is out of the question in reality. [...] Our releasing process can only initiate if this disposition happens to have been attained. It is this circumstance which makes the photoelectric flow, as we presume, constant over time.[92]

4.8   Debye, von Laue and Schrödinger on Wave Packets

Around 1910 many other theoretical physicists besides Max Planck, Peter Debye and Arnold Sommerfeld had strong doubts about Einstein's light-quantum hypothesis and preferred to continue to describe light and electromagnetic fields by classical electrodynamics without any energy quantization. How did these opponents of light quanta model processes in which light interacts with matter within finite and limited time and space? Experiments on the photoelectric effect showed that the photoelectrons triggered by electromagnetic radiation have an energy distribution. Together with the proportionality between energy E and frequency ν also found since 1914 from these experiments, one may conclude that this radiation triggering the photoelectric effect is not monochromatic but manifests a spectral distribution.[93] Beams of rays with such a distribution of different frequencies can be interpreted mathematically as wave packets. Peter Debye and Planck's pupil Max von Laue (1879–1960)

---

[91] Millikan (1913) pp. 123–124 (original emphasis).

[92] Debye and Sommerfeld (1913) pp. 927–928.

[93] See Sects. 3.4, 3.5, 3.6 above and, e.g., Ramsauer (1914).

delivered important mathematical contributions toward a consistent description of
such wave packets around 1910 (see Fig. 4.2).[94]

A wave packet $\psi(x, t)$ propagating along the x-axis and in time t can be regarded
mathematically as the sum of plane waves:

$$\psi(x, t) = \sum_j C_j \cdot e^{i(\omega_j \cdot t - k_j \cdot x)},$$

where the amplitudes C_j of each individual wave can be chosen freely and summed
together determine the special form of the wave packet. Each contributing wave j
has its own frequency ν_j or angular frequency $\omega_j = 2\pi\nu$ and wave number $k_j = \omega_j/v$. Monochromatic waves have only *one* propagation speed v, what is known as
the phase velocity $v_p = \lambda \cdot \nu = \omega/k$. The decisive magnitude for the dispersion of
a signal and the transmission of energy and momentum in the case of wave packets is
the so-called group velocity v_g. For electromagnetic waves v_p and v_g in vacuo both
are equal to the speed of light c. In matter, however, the phase velocity of waves is
generally dependent on the frequency ν_j of the jth component. That is why the wave
packet becomes broader over time and 'dispels.' One also uses the term dispersion.[95]

According to Debye, beams of radiation are analogous to superposed plane waves;
each on its own would have traversed the entire space. A continuum of such wave
trains is superposed in that the wave normal is varied within a volume angle and the
contributions from all the frequencies are summed. The result of this superposition
is then a spatially limited beam of rays; the edges beyond its enclosing double cone are
almost entirely destroyed by destructive interference. At the center of this beam itself,
constructive interference is observable. "By suitable distribution of the amplitudes
and phases [one can] limit the excitation of one area that is relatively small also in
the longitudinal direction. That is how one obtains the analytical representation of
an 'energy packet' of relatively small dimensions, that travels on at the velocity of
light or, if dispersion is present, at the group velocity."[96]

Erwin Schrödinger recognized in his "Wave Mechanics" of 1925–26 that matter
can also be described as "matter waves," which just like electromagnetic waves satisfy
a wave equation that later took his name. He turned his mathematical examination
to existing papers about wave optics as far as they allowed. For matter, too, "such
wave groups can be constructed, namely by the very same design principle that
Debye and von Laue used in regular optics to solve the problem of showing the exact
analytical representation of a cone or beam of rays."[97] In a semi-popular article
Schrödinger also drew a diagram of such a spatially and temporally limited "wave

---

[94]See Debye (1909), von Laue (1914) as well as, e.g., the Wikipedia article on wave packets and
Brandt and Dahmen (1985) pp. 20 ff. on their visualization.

[95]On this point see the exchanges of letters: H.A. Lorentz to E. Schrödinger 27 May 1926 and
Schrödinger to Planck 31 May 1926 and to Lorentz 6 Jun. 1926, both exchanges reproduced in
Przibram (1963), esp. pp. 9, 43–45 and 54.

[96]Schrödinger (1926b) p. 501, with reference to Debye (1909) and von Laue (1914).

[97]Schrödinger (1926b) p. 500, citing Debye (1909) and von Laue (1914).

group," which he proposed in his book "Wave Mechanics" from 1926 not only as a wave packet of the Maxwellian continuous field theory of electromagnetism but also as a mental model for matter. He regarded the extension or spread of the wave group as correlated with the "thickness of the point mass"[98] assigned to this wave group in his "Wave Mechanics." This spread (in Figure 4.2 from +10 to −15 in the x-direction) and the shape of this wave packet remained unchanged as long as there was no dispersion. The hope of Schrödinger and his supporters that these matter waves could get a simple realistic interpretation as waves in three-dimensional space was disappointed, though. The Schrödinger equation was a differential equation of second order just like the classical wave equation, but its solutions were state amplitudes ψ in complex space. Only their squared real values were empirically interpretable as spatial probabilities.[99]

This mental model of wave packets was an extremely powerful notion for the quantum-theoretical description actually intended by Schrödinger as well as for interpreting electromagnetic radiation.[100] One advantage that Schrödinger himself, along with other physicists, believed he saw in his wave mechanics in 1925–26 was the "gradual transition from micromechanics to macromechanics."[101] The greatest advantage of his mental model of wave packets was their compatibility with wave-particle duality and Heisenberg's uncertainty relation. For, the temporal breadth Δt and the energetic uncertainty ΔE of a given wave packet could be adjusted, depending on the choice and weighting of the frequencies included in the superposition. Heisenberg's uncertainty relation, according to which $\Delta E \cdot \Delta t \geq \hbar$, is quite automatically satisfied for electromagnetic waves in wave packets. The more precisely a wave packet is temporally constrained (the smaller the temporal imprecision Δt is), the greater is its energy imprecision ΔE and vice versa: If a wave packet is energetically very precisely defined, then its spatial and temporal indeterminacy is very large.[102] Here we have a strong accord with semantic layer 8 of our conceptual development (Sect. 3.8), which developed out of Einstein's and Louis de Broglie's considerations about wave-particle duality. This is a major reason why this mental model is still very frequently used because it can offer quite excellent explanations for many experimental situations.

---

[98] Schrödinger (1926a) pp. 59–60 and Schrödinger (1927b) on wave mechanics. Compare further Brandt and Dahmen (1985) Chaps. 3–4.

[99] See Schrödinger (1926b) and Schrödinger's letter to Planck, 11 June 1926, in Przibram (1963) p. 14. The probability interpretation of the Ψ-function comes from Max Born (1926).

[100] The powerful influence of Schrödinger's wave mechanics on the history of quantum mechanics generally is covered by Rechenberg (1982ff.) vol. 5, Darrigol (1986, 1992), also citing sources.

[101] Thus the programmatic title of Schrödinger (1926a).

[102] Compare Brandt and Dahmen (1985) pp. 40–49.

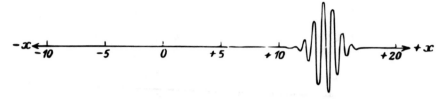

Fig. 4.2 A wave packet according to Schrödinger 1926. Source: Schrödinger (1926a) p. 60.
Reprinted by permission of Verlag Julius Springer© 1926

4.9 Gilbert N. Lewis's Mental Model of Photons 1926

We already had occasion to mention the American physical chemist Gilbert Newton
Lewis and the motivation behind his coinage of the neologism 'photon' (in Sect. 2.6).
The context of his reflections was the quest for a rigorously temporally symmetric
description of the transmission of electromagnetic radiation between emitter and
absorber. It is worthwhile to return briefly to the mental model behind this transmission process and the carriers of those energy packets. The conventional picture
of this process was a gradually initiated emission of electromagnetic radiation by
an oscillating electrically charged particle (e.g., an electron still bound to an atom's
nucleus). This particle performs some hundreds of thousands or some millions of
such vibrations and in the process emits an electromagnetic wave into the surrounding space that first gradually builds up, then stabilizes at a frequency ν, and finally
fades away again. This wave at some point meets another object that is able to oscil-
late at this frequency ν (e.g., another electron bound in the same way) and excites it
to gradually augment its oscillations, then stabilize at that frequency and thereby be
absorbed by the absorber, i.e., a second oscillating particle and its environment. The
process required a certain finite amount of time.[103] The new scattering experiments by
Compton 1922–23, Bothe and Geiger 1924–25, among others, suggested an entirely
different picture of the energy transmission of very hard x-rays and γ-rays—namely,
the mental model of very sudden pulses of virtually point-shaped sharply defined
particles or quasi-particles, just like Einstein's elusive light quanta.

> Whatever view is held regarding the nature of light, it must now be admitted that the process
> whereby an atom loses radiant energy, and another near or distant atom receives the same
> energy, is characterized by a remarkable abruptness and singleness. We are reminded of the
> process in which a molecule loses or gains a whole atom or a whole electron, but never a
> fraction of one or the other. [...] Had there not seemed to be insuperable objections, one

---

[103] Einstein (1927) p. 546, for instance, mentions "Hunderttausende oder Millionen von Schwingun-
gen" as being necessary to generate the wave. Heisenberg (1927) showed that the relation between
the duration Δt and the breadth of the energy spectrum ΔE connected with his uncertainty relation,
$\Delta E \cdot \Delta t \geq \hbar$, demands very large times t in order to make the energetic uncertainty ΔE and thus
also the frequency uncertainty $\Delta \nu$ sufficiently small.

might have been tempted to adopt the hypothesis that we are dealing here with a new type of atom, an identifiable entity, uncreatable and indestructible, which acts as the carrier of radiant energy and, after absorption, persists as an essential constituent of the absorbing atom until it is later sent out again, bearing a new amount of energy.[104]

These quasi-atoms of radiation were what Lewis described at the end of 1926 by his new term 'photon.' It was purposefully named in analogy to the artificial minting 'electron' from the end of the nineteenth century. Lewis considered electrons and photons as equally indestructible elementary material building blocks of the atom which are normally confined inside it. It was possible for them to be released by scattering, radioactive decay or other processes when they could leave the atom at very high velocity—in the case of x-rays and γ-rays as well as emitted optical spectral lines, at the speed of light; in the case of heavier particles, such as electrons (i.e., particularly β-rays and cathode rays) at lesser but still very high speeds. Other suitable, similarly built atoms could recapture these quanta of radiation under favorable conditions in a kind of resonance process. During this transition these packets of radiation transporting energy and momentum remain indestructible units emitted and re-absorbed only as a whole, as closed packets. Lewis viewed photons as sealed electromagnetic packages of energy and momentum, in the form of quasi-particles that differed from other particles by propagating with the velocity of light and satisfying the ultrarelativistic limiting formula for collision processes such as Compton scattering, for which Compton had provided good experimental proof.

In this short article in *Nature*, which appeared on 18 December 1926, Gilbert Lewis already recognized that the notion of thorough conservation of photons and their total numbers would lead to major problems in interpreting spectroscopic findings. If an atom goes from an excited state A through two intermediary levels B and C into a final state D, according to his conception three photons would be emitted; whereas in a direct transition from A to D only one would be released. In order not to produce contradictions with his postulated 'law' of the conservation of photon number, when observing many such transition events occurring in parallel at the same time, one would have to assume continual exchanges of very many lower-energy photons occurring in the background:

> The puzzle that one, and only one, photon is lost in each elementary radiation process, is far more rigorous than any existing selective principle, and forbids the majority of processes which are now supposed to occur. To account for the apparent existence of these processes, it is necessary to assume that atoms are frequently changing their photon number by their exchange of photons of very small energy, corresponding to thermal radiation in the extreme infra-red.[105]

---

[104]Lewis (1926c) p. 874. On Compton's scattering experiments and Bothe & Geiger, see here pp. 126ff. and 131f.

[105]Lewis (1926c) p. 875.

Paradoxically, Max Planck had originally come forward in 1900 to prevent precisely such interminable subdivisions of electromagnetic field energy into ever smaller packets of energy. Planck's energy quantum $E = h \cdot \nu$ set a lower spectral limit $\nu = E/h$ for those energy packets. Be this as it may, these paradoxical implications of Lewis's hypothetical (and as we now know, wrong) conservation law of photons motivated him to abandon this theory soon afterwards and never return to it. His short and succinct 'photon' would quickly establish itself as a term nonetheless.[106]

---

[106] See Sect. 2.6 and further references there about its history and statistics.

Chapter 5
Early Reception of the Light Quantum

5.1 Initial Skepticism

If Einstein himself could not fully cope with his concept of light quanta well into the 1920s—indeed up to the end of his life—it is no wonder that his contemporaries would not enthusiastically rally behind it from the start, either. During its first decade from 1905 to 1915, skepticism if not outright rejection predominated. Max Planck was among these skeptics even though he was the motor behind Einstein's appointment in 1913 as director of the new *Kaiser-Wilhelm-Institut für physikalische Forschung*, a research institute that for a long time existed only on paper (without a building of its own until 1937) with funding by the Koppel Foundation. He wrote in 1909:

> It seems to me that the greatest caution would be due as regards Einstein's new corpuscular theory of light [...] The theory of light would be thrown back not decades but centuries, to the time when Christian Huygens dared to wage his battle against Newton's overwhelming emissive theory [...] And all these accomplishments, which are among the proudest successes of physics—indeed, of science generally, are supposed to be relinquished for the sake of some still quite contestable observations? Much heavier guns would need to be run out to make the now very firmly founded structure of the electromagnetic theory of light begin to quake.[1]

In his proposal for Einstein's membership in the Prussian Academy of Sciences in 1913, Planck could not refrain from adding the remark: "That he might sometimes have overshot the target in his speculations, as for example in his light quantum hypothesis, should not be counted against him too much."[2] Clearly, in 1913 strong skepticism if not open opposition still prevailed about the concept of light quanta even among Einstein's strongest promoters and mentors such as Planck and his pupil Max von Laue (1879–1960). Einstein's considerations about quantizing the energy of the electromagnetic field seemed too radical a break with the very foundations of

---

[1] Planck (1910b) esp. pp. 763f., reprinted in Planck (1958), quote on pp. 242ff.

[2] See Kirsten and Treder (1979) vol. 1, p. 96 and CPAE, vol.5, doc. 445, p.527 (transl. pp. 337–338).

© Springer International Publishing AG, part of Springer Nature 2018
K. Hentschel, *Photons*, https://doi.org/10.1007/978-3-319-95252-9_5

Table 5.1 Comparative listing of supporters and opponents of a discontinuous model of light and other electromagnetic radiation prior to the publication in 1923 of Compton's experiments. Modified from Brush (2007) p. 232, who specified "converts" (cv) to the particulate model before 1923, and "backsliders" (bs) who retracted their support after 1923. The age of each in 1920 is indicated in parentheses. A third column is added here to distinguish between the notions of post-Newtonian particles and Einstein's light quanta (left and middle)

Particle model	Light quanta	Opponents of light quanta
William H. Bragg (58)	Albert Einstein (41)	Alfred Berthoud (46)
William L. Bragg (30)	Louis de Broglie (28)	Niels Bohr (35)
Daniel F. Comstock (37)	Maurice de Broglie (45)	Leon Brillouin (31)
C.D. Ellis (25)	Norman R. Campbell (40)	A.H. Compton (28) cv
Arthur L. Hughes (37)	James A. Crowther (37)	Peter Debye (36)
Oliver Lodge (61)	A.S. Eddington (38) bs	William Duane (48)
D.V. Mallik (54)	Paul Ehrenfest (40)	Franz S. Exner (71)
G.W.C. Kaye (40) cv	James Jeans (43)	G.W.C. Kaye (40) cv
Johannes Stark (46)	Abram Joffe (40)	Max von Laue (40)
J.J. Thomson (64)	Arthur Haas (36)	H.A. Lorentz (67)
Robert W. Wood (52)	H.A. Kramers (26) bs	Robert A. Millikan (52)
	Rudolf Ladenburg (38)	J.W. Nicholson (39)
	Walther Nernst (56)	Max Planck (62)
	Leonard T. Troland (31)	O.W. Richardson (41)
	Fritz Reiche (37)	Arnold Sommerfeld (52)
	Erwin Schrödinger (33) bs	Siegfried Valentiner (42)
	Mieczyslaw Wolfke (37)	

Maxwell's electrodynamics. The latter modeled fields as an energetic continuum in space and time and had been accepted for some fifty years.

Numerically seen, by 1923 a tie existed between outright opponents of any quantization of the radiation field and professed supporters of Einstein's concept of light quanta (see Table 5.1). The names of some of the most respected senior scientists (in the left column) appear among the advocates of various forms of a return to post-Newtonian particulate concepts of high-energy radiation. Among Einstein's supporters, by contrast, we find the names of many junior members of the field (ten of them under forty years of age). This provides some evidence in favor of Max Planck's cynical proposition that the backers of a dominant theory first have to die out before new theories can become established.[3] Very few of the persons listed in this table converted after 1923 to the contrary persuasion. Surely most notable among these is Arthur Holly Compton, who published his change of heart in 1923 on the evidence offered by his own scattering experiments. One of these scientists (G.W.C. Kaye)

---

[3] See Planck (1948) p. 22.

vacillated indecisively between the two alternatives. The overwhelming majority of physicists, however, had not yet committed themselves by then and followed the debate without personally intervening.

5.2 The Compton Effect as a Turning Point 1922–1923

The vanguard of hard-core opponents to any discontinuous model (in the right-hand column of Table 5.1) began to fall apart only when further experimental support of the light-quantum hypothesis began to accumulate in the decades that followed. Remarkably, this support was provided by two Americans who certainly did not number among the inveterate defenders of the 'new physics.' One of them had even explicitly set out to refute Einstein's quantum hypothesis. In 1916 Robert Andrews Millikan (1868–1953) landed up confirming Einstein's phenomenological formula, $E = h\nu - W_A$ with the further spin-off that Planck's action quantum h became very accurately measurable.[4] Einstein had predicted that scattering off an electron would diminish the frequency ν of a light quantum and this was confirmed experimentally by Arthur Holly Compton (1892–1962) in 1922 from the scattering of high-energy x-rays off graphite. Peter Debye, who was conducting parallel research, also arrived at a similar deduction at the beginning of 1923.[5] The historian of physics Roger Stuewer warned against drawing a direct interpretational link here from Einstein's theoretical prediction to clear experimental confirmation by Millikan. As he pointed out, this historical connection appears to be linear only in retrospect:

> Millikan and his students finally confirmed the predicted linear relationship between the frequency of the incident radiation and the maximum energy of the ejected photoelectrons. However, Millikan categorically rejected Einstein's light-quantum hypothesis as an interpretation of his experiments—despite his own words to the contrary in his later, self-aggrandizing autobiography.[6] Compton in fact began his postdoctoral career in 1916 in an atmosphere of virtually universal skepticism toward Einstein's light-quantum hypothesis. [...] Then, as Compton's own X-ray scattering experiments progressed, he rejected one interpretation after another, misread his experimental data, then read it correctly, and in general struggled on his own to the extent that Einstein's name does not appear once in Compton's published papers. In the end, moreover, Compton was nearly scooped in his discovery by Peter Debye, who in contrast was directly influenced by his knowledge of Einstein's light-quantum hypothesis.[7]

Compton's experimental setup in 1922 (see Fig. 5.1, top left) was as follows: A water-cooled, very high-performance tube generated hard x-rays that were sent through lead screens with a very narrow slit of 0.1 mm width and a filter to focus it into an approx-

---

[4]See Millikan (1916a, b) and Sects. 3.6–3.7 above.

[5]On this episode see Compton (1922, 1923a, b, 1927), resp., Debye (1923); cf. Stuewer (1975a, 1998) and further sources mentioned there.

[6]Millikan (1950) pp. 101–102, also cited above, on p. 48. Millikan's rewriting of history is unreservedly criticized by Holton (2000) and Stuewer (1998, 2014) p. 143: "Millikan's philosophy of history: if the facts don't fit your theory, change the facts."

[7]Stuewer (1998).

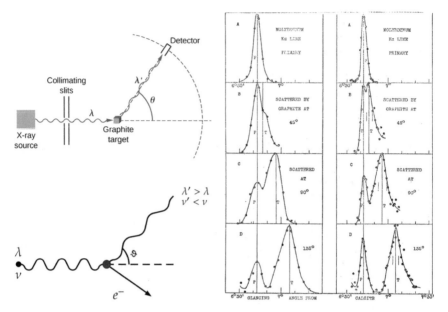

Fig. 5.1 Schematic setup and results of Compton (1922). Compton's observations showed (on the right) that not only the original wavelength λ from the x-ray tube occurred in the scattered radiation but also a larger wavelength λ' that continued to increase with increasing angle ϑ. *Sources* (right) Compton (1923b) p. 411. Reprinted by permission of the American Physical Society © 1923; (top left) https://cnx.org/resources/43f60758c3fff3fdb7cf96a50746646ba509ffcd/CNX_UPhysics_39_03_compton1.jpg

imately monochromatic beam that was directed onto a cylindrical scattering target made of graphite. The electrons loosely bound to the graphite atoms were dislodged by this 'needle radiation' and catapulted out in a statistically broad scattering of unknown direction. The x-rays, which are also scattered as a function of the scattering angle ϑ, went through more lead screens and were deflected off a rotatable adjustable calcite crystal into a scintillation counting tube for detection. Compton's apparatus permitted a variation of the scattering angle ϑ from 0° to well beyond 90°. The rotating crystal positioned in front of the scintillation counting tube made it possible to determine the wavelength of the scattered radiation λ' at the same time from the diffraction of these x-rays off the crystal lattice.

The observations (see Fig. 5.1, right) showed that in addition to detecting the original wavelength $\lambda = 0.711$ Å for the K_α-line of molybdenum emitted from the x-ray tube,[8] larger wavelengths λ' were also detectible that increased as the scattering angle ϑ increased. At that time, Compton estimated his measurement precision at ± 0.0001 Å, which corresponded to a margin of error of only 1.5 ‰. Because of the de Broglie relation $\lambda = h/p$, this meant that the larger this scattering angle was, the more

---

[8]Molybdenum was an element of the tube supplied by General Electric to Compton. See Compton (1923b) pp. 410, 413.

energy and momentum the x-radiation lost. Qualitatively speaking, this was similar
to an analogous classical collision of elastic balls. However, the balance of energy and
momentum here was different because x-rays are a kind of electromagnetic radiation
that propagates at the velocity of light and is subject to a dynamics that is not classical
but ultrarelativistic. For classical particles with non-vanishing resting mass m, it
follows that $E = [(mc^2)^2 + (mv)^2c^2]^{1/2}$. Light and other electromagnetic waves,
however, do have a vanishing rest mass; hence $E = pc = h/\lambda$, and analogously for
scattered x-rays: $E' = pc = h/\lambda'$. In the case of Compton scattering of electrons of
mass m_e, it therefore followed from the law of conservation of energy and momentum,
after a few more reformulations: $\Delta\lambda = \lambda' - \lambda = \frac{h}{m_e c} \cdot [1 - \cos(\beta)]$.[9]

This experiment demonstrated irrefutably that hard x-rays carry energy *and*
momentum[10] and that interactions between this radiation and matter can be reduced
to interaction zones of the size of a single electron. In these zones the x-ray quanta
undergo collision and scattering processes the way particles do and these processes
can be described to great precision by ultrarelativistic dynamics. Thus Compton's
experiments confirmed Einstein's light-quantum hypothesis and simultaneously also
his special theory of relativity. Initially, others including William Duane (1872–
1935) at *Harvard University* had difficulty repeating Compton's experiments but
after Peter Debye at Zurich succeeded in March 1923 along with other North American experimenters at the end of 1924, more physicists began to regard Einstein's light
quantum as a hypothesis to be taken into serious consideration. Arnold Sommerfeld,
for instance, promptly included the Compton effect in the fourth edition of his handbook *Atombau und Spektrallinien* in 1924. Arguing in favor of the "fundamental
importance of Compton's result," he launched the following broadside against the
interpretative supremacy of the wave theory:

> A ray in which energy and momentum are localized in a point shape does not essentially
> differ from a corpuscular ray; we have revived Newton's corpuscles. […] If the observation
> of x-rays does in fact result in a change in wavelength, then it seems that the wave theory is
> done for! Then it would hardly be able to uphold its position anymore in the optically visible
> range either; it must cede to quantum theory not only the phenomena of the generation,
> destruction and conversion of light but also the facts of reflection, refraction, dispersion and
> scattering; and only the processes of pure phase relations, interference and diffraction are
> left standing. In this way, the scope of validity of quantum theory is expanding from year to
> year and is narrowing down that of the wave theory.[11]

Both of Millikan's publications from 1916 and Compton's from 1923 count among
the most frequently cited physical papers from the 1920s and 1930s—the same
applies to their famous and multiply reprinted speeches delivered in 1923 and 1927,
resp., at the awards of their Nobel prizes. The paper published by Compton (1923a)
in the *Physical Review* even takes first place among the most frequently cited papers

---

[9]The full derivations are now textbook knowledge and are also called the Compton wavelength in
the form $\frac{h}{m_e c} = \lambda_c = 2.42610^{-12}$m. See, e.g., Compton (1922, 1923a); Debye (1923).

[10]See, e.g., Albert Einstein to Michele Besso, May 24, 1924, in Speziali (1972), p. 202.

[11]Sommerfeld (1919)c pp. 57–59; Michael Eckert discussed six editions of this standard textbook
of quantum theory: http://edition-open-access.de/studies/2/7/index.html.

from 1920 to 1929.[12] However, frequent citation does not necessarily mean that the quoted article was believed one hundred percent. It was only indisputable that the transfer of energy and momentum from a light quantum to an electron does in fact occur. Robert Millikan's formulation of his standpoint at the time still sounds very skeptical:

> After ten years of testing and changing and learning and sometimes blundering [...] this work resulted, contrary to my own expectation, in the first direct experimental proof [...] of the exact validity [...] of the Einstein equation and the first direct photo-electric determination of Planck's h. [...] the general validity of Einstein's equation is, I think now universally concluded, and to that extent the reality of Einstein's light quanta may be considered as experimentally established. But the conception of localized light quanta out of which Einstein got his equation must still be regarded as far from being established.[13]

5.3   The BKS Theory 1924 and Experiments by Bothe and Geiger

Compton's experiments also had an impact on the question to what extent longstanding principles such as the law of conservation of energy in classical physics could remain unaltered in the new quantum theory and quantum mechanics. The Dane Niels Bohr assumed a particularly radical position. For the Bohr-Sommerfeld model of the atom (see here pp. 31 and 77) he had been willing to dismiss Larmor's theorem of classical electrodynamics *per fiat*, according to which circularly accelerated charges radiate energy. Many physicists were initially shocked about this audacity but eventually got somewhat used to it, because the Bohr-Sommerfeld model was empirically so successful. In January 1924 confronted with Compton's results, Bohr presented another bold theory. His long paper on the quantum theory of radiation coauthored with his close collaborator Hendrik Anthony Kramers (1894–1952) and an American guest researcher John Clarke Slater (1900–1976) gained fame as the Bohr–Kramers–Slater (BKS) theory.[14]

The great reservations that these three authors had about Einstein's light-quantum hypothesis are already evident in the paper's introduction. They argued that discontinuities in the interaction between radiation and matter had led to the introduction of the light-quantum hypothesis, which in its most extreme form demolished the wave-like nature of light. However, they continued, this quantum theory was still purely formal because it offered no concrete model about what really occurred during the transition between stationary states inside the atom. This the three authors

---

[12] See Small (1986) pp. 144–145. By the end of 1929 Compton (1923a) had been cited seventy-eight times in twenty leading international journals of physics; based on the *Physics Citation Index 1920–29* accessible to Small (1981). Cf. Brown (2002) on Compton's later influence.

[13] Robert Millikan (1924) pp. 61–62, see also p. 64.

[14] On the following see Bohr et al. (1924a) pp. 785ff. and Mehra and Rechenberg (1982) vol. 1, pp. 532–554, Pais (1991) pp. 232–238, Kragh (2009); on Kramers see Dresden (1987); on Slater see Morse (1982).

proposed to change. Slater had the idea that a virtual guiding field prepared such a transition by sending out feelers, so to speak, about which second orbit (or available energy level E_2) an electron could jump onto if it were to leave its present orbit (at energy level E_1). This seemed to be the only way for it to be certain that upon leaving the first orbit the emitted light quantum would really have the suitable energy $\Delta E = E_2 - E_1 = h \cdot \nu$ and frequency ν.[15]

The authors even sometimes set the term 'light quantum' within quotation marks to distance themselves from the notion of "entities, each of which contains the energy $h\nu$, concentrated in a minute volume." They granted that it might have some heuristic value, for instance, in interpreting the photoelectric effect, but in their eyes it evidently offered no satisfactory solution to the problem of the spreading of light and was not able to explain interference phenomena either.[16] Pursuant to the so-called correspondence principle of quantum theory, which Bohr and Kramers had successfully applied on many occasions, for large quantum numbers it must be possible to draw a link between quantum-theoretical expressions and those from classical mechanics and electrodynamics. This only regarded a boundary transition in the case of very many transitions, however, and said nothing about an individual jump between two energy levels. Hitherto it had always been presumed that the lasting laws of mechanics, particularly the law of conservation of energy, also applied to individual atomic processes—but who guaranteed this? Planck's theory of black-body radiation and even Einstein's theory of spontaneous and induced emission showed that further advancement in quantum theory was only possible if a connection was made to probabilistic considerations. Thus it seemed suggestive to make the assumption that the individual quantum jump was governed not by deterministic laws but by statistical laws. Each *individual* process would not necessarily be compelled to satisfy the law of conservation of energy strictly. Only the statistical sum of them would have to guarantee this law, since this total was all that macroscopic measuring instruments were able to establish. The question that automatically posed itself was: How could this statistical upholding of the conservation of energy and momentum be guaranteed? The three authors had an answer ready: by the virtual radiation field which Louis de Broglie and Slater had postulated for other reasons. This field induced each transition and influenced those probabilities in a way that overall "the observed statistical conservation of energy and momentum" resulted.[17]

After hearing about these ideas by Bohr and his collaborators, Einstein wrote in one of his famous letters to Max Born (1882–1970) in Göttingen, whom he regularly contacted about the latest developments in quantum theory:

---

[15]Bohr et al. (1924a) Sect. 1, p. 786 and Sect. 2, p. 793, further Slater (1924). Rutherford had originally presented this problem with causality in quantum jumps in 1913. The Bohr-Sommerfeld atomic model had always deferred it as irresolvable: see Hentschel (2009b) for evidence and further literature. Similar ideas about guiding fields had already been advanced by Louis de Broglie (1922, 1923).

[16]Ibid., p. 787. Here they were implying the objections raised by Lorentz (see Sect. 4.3).

[17]Ibid., p. 793. This last assumption was criticized as *ad hoc*. In the philosophy of science, the BKS theory would later even become a model example of an unsatisfactory *ad-hoc* theory.

Bohr's opinion of radiation interests me very much. But I won't be driven into abandoning strict causality before entirely different defenses have been tried against it than hitherto. The thought that an electron subjected to a ray chooses *of its free volition* the instant and direction in which it wants to bounce away is intolerable to me. If so, then I would rather be a cobbler or even a casino employee than a physicist. My attempts to give quanta tangible form have nevertheless always failed, but I haven't given up hope yet, not by a long shot.[18]

As the old English saying goes: "The proof of the pudding is in the eating." The only way for this very bold BKS theory to gain plausibility was by proving its merit in experimental application. The BKS paper from Copenhagen had alluded to some consequences of this theory which were assumed to perhaps be verifiable in principle. One consequence in particular was a statistically broader distribution of energy and momentum for light quanta. In scattering processes of light quanta involving electrons, a proportionately broader angular distribution of scattered electrons should occur, considerably broader than would be suggested by the classical picture of a collision process resembling billiard balls with a definite result.

This is where Walther Bothe (1891–1957) and Hans Geiger (1882–1945) saw their point of departure. In June 1924 they proposed to conduct a more precise Compton-scattering experiment in which not only the direction of the scattered electron be measured exactly but also the direction of the scattered x-rays (see Fig. 5.2). The validity of the law of conservation of energy and momentum could thus be checked for individual cases.[19] The development of a point counter tube (a precursor of the later Geiger-Müller counter) had meanwhile brought very accurate coincidence measurements within the range of experimental feasibility. Bothe and Geiger devoted the remainder of 1924 to the scattering experiments at the bureau of standards (PTR) in Charlottenburg near Berlin. They became the first to evaluate coincidences for both parties in a scattering event. The result confirming the exact validity of the conservation of energy and momentum was published in April 1925. The BKS theory was thereby experimentally proven false.[20]

A.H. Compton also conducted his own experiments on electrons scattered by x-rays in 1925. He positioned his interaction zone in the interior of a Wilson cloud chamber and generated very brief pulses of light by igniting an explosive tungsten filament. The almost 1,300 stereoscopic exposures taken this way depict thirty-eight traces of collision pairs, allowing one to deduce their site of interaction and their momenta. Eighteen of these exposures were interpretable as photographs of a single collision process and showed that the conservation of energy and momentum was obeyed in all these cases. They provided new evidence against the hypothesis by Bohr, Kramers and Slater that for subatomic processes the conservation of energy might only be satisfied on statistical average. These photographically documented

---

[18] A. Einstein to M. Born, 29 Apr. 1924, first published in Born et al. (1926) pp. 118f.; reprinted in CPAE vol. 14 (2015) doc. 240, p. 371 (transl., p. 237).

[19] See Bothe and Geiger (1924) as well as Bothe (1924) on quantitative estimates in detail.

[20] See Bothe and Geiger (1925) as well as on the experimental methods Trenn (1976), Galison (1991) pp. 440f. The philosopher of science Karl R. Popper regarded this experiment by Bothe and Geiger as a paragon *experimentum crucis*, which makes possible a clear decision between deterministic and merely stochastic laws of nature. See Popper (1934, 1935) Sect. 77, p. 179.

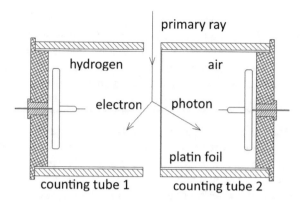

Fig. 5.2 Experimental setup by Bothe (1924); Bothe and Geiger (1925). A primary beam of x-radiation entering into the left half of a hydrogen atmosphere through a small gap between two point counters is scattered by an electron. The thin platinum foil mounted in front of the point counter in counter tube 2 absorbs electrons but allows photons with wavelengths within the x-ray range to pass through. Thus coincidences between the detection times of an electron in counter tube 1 and a photon in counter tube 2 within a period of 10^{-3} s can be associated with individual scattering processes. *Source* https://de.wikipedia.org/wiki/Datei:Koinzidenzmessung.png

collision processes also showed "that at least a large proportion of the scattered x-rays proceed in directed quanta of radiant energy."[21]

Compton revealed his own mental model of these scattering experiments in a popular report about his latest findings for *Scientific American*: "Recent experiments show that light rays act like projectiles in knocking electrons out of matter. [...] We thus have a new picture of light as consisting of streams of little particles [...] light bullets."[22] The illustrations attached to this article underscore this model. They depict a pattern of many tiny points drawn under a wave in the shape of a sinus curve. Wherever there is an extreme point on the wave the point density is high, and in the nill areas it is very low. The new mental model Compton attached to light was that of a fine stream of very small quasi-Newtonian corpuscles. Thermal heating of bodies from incident light was reconceptualized as analogous to a "stream of bullets from a rapid-fire striking a target."[23] As we shall see in Sects. 9.1–9.3, this conception of photons as precisely localizable and individually disparate particles would prove to be wrong and misleading, even though it apparently still continues to haunt many physicists to this day. A consistent interpretation of the light quantum—from 1926 on, English speakers referred to them as 'photons'—as Bragg, Compton, Sommerfeld and Millikan 1924–25 demanded, was long in coming. And we are perhaps even still waiting for it now (for newer evidence, see Chaps. 8–9).

---

[21]Compton and Simon (1925) pp. 289, 299.

[22]Compton (1925) p. 246; cf. further Silva and Freire (2011).

[23]Compton (1925) p. 246. This matches the caption he wrote under a portrait photo of Einstein: "Professor Albert Einstein. He revived the old Newtonian idea of light corpuscles in the form of quanta."

Paradoxically, later reanalyses of the Compton effect showed very clearly that it was entirely possible to calculate x-ray scattering semiclassically equally well from the balance between energy and momentum, hence without making any assumptions about a quantization of the field. These reassessments began in 1924 with Otto Halpern, then Guido Wentzel and Guido Beck, Erwin Schrödinger and Paul A.M. Dirac followed, until the two joint papers by Oscar Klein and Yoshio Nishina appeared 1928–1929. Thus—strictly speaking—Compton scattering is *not* the irrefutable proof of the light-quantum hypothesis that it is often held to be—any more than is the photoelectric effect. Nevertheless it was a fork in the road, or as Roger Stuewer (1975a) put it: "a turning point in physics". It pointed the way toward acceptance of the light quantum among physicists. Compton's, Bothe and Geiger's experiments proved, as Heisenberg put it, "the reality of light quanta"; Pauli wrote to Kramers on July 27, 1925: "It now can be taken for granted by every unprejudiced physicist that light quanta are as much (and as little) physically real as electrons. Classical concepts in general, however, should not be applied to either."[24] Those hefty debates about the angular distribution of the intensity of scattered radiation in the Compton effect also shed guiding light on the development of quantum electro-dynamics from 1927 on.[25]

---

[24] All three former quotes from Stuewer (1975a) pp. 287, 303–304.

[25] On semiclassical derivations of the Compton effect and on its importance for the genesis of QED, see Brown (2002) and primary literature analyzed there, esp.: Halpern (1924), Wentzel (1925, 1927), Beck (1927), Schrödinger (1927a), Dirac (1927a), Klein and Nishina (1928, 1929) Halpern and Thirring 1928/29a, b, pp. 441–444, furthermore Dodd (1983) and Strnad (1986b).

Chapter 6
Light Quanta Reflected in Textbooks and Science Teaching

Our attention so far has been concentrated on 'light quanta' or 'photons' within the scientific contexts of physics and neighboring fields. This means that we have stayed within the *research* scene of scientists personally involved in these discoveries. Nowadays, the history of science also looks beyond these inner-science contexts to the *teaching and learning* of scientific subjects. The reason is that science pedagogy and the history of education can better explain how successive generations of scientists digested all of this material. Even when one's primary focus is on the internal development within a science, a glance further to the instruction of scientific content is indispensable, especially if one wants to find out when such material entered into the curriculum or how it was grasped and received by college students and later by secondary-school pupils. More recent historiographic treatments of scientific instruction and pedagogical transmission of difficult material are provided, for instance, by David Kaiser's examinations of post-World War II teaching with Feynman diagrams, or Kathryn Olesko's analyses of instruction at the Koenigsberg seminar for physics.[1] A dozen of the most influential textbooks specifically devoted to quantum theory and quantum mechanics have been examined in a collection of studies edited by Massimiliano Badino and Jaume Navarro.[2]

Evidently, within the scope of our history of the early quantum theory, disturbingly many historical distortions, misleading accounts, errors and myths are in circulation about Einstein's light-quantum hypothesis, his experimental predictions and retrodictions and diverse experimental tests of those predictions. This is shown by a comparison of textbooks, practical laboratory instruction manuals and other didactic material in physics against microhistorical analyses of the related episodes. In many cases, the actors themselves spread such pseudohistories in order to conceal earlier

---

[1] See Kaiser (2004) and Olesko (1991). Historiographic overviews along these lines are found in Kaiser (2005) and Simon (2012, 2013).

[2] See Badino and Navarro (2013), also available on line at http://edition-open-access.de/studies/2/index.html

© Springer International Publishing AG, part of Springer Nature 2018
K. Hentschel, *Photons*, https://doi.org/10.1007/978-3-319-95252-9_6

errors or else to distract attention away from how very differently things had turned out from what they had once prognosticated. Other skewed accounts originated from sloppy utilization of the facts in contemporary reports or commemorative essays; yet others came from well-meaning attempts to simplify complex historical events in textbooks (presuming it was not the result of sheer ignorance about the primary sources, which in the case of too many textbook authors unfortunately cannot be excluded). The distortions and **historical myths** most difficult to eradicate include:

1. **Heinrich Hertz**—not Wilhelm Hallwachs—**being the discoverer of the photo-electric effect** (cf. here p. 57 on the true context of its discovery)
2. **Lenard being the supposed discoverer of the proportionality between energy and frequency in the photoelectric effect** (cf. here pp. 58ff. for evidence that Lenard had *not* been looking for any frequency dependence in the photoelectric effect but, on the contrary, had even explained it away with his 'trigger hypothe-sis')
3. **Planck as discoverer of energy quantization, or even of light quanta** (cf. here Sect. 2.1 on his half-hearted conclusions and the timidity that over-whelmed him soon after the appearance of his bold papers from 1900–01; or Sect. 2.4 about his second quantum theory from 1909–13, in which the radia-tion field is again modeled as a Maxwellian continuum, and his opposition to Einstein's light quantum)
4. **Einstein as the proclaimer of an ontological interpretation of light as a stream of particles** (cf. Sects. 2.3 and 2.4 on Einstein's exceedingly cautious introduction of the light-quantum hypothesis as a "heuristical point of view," and Sects. 4.2 and 4.3 on his mental modeling and temporary doubts)
5. Reducing **Einstein's article from 1905 exclusively to theorizing about the photoelectric effect** is equally misleading (cf. here Sect. 2.2 on Einstein's much broader agenda, which included far more empirical effects and aimed at reinter-preting Planck's energy quantization rather than only explaining the photoelectric effect)
6. **Millikan as self-assigned verifier of Einstein's predictions by experiment** (cf. here pp. 63ff. on Millikan's original intention to disprove what he considered to be an absurd theory and his later 'rewriting' of history.[3]
7. **Compton as decided 'confirmer' of Einstein's predictions** (cf. here pp. 126ff. on his equally skeptical initial expectations and serious problems with interpreting his own data)
8. **The photoelectric effect, Compton scattering and other early experiments as purportedly irrefutable evidence of Einstein's light quanta** (cf. here p. 145 about possible semiclassical interpretations of these early experiments and the otherwise extremely hesitant attitude that prevailed in the scientific community,

---

[3] See further Holton (2000), Stuewer (1998, 2014) p. 143: "Millikan's philosophy of history: if the facts don't fit your theory, change the facts."

a great majority of its members still not being ready to consider light quanta as extant even *after* Compton's experiments, preferring at best to tolerate them as a perhaps useful fiction)

Concerning these eight specific historical episodes there remains the question of whether physics textbooks draw an accurate overall picture of physical research, for instance, about the interplay between experiment and theory formation.[4]

I dispense with recapitulating each of these eight episodes here and providing full citations from the various textbooks, considering that the former has been presented in foregoing chapters and the latter has luckily already been closely surveyed by a group of American, Canadian, Danish and Venezuelan physics instructors and pedagogues on the basis of 103 physics textbooks and laboratory instruction manuals in the English language (published from 1937 into the first decade of the present century).[5] Allow me to skip to the statistical results of this critical survey of the literature for six of the eight myths listed above, which had also caught the notice of these physics instructors. These examples speak volumes about the faulty transmission of historians' research findings to physics textbook material and hence also about its instruction in upper schools, colleges and universities in Great Britain and USA. I would doubt very much whether these results would improve substantially if the analyzed texts were by German or French-speaking authors. It would nevertheless be interesting to see whether such a critical survey of the more recent literature in other linguistic spheres would lead to a similar result in all countries or whether the numerous well-researched academic publications in the history of science have in fact led to some betterment, after all, particularly since the Einstein centenary 2005.[6]

Stephen Klassen at the *University of Winnipeg* in Canada and his Canadian and Venezuelan collaborators studied six historical episodes which they categorized as decisive for a deeper historical and physical understanding of the photoelectric effect (numbers 2 and 5–8 in the above list). For each of these episodes, they drew up a detailed list of four to six important points. For a given textbook account to earn the predicate 'excellent,' it had to have mentioned or taken into account all of these points. If more than two points were mentioned, the textbook got the predicate 'satisfactory'; if only one or two points were mentioned in passing, the predicate 'mention' was assigned, and a total absence of all points got the predicate 'no mention.' It is not surprising that in five of these six historical episodes not a single one of these over a hundred analyzed textbooks could be ranked as 'excellent' overall. The purpose of a textbook is quite different from a historical text, of course. The almost equally bad

---

[4]Niaz, Klassen and Metz (2010) p. 922 include this as their sixth and last evaluation criterion: "Criterion C6: The historical record presented and its interpretation within a history and philosophy of science perspective." On the interplay of instrumentation, experiment and theory formation, cf. Hentschel (1998a, 1960b).

[5]On the following see Klassen (2008), which is identical with Klassen (2011), as well as Niaz et al. (2010) on physics textbooks, Klassen et al. (2012) on laboratory instruction manuals. More generally, see Kragh (1992) and Passon and Grebe-Ellis (2016).

[6]I am pessimistic, though. The results of a generally oriented critical survey of popular articles for the centenary of Einstein's famous discoveries, conducted by Jonas Keck as a bachelor's thesis under my supervision in 2016, are not very encouraging, at least.

Table 6.1 Statistical analysis of 103 physics textbooks for a correct rendition of six historical episodes (E1 to E6). The bracketed numbers refer to the eight typical myths about the photon from my list above. The column (total) preceding the averages for each row refers to the sixth on the historical record, as specified here in footnote 4. From Niaz, Klassen and Metz (2010) p. 919

Category	E1 = (2)	E2 = (5)	E3 = (8)	E4 = (7)	E5 = (6)	Total	Average
No mention	84	14	84	69	95	91	73
Mention	16	39	11	16	2	7	15
Satisfactory	0	48	2	16	3	2	12
Excellent	0	0	3	0	0	0	0

results (with few exceptions) for the next two categories is much more disappointing, though (see Table 6.1). On average, only 12% were 'satisfactory' and only 15% considered any mention of one of these points necessary. Almost three quarters (73%) took no account at all of any historical context, apart from one exception to be discussed below, ranging on average between 69 and 95%.

Astonishingly enough, three textbooks in English also took properly into account my eighth point in the above list of myths (Klassen's episode 3). According to the evaluation by Klassen et al., these three textbooks covered the long-lasting rejection of the light-quantum hypothesis by virtually the entire physics community, including those among its members who in so many other accounts were wrongly identified as early confirmers and backers of Einstein—namely, von Laue, Planck, Bohr, Millikan, etc.[7] But otherwise, that's about it. A meager two other textbooks fall under the next category 'satisfactory' for this episode. It is not that surprising that our fifth myth (Klassen's second episode) got at least one mention particularly frequently and also got fuller treatment considerably more often than all the others. That is because Einstein's explanation of the photoelectric effect is the basis of any student replication with evaluation as a laboratory exercise. All in all, the table of all these findings (Table 6.1) shows how little the contextual relations developed by historians of physics have been taken into account for decades.

An analysis of textbooks by a temporal series of 10-year publication intervals did not lead to any recognizable trend of improvement between the 1950s and the first decade of the 2000s, either.[8] Altogether, Klassen and his coauthors come to the dismal conclusion: "most of the textbooks were largely deficient of any material on most of the criteria. [...] If an overall characteristic of the 103 textbooks could be identified, it might be the sporadic nature of the inclusion of the relevant historical information."[9] They do at least see a few rays of hope in the form of a few isolated pertinent observations: "Despite not approaching the photoelectric effect in an overall

---

[7]Two of these texts were authored by Eugene Hecht, published in 1998 and 2003, the other one was the much-read elementary physics textbook by D. Halliday and R. Resnick from 1981.

[8]Niaz et al. (2010) pp. 923f.

[9]Ibid., pp. 915–916; cf. ibid. pp. 917–920 for the detailed evidence in tabular form.

Table 6.2 Statistical analysis of 38 laboratory instruction manuals for a correct rendition of a total of four historical episodes, which for ease of comparison with the foregoing table I refer to here as E2 to E5, with the bracketed numbers referring to one of the eight typical myths about the photon from my above list. From Klassen et al. (2012)

Category	E2 = (5)	E3 = (8)	E4 = (7)	E5 = (6)	Average
No mention	61	100	92	100	88
Mention	18	0	8	0	7
Satisfactory	21	0	0	0	5
Excellent	0	0	0	0	0

satisfactory manner from the point of view of our six criteria, some textbooks did offer good incidental insights."[10]

It doesn't look much better if we look at the laboratory manuals evaluated by Stephen Klassen and Barbara A. McMillan by the same method.[11] In 2012 they published the results of their analysis of a total of thirty-eight practical laboratory manuals published in the English language between 2001 and 2010 that were available on line. They used the same method as before but only examining four of the former six historical episodes (E2–E5, see Table 6.2). The total result of this survey of relatively recent texts was again sobering: "in general, the manuals ignored the historical context and the difficulties involved in understanding the experimental data that led to alternative interpretations."[12]

It may well be that physics textbooks and laboratory manuals are part of the context of Kuhnian 'normal science,' within which pupils and students are efficiently taught by means of paradigmatic examples the scientific norms and model solutions to use during their studies and later professional practice. It has often been asserted, even by Thomas S. Kuhn himself, that—as opposed to scientific research—this beginner science is characterized by "oversimplification and dependency on textbooks" and that it "tends to a lack of context, imagination, and engagement."[13] It may well be doubted whether this purely formal or result-oriented and ahistorical form of presentation really is the best way to awaken interest and generate an integrative overall comprehension by advanced pupils, college students and beginning physics majors at university. Physics instructors just as physics historians have often argued against it. Mentioning the historical course of developments certainly makes the subject matter more lively but also more complex, especially when it does not construe linearized plots but reveals cognitive and social resistances and conflicts. At the same time it also makes it more understandable why it took so long for many central ideas to be developed and become accepted. A figure like Albert Einstein becomes even more

---

[10] Ibid. p. 915; for our case, for instance, the following one-liner about the above myth (1) in Weidner and Browne (1997) p. 867: "how the photoelectric effect was discovered was an irony of history."

[11] On the following see Klassen et al. (2012).

[12] Klassen et al. (2012) p. 739.

[13] Quotes from T.S.Kuhn (1962) cited by Niaz, Klassen and Metz (2010) p. 924; cf. also Brush (1974), Jones (1991).

fascinating when one knows that he struggled cognitively with his own concept of light quanta from 1905 all the way up to the end of his life. Also, students encountering difficulties in attempting to replicate the photoelectric effect by experiment during one of their practical sessions will be less discouraged if they learn that during the first decades of the twentieth century, even the greatest theoreticians and experimenters were having trouble with the effect. As we saw in Sects. 3.5–3.6, it took at least ten years (from 1906 until 1916) before it became certain that energy and frequency were in fact linearly related in the way Einstein had predicted. Nowadays, that practical experiment takes no more than 20 minutes. Is this sufficient to lodge this finding and its implications in a beginner's mind? No glory is lost to the 'heroes' of physics, such as the Nobel laureates Compton, Millikan and Planck, if it becomes better known that these grand physicists all had erred on some point. On the contrary, such episodes may well help the general public to treat expert opinions and findings of scientific research more maturely and securely, in better awareness of their professionalism and historicity. By contrast, another Nobel laureate in physics, Johannes Stark was—entirely justifiably—sharply criticized for his exaggerated addiction to visualizable intuition and particularly for his complicity in the Nazi "Aryan physics" movement. It is a historical paradox that this figure, who had once been among the earliest and most passionate defenders of Einstein's light-quantum hypothesis, later became one of Einstein's most aggressive opponents, lambasting him as a "Jewish propagandist."

The many clichés proliferated by sloppy 'quasi-history' of science are so wrong and misleading that it is never too soon to wean aspiring physicists and anyone else from them. It almost doesn't matter anymore what the reason was for such historical negligence or distortion, whether it be indolence about procuring the primary sources, which have meanwhile often become quickly and easily available digitally, or about at least reading reliable secondary literature in the history of science; or whether it be "a rather misguided desire for order and logic, as a convenience in teaching and learning."[14] At the end of the day, this myth-ridden quasi-history, which continues to reappear in forewords and paratexts for textbooks, which follow a sorely ahistorical approach in other respects as well, produces a very distorted image of scientific research. Complex processes of discovery are either trivialized as purportedly purely logical conclusions drawn from foregoing experiments or are mystically presented as lucky strikes out of nowhere. Decades of debating about experiments, concepts and theories are ignored, in which good and noteworthy arguments were put forward by both sides as they grappled for the truth. The intricate interplay between scientific instrument design, experimental practice and the interwoven scientific theories is artificially linearized. Historical dead ends and false tracks, alternative interpretations and profound problems of interpretation are simply skipped. Figures like Einstein or Heisenberg, Compton or Millikan are stylized as heroic luminaries, their scientific opponents as simpletons or malevolent demons.

---

[14]Whitaker (1979) p. 239 with good examples from the literature about physics textbooks.

Consequently, we can only applaud the conclusion reached by Klassen and his collaborators: "We recommend that historical presentations should be an integral part of the presentation of the photoelectric effect in textbooks. [...] the history of physics is 'inside' physics."[15]

---

[15]Niaz, Klassen and Metz (2010) p. 924, themselves citing Fabio Bevilaqua. Further complaints about 'quasi-history' along very similar lines: Holton (1973), Brush (1974), Whitaker (1979), Simonsohn (1979, 1981) and Kragh (1992) or Weinmann (1980), Niedderer (1982), Tarsitani (1983), Rahhou et al. (2015) about the constructivist role of the history of science in the transmission of science.

Chapter 7
The 'Light Quantum' as a 'Conceptual Blend'

Up to this point we have been practicing a kind of retrospection. Semantic accretion resulted from a step-wise, nonlinear enrichment of levels of signification. We took this retro perspective, departing from the developed concept, to establish which semantic layers were added when, or at what point which other ones dropped out or were altered. I find it important to reiterate here that it is not a matter of simple accumulation but of complex and nonlinear 'accretion.' The advantage of this form of consideration is a definite referencing of the term whose development is of interest. The selection criterion for which layer is relevant in the formation and semantic accretion of the later fully formed concept is clear.

We could also take a forward-looking diachronic perspective, though. Mark Turner's interpretational approach to terminological development, 'conceptual blending,'[1] makes do without referencing later developmental stages and inquires about when which concepts are interconnected and how. The strength of this form of observation is greater openness. Its weakness is that the clarity about the selection criterion is gone.[2] The idea behind Mark Turner's 'conceptual blending' is a kind of conceptual melding of two formerly completely different semantic base areas. Let us first look at a simple example to understand this basic idea: The notion of a 'black hole' is the conceptual melding of the known everyday phenomenon of objects almost automatically falling into a hole, combined with the mathematical concept of a singularity. Take, for instance, a mass-point with the gravitational force F_G. This gravitational force $F_G \sim m/r^2$ at the distance r around a spatially singular mass-point of mass m, which is a well-defined expression for all finite distances r, diverges at $r \to 0$. Combined in a process of conceptual blending as the two meld together, the outcome is the new and radical concept of a 'black hole' with its inescapable force of attraction into which all bodies with mass must fall at a greater or lesser speed. This is the

---

[1] See on the following Turner (2006) and Fauconnier and Turner (2002).

[2] De facto, though, here too the strands that are blended at a given time are identified only in historical retrospect.

© Springer International Publishing AG, part of Springer Nature 2018

K. Hentschel, *Photons*, https://doi.org/10.1007/978-3-319-95252-9_7

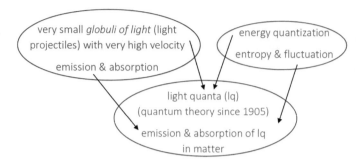

Fig. 7.1 Light quanta as the outcome of simple conceptual blending. My diagram illustrates schematically the idea of a superposition of semantic layers from two entirely different base areas, by which a new concept forms

basic idea of 'conceptual blending' introduced by Mark Turner (∗1954) and Gilles Fauconnier (∗1944).

Adapting this to our case, now, we reinterpret the light quantum as a conceptual blend, that is, as a conceptual fusion or superposition of Newton's "globulus of light" and indispensable parts of quantum theory from 1900 onwards:

A single blending (as in Fig. 7.1) is obviously far too coarse-grained as a model, because in order to end up with the complex concept of a 'light quantum,' many conceptual integrations—both historical and semantic ones—must be stacked on top of each other. That is why it is better to speak of iterated steps of conceptual blending; for example: Einstein's 'light energy quantum' from 1905 as a first step followed by the light quantum, and wave-particle duality from 1909 as the "second blend" (see Fig. 7.2).

If this is still deemed too coarse, one can consider whether a so-called "mega-blend" of two overlain blends might fit the historical development better. Then (see Fig. 7.3) the first blend is Einstein's 'light energy quantum' (1905) and the second incorporates Einstein's fluctuation analysis (1909), hence the light quantum and wave-particle duality taken together.

From Turner's perspective, quantum electrodynamics (QED; cf. Sect. 3.12) appears as the outcome of such a second-step conceptual blending. Quantization of the electromagnetic field is achieved in QED by a kind of virtualization of the photon, which is reinterpreted as a massless exchange particle in electromagnetic interactions. This explains why, different from the weak interaction which is borne by exchange particles with mass, this interaction in principle has an infinite range, although it decreases for distance r by the relation $1/r^2$. We have already become acquainted with the famous Feynman diagrams (in Sect. 3.12). They visualize these exchange relations and at the same time allow one to classify the various arrangements of the processes

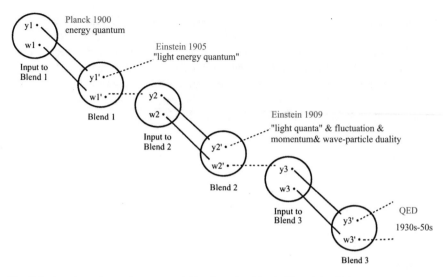

Fig. 7.2 A twice-reiterated conceptual blending. 1st blend: Einstein's 'light energy quantum' (1905); 2nd blend: Einstein's 'light quantum' with fluctuations and wave-particle duality (1909); 3rd blend: QED of the 1930s. My modification of a figure for a different example in Fauconnier and Turner (2002) p. 158

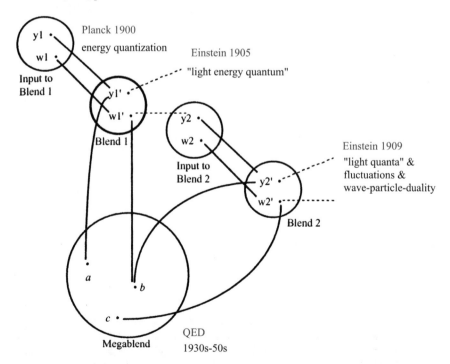

Fig. 7.3 A 'megablend' of two conceptual superpositions—again a greatly simplified schematic modification from Fauconnier and Turner (2002) p. 159

of interaction as well as to estimate the magnitudes of the interaction energies.[3] Each individual one of these exchange processes is, in QED, a purely virtual one—a mere conceivable possibility so to speak, which can be conveniently inventorized and summed. QED photons are partially virtualized, whereas they really exist in propagating light. Thus we have come far beyond the horizon of thinking of Einstein and his generation, which we have been reconstructed up to now.[4] We shall devote another chapter (Chap. 8) to these more recent developments.

---

[3] Each vertex, at which a photon couples to a charged particle, enters into the calculation as a factor $\alpha \simeq 1/137$, for the so-called fine-structure constant. Therefore, processes of higher order have less probability. However, higher-order scattering processes are certainly not negligible because of their combinatorially rapidly increasing number. They can add up to nonnegligible totals (the so-called renormalization problem of QED).

[4] On later developments, see Kidd et al. (1989) as well as here Sect. 3.12 on QED.

Chapter 8
Quantum Experiments with Photons Since 1945

From the historical perspective, many of the experiments that we have discussed undoubtedly helped bolster Einstein's light-quantum hypothesis. Strictly speaking, though, they are not compelling 'proofs' of Einstein's radical interpretation of the light quantum, even if many of his contemporaries may have thought so. The older experiments on the photoelectric effect, the Compton and Raman effects, the law of photochemical equivalence and many other central links in the chain of 'evidence' have all been shown to be semiclassically explicable.[1] These 'semiclassical' theories make the assumption that restrictions are only imposed on material systems and their oscillations in discrete energy levels, whereas the radiation field surrounding them can be described semiclassically by classical Maxwellian electrodynamics. To this extent these 'old' experiments of quantum theory (between 1900 and 1924) and within the subsequent period of early quantum mechanics (from 1925 on) only enforce a quantization of matter but not any quantization of the electromagnetic field. From 1927, the emergent quantum electrodynamics actually took this very next step: a "second quantization" of the radiation field itself, not just of matter (see Sect. 3.12). Until 1945, not a single experiment necessarily presupposed this radical second step. This changed in 1950 when increasingly sophisticated experiments with rapid-response photon detectors and from 1960 on with lasers closed more and more of the gaps for semiclassical interpretations to sneak in. Afterwards the predictions of quantum mechanics and quantum electrodynamics, some of which were counterintuitive, could be seriously taken on.[2]

---

[1] See, e.g., Beck (1927), Wentzel and Beck (1927) on the photoelectric effect as well as Schrödinger (1927a) on the Compton effect; further Sudarshan (1963), Lamb and Scully (1969), Crisp and Jaynes (1969), Jaynes (1973), Henderson (1980). On variants of these experiments on single-photon states or correlated photon pairs, which are not classically describable, see Clauser (1974).

[2] Good bibliographic surveys on this are found in Paul (1985, 1986), Meystre and Wall (1991), Scully and Zubairy (1997), Sulcs (2003), Zeilinger et al. (2005), Chiao and Garrison (2008).

© Springer International Publishing AG, part of Springer Nature 2018   145
K. Hentschel, *Photons*, https://doi.org/10.1007/978-3-319-95252-9_8

8.1   'Photon Clumping': Hanbury Brown and Twiss (HBT) 1955–57

In the mid-1950s, a British radio astronomer employed since 1949 at the *Jodrell Bank Experimental Station* of the *University of Manchester*, Robert Hanbury Brown (1916–2002), had the idea of applying to optical astronomy the baseline interferometry in regular use in radio astronomy with its arrangements of many detectors set larger distances apart.[3] Michelson interferometers essentially rely on the phase of the interfering waves. Hanbury Brown and his collaborator Richard Quentin Twiss (1920–2005) designed a so-called intensity interferometer (Fig. 8.1, left). The light from Sirius was cast by two concave mirrors M_1 and M_2 onto photocathode detectors P_1 and P_2 a few meters apart and converted into an electrical signal that was amplified and analyzed. After successfully testing the arrangement with arc lamplight and other terrestrial sources, Hanbury (as his friends called him) thus managed to analyze the starlight of one of the brightest fixed stars by this method during the clear nights of the frigid winter of 1955/56. Suitable adjustments and the elimination of time delays between the signals in the left and right arms of the apparatus (with detector intervals set between two and nine meters) allowed him to find a positive correlation for the light of Sirius and, as anticipated, to establish that with increasing distance there is a rapidly decreasing positive cross-correlation between the two detectors (see the four data points and their error bars in Fig. 8.1, right). This could be theoretically modeled quite well by ascribing to Sirius an apparent magnitude that corresponded to a minuscule angle of observation of 0.0068″ (with an uncertainty of 0.0005″).[4] This apparent angular size agreed in order of magnitude with astrophysical models for such stars in the spectral class A1 and with estimated surface-temperatures of a little over 10,000 °C. Control measurements showed that no such correlations could be established when two different sources of light were used for the irradiation. Cosmic rays or other sources of error could be excluded as well.

This finding of a clear positive correlation (+0.85, over distances of about two meters, therefore relatively close to the maximum correlation coefficient of +1) initially encountered considerable resistance in the scientific community. As Hanbury Brown himself wrote:

> The most common objection to our work was that the time of arrival of one photon at a detector cannot conceivably be correlated with that of another because individual photons are emitted at random times and must therefore arrive at random times. If our system was really going to work, we would have to imagine photons hanging about waiting for each other in space![5]

---

[3] Originally trained as an engineer at *Brighton Technical School*, Hanbury Brown continued to study physics at *Imperial College* and worked on radar development during World War II. On his life and career, see his autobiography Hanbury Brown (1991); furthermore Lovell (2002) and Davis and Lovell (2003). Twiss was educated at Cambridge, England, and served during the war as a radar researcher for the British *Admiralty*, then was employed by the *Naval Service*, the *Armed Forces* and the *Division of Radiophysics* in Sydney before coming to Jodrell Bank. See Tango (2006).

[4] See Hanbury Brown and Twiss (1956a) on the apparatus and (1956b) on the results for Sirius.

[5] Hanbury Brown (1991) p. 121. The reception of HBT: Bromberg (2010), Silva and Freire (2013).

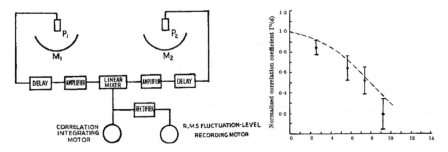

Fig. 8.1 Experimental arrangement (left) and measurement data in stellar intensity interferometry for Sirius. *Source* Hanbury Brown and Twiss (1956b) pp. 1046 and 1047

Because photons were known to propagate at the velocity of light, such an 'explanation' was obviously impossible. But how else could this apparently paradoxical finding be explained? How could two photons from such a faraway source be so intimately correlated with each other? Hanbury Brown and Twiss themselves estimated that on average at any instant just one photon was inside their interferometer. This interpretation presumed that the intensity of such a light signal was so low that this single photon had split itself into two halves to enter into both detector arms. Considering that photons are indivisible, this conclusion was equally absurd.

Initially, the opposition to their results was formidable and these two—rather inexperienced experimenters in quantum theory—were completely unprepared for it. Worse still, two other teams were unable to establish any such photon correlations in similar types of experiments: Eric Brannen with his graduate student Harry I.S. Ferguson at the *University of Western Ontario* in Canada and Lajos Jánossy (1912–1978) and his collaborators at the Central Physics Research Institute (*Központi Fizikai Kutatóintézetének*) in Budapest. In mid-1956 Brannen and Ferguson, for example, found a correlation of just 0.01% in their coherent light. But Peter Fellgett and Hanbury Brown and Twiss demonstrated in 1957 that the detectors used by both these null-result teams were simply not sensitive enough.[6]

Edward M. Purcell (1912–1997) had been able to show at the end of 1956 that this finding by Hanbury Brown and Twiss was not physically impermissible at all. It was rather explained as statistically expected fluctuations in light intensity (or put in terms of the particle picture: as fluctuations in photon emissions). If one were to observe just one emitter and just one detector, one would note a "tendency for the counts to 'clump.' From the quantum point of view, this is not surprising. It is typical of fluctuations in a system of bosons. [...] this extra fluctuation in the single-channel rate necessarily implies the cross-correlation found by Brown and Twiss."[7] Purcell,

---

[6]Brannen and Ferguson (1956) p. 482 and Jánossy and Zs. Náray (1957) versus Fellgett (1957) p. 956, Hanbury Brown and Twiss (1957) p. 1448, and finally Fellgett et al. (1959). On the methods and contexts of both teams cf. Bromberg (2010) pp. 11f., (2016) pp. 245f., Silva and Freire (2013) pp. 468–471.

[7]Purcell (1956) p. 1449. As is shown by Silva and Freire (2013) pp. 472–474, Purcell's decision to take sides in favor of HBT was crucial for the acceptance of their result.

Fig. 8.2 Bunching (**a**), random distribution (**b**) and antibunching (**c**) of photon-count rates for a correlation period τ_c (visualized at the very top). Temporally very closely lying events. Frequently in bunching which occur especially are suppressed in the antibunching graph. *Source* Scully and Zubairy (1997) p. 136. Reprinted by permission of Cambridge University Press © 1997

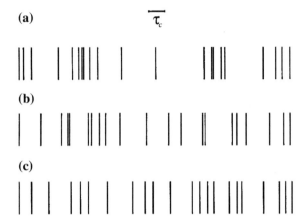

a Nobel laureate in physics at *Harvard University*, viewed the correlation as strongly indicative of the quantum trait of "clumping of the photons" from Bose–Einstein statistics. He and other theoretically trained experts in quantum optics regarded the HBT experiment, interpreted "from a particle point of view," as "a characteristic quantum effect" and discouraged approaching it with the naive classical image of photons as disjunct and indivisible particles lacking any explanation for such correlations.[8] The basic derivation of this effect in quantum optics was later called "photon bunching" (that is, the tendency of photons to cluster together, cf. Sect. 8.6 below). It is most easily understood[9] when one compares the classical and the quantum-mechanical derivations of the intensity measurements by two detectors A and B of two light sources a and b. Let us assume that both A and B can receive the light coming from a as well as from b, and additionally assume that both emitters radiate equally intensely at the amplitude T and that both detectors are equally sensitive. Then the classical total result is the sum of the four detection possibilities: a–A, a–B, b–A and b–B at $I = 4T^2$. Quantum mechanically, the probability amplitudes for these four different processes must *first* be summed together and *then* squared, which gives: $I = T_{aA}^2 + T_{bB}^2 + (T_{aB} \pm T_{bA})^2$. For bosons the plus sign applies in the interference term, for fermions the minus sign. Thus, for equal $T_{ij} = T$, it follows that: $I = 6$ for bosons and $I = 2$ for fermions. In other words, bosons manifest a tendency to bunch, whereas fermions to antibunch (Fig. 8.2).[10]

Latest with the award of the Albert Michelson Medal to R. Hanbury Brown and R.Q. Twiss by the venerable *Franklin Institute* in Philadelphia in 1982, the HBT

---

[8]Ibid., p. 1450. This was rendered in German as *Photonenklumpen* by Paul (1985) p. 142 and (1995) p. 171 or for the tendency, *Anhäufelung*.

[9]See Loudon (1973b) Chap. 3, pp. 111ff. and Scully and Zubairy (1997) pp. 110–136 on the details about the HBT experiment. The following simplified description is based on Sillitto (1957) and more completely: Fano (1961) and Paul (1986).

[10]Experimental proof of antibunching for fermions only arrived at the end of the 1990s among others by Henny et al. (1999), Kiesel et al. (2002) and Spence (2002).

effect of *photon clumping* or *photon bunching* as it was mainly being called by then, had become generally known. One important effect of these debates surrounding the HBT effect of the late 1950s was heightened awareness in the scientific community as a whole about problems in quantum optics (e.g., the difference between coherent and incoherent radiation) and the multifarious interpretations of photons within this context. The earliest attempts to explain the HBT effect were still based on semi-classical approaches.[11] The American pioneer in quantum optics Roy Jay Glauber (∗1925) from the *Lyman Laboratory* at *Harvard University* worked out a complete quantum theory of optical coherence and insisted: "There is ultimately no substitute for the quantum theory in describing quanta."[12] Richard M. Sillitto (1923–2005) from the *Department of Natural Philosophy* at the *University of Edinburgh* agreed with him in 1960: "It is one of the interesting features of [the HBT] result that it cannot be understood in terms of the crude—too crude!—model of a beam of light as a stream of discrete, indivisible, corpuscular photons." But unlike Glauber he arrived at a completely different conclusion about further usage of the photon concept. He advised: "we should be very hesitant about accepting arguments which rely on the 'corpuscular photon' model—at any rate beyond the point where the photons are required to do more than obey the laws of conservations of energy and momentum. In fact, I think that if we abolished the word 'photon' from our vocabulary for ten years, we should find that we could get on perfectly well without it."[13] Obviously, no one heeded this call to impose a ten-year moratorium on the term. Quite the contrary. Its usage intensified immensely, notably with the emergence of lasers from 1960 on. In 1962 interference between beams generated by two independently operating lasers first became feasible. One expression of this was the discovery of 'quantum beats,' an extenuated kind of rising and falling of intensity at a beat set by the difference between the two laser frequencies.[14] Laser interference experiments such as these also revealed nonclassical effects. The full arsenal of mathematical tools and concepts of QED, including wave-particle duality, had to be engaged for their interpretation.[15] Paul Dirac's frequently quoted dictum: "Each photon interferes only with itself. Interference between two different photons never occurs," was thus quite

---

[11] See, e.g., Purcell (1956) on methods of microwave perturbation theory or Kahn (1958), Fellgett et al. (1959) as well as Mandel and Wolf (1961) based on a model of light as a "Gaussian beam," which is subject to "shot-gun noise," i.e., classically stochastic fluctuations; cf. Bromberg (2016).

[12] Glauber (1963a) p. 85, likewise (1963b) p. 2529: "largely outside the grasp of classical theory" and (1963c) p. 2788 "an intrinsically quantum mechanical structure and not derivable from classical arguments." Glauber got the Nobel prize in physics for 2005, cf. Glauber (2005), Bromberg (2010).

[13] Sillitto (1960) p. 131, resp., p. 134; analogous arguments appear later again in Cohen-Tannoudji (1983) p. 198.

[14] See Javan et al. (1962), Magyar and Mandel (1963), Pfleegor and Mandel (1967), Paul (1985) pp. 111ff., (1986), Louradour (1993) and Wallace (1994) as well as Kuhn and Strnad (1995) pp. 173f.

[15] See, e.g., Ghosh and Mandel (1987) and Mandel (1986) for a bibliographic overview. Comp. further Wallace (1994) versus Louradour et al. (1993) on the interpretation of these findings.

definitely falsified.[16] It was very rare, by the way, for Paul Dirac to ever make any wrong prediction.

8.2   Single Photons and a Semitransparent Mirror: Campbell (1909) and Clauser (1974)

What happens when light hits a semitransparent mirror? According to the classical theory of light, the probability of finding light at a particular location is directly proportional to the square of the light intensity at that location. The wave theory of light allows very accurate predictions about the spatial propagation and temporal development of this light intensity and is extremely useful in diffraction and interference experiments, for example. But what happens if we allow the light intensity to drop? According to the classical theory, exactly 50% of the total intensity will still be distributed between the two partial beams. If we continue to drop the intensity far enough down that instead of $n \gg 1$ applying in the equation $E = n \cdot h \cdot \nu$ we have $n \simeq 1$, then the quantum theory of this very low-intensity radiation arrives at a different prediction than the classical theory. The classical theory predicts that a pair of highly sensitive detectors, positioned one in each of the two possible paths for the light behind the semitransparent mirror, will occasionally also be hit simultaneously whereas if only *one* quantum of light has hit the mirror strict quantization must require that such a coincidence must *never* occur. The reason is that this single light quantum is indivisible and can only propagate in *one* of the two directions. Which one can only be calculated probabilistically. The quasi-corpuscular interpretation of light asserts, though, that it definitely cannot travel in both directions at the same time. A wave of light *is* divisible; an individual photon simply *is not*. The first person to want to test this idea by experiment was an English physicist and philosopher of science in 1909. Norman Robert Campbell (1880–1949) was a fellow at *Trinity College* in Cambridge at the time as well as research assistant to J.J. Thomson at the *Cavendish Laboratory*. Campbell wanted to decide between Planck's quantum theory of radiation, according to Johannes Stark's more radical interpretation of 1908, and classical electrodynamics of a continuum. But he was unable to overcome the experimental difficulties in producing conditions with low enough numbers of light quanta.[17]

A single-photon experiment was successfully completed for the first time in 1973 by John Francis Clauser (∗1942) in *Lawrence Berkeley Laboratory* at the *University of California*. Twelve years later the quantum optics expert in Orsay near Paris repeated this experiment under more stringent conditions.[18] The predictions in

---

[16]See Dirac (1930c) Sect. I.3; cf. Paul (1986) pp. 209 and 230, Chiao and Garrison (2008a) p. 315, Bromberg (2010) pp. 8f. on the lasting impact of this pronouncement.

[17]See Campbell (1909); this is one of the few publications conceding its experiment as a failure.

[18]See Clauser (1974) resp. Grangier et al. (1986). The setup of the latter is discussed in Sects 8.3–8.4 below; furthermore Meystre and Walls (1991), Sect. V, Sulcs (2003) pp. 371–374, Zeilinger (2005b)

quantum mechanics of an anticorrelation between the photons in the two detectors were confirmed every time whereas the predictions of the (semi)classical theories were statistically significantly violated. Clauser concluded: "The results, to a high degree of statistical accuracy, contradict the predictions by any classical or semiclassical theory in which the probability of photoemission is proportional to the classical intensity. [...] So far no [experiment] has uncovered any departure from the quantum-electrodynamic predictions, but severe departures from [semiclassical] predictions have been found. The classical (unquantized) Maxwell equations thus appear to have only limited validity."[19]

8.3   Single-Photon Interference Experiments: Taylor (1909) to Grangier et al. (1986)

The next group of interference experiments also lies within the range of extremely low light intensities. Strictly speaking, among the experiments discussed in this chapter this group goes the farthest back in time. Fresnel's wave theory of light predicted the existence of interference patterns. The English physicist at the *Cavendish Laboratory* in Cambridge, Geoffrey Ingram Taylor (1886–1975), was the first to think of testing this prediction experimentally for very low-intensity light in 1908. His purpose was certainly not in order to test Einstein's quantum theory of 1905. J.J. Thomson's quasi-corpuscular model was what interested him. It expected there to be an inhomogeneous distribution of the energy along the wave front. Thomson asked Taylor to examine whether at very low intensities of light modifications to the interference pattern appeared or else noticeable fluctuations, at least.[20]

In Taylor's pioneering experiment, the light of a gas flame was attenuated through a sooted plate and then cast very close by the point of a needle. The light diffracted in this way then fell onto a photographic plate for longer durations for recording. The exposure times varied between a few minutes and about three months. The shortest exposure times produced a coarse-grained pattern on the photographic plate, which indicated a few punctiform processes of absorption that were statistically nearly completely evenly distributed. Thus it suggested a corpuscular structure of light rather than a spatially more even spread. Longer exposure times led to the formation of the same interference pattern as was generated by short exposure to high-intensity light: "in no case was there any diminution in the sharpness of the pattern," Taylor wrote.[21] The wave theory of light predicted exactly such an interference pattern, as regards positions and breadths of the interference stripes. However, (as shown in

---

pp. 275ff., Zeilinger et al. (2005) pp. 230–232. On John Clauser, see Whitaker (2012) pp. 149ff. and Freire (2013).

[19]Clauser (1974) pp. 853 and 856.

[20]Cf. Sect. 4.5 above on J.J. Thomson's mental modell. On Taylor, cf. further Batchelor (1996) and Sillitto (1960) p. 129.

[21]See Taylor (1909) p. 119.

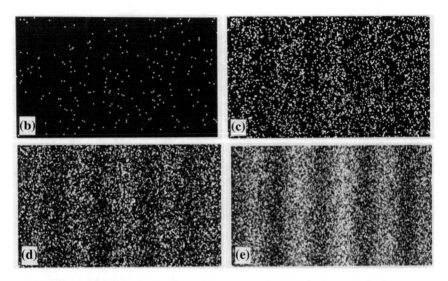

Fig. 8.3 Gradual build-up of an interference pattern from separate points at very low light intensity in Taylor's experiment (in the temporal sequence b–c–d–e). From Taylor (1909), produced in an analogous experiment by Tonomura, Belsazar 2012 with matter waves of electrons. Figure in the public domain from https://de.wikipedia.org/wiki/Datei:Double-slit_experiment_results_Tanamura_four.jpg (accessed 4 Feb. 2017)

Fig. 8.3) this interference pattern was always only composed by the superposition of many individual points. They supported the wave theory only in the way they were distributed. As individual points, however, they suited the particle model of light. As such, waves *and* particles seemed to occur together in this pioneering experiment, not complementarily, as an either/or option in the way already familiar to us from wave-particle duality (Sect. 3.8).

It later became evident, however, that the intensity of the light examined in Taylor's original experiment in 1909 as well as in its repetitions in 1927 and 1957 had not been diminished enough to exclude semiclassical interpretations completely.[22] Rodney Loudon pointed out in his textbook on the *Quantum Theory of Light* that "chaotic light" (with photon numbers $n \gg 1$) cannot be sundered from classical fluctuations of second-order quantum fluctuations.[23] More recent variants of this experiment performed by quantum optics specialist Alain Aspect (∗1947) and his two doctoral students Philippe Grangier (∗1957) and Gérard Roger[24] in a laboratory at the *Institut d'Optique Théorique et Appliquée* in Orsay near Paris in association with the *Centre*

---

[22] A survey of the literature with a critical perusal of all the earlier experiments on what were then considered single-photon interferences, including Dempster and Batho (1927) and Jánossy and Zs. Náray (1957), is offered by Sillitto (1960) and Pipkin (1978) with primary-source citations.

[23] See Loudon (1973b) Chap. 3, pp. 82–111.

[24] See Grangier et al. (1986). Alain Aspect and his collaborators are mentioned by Whitaker (2012) pp. 191ff. See also https://fr.wikipedia.org/wiki/Alain_Aspect.

National de la Recherche Scientifique (CNRS), demonstrated that the interference pattern (this time, interference through a double-slit, not off the tip of a needle) is indeed composed of individual points. Their distribution supported the wave theory of light but each individual point supported the particle theory. Quantum mechanics predicts that this interference pattern must immediately break down if the apparatus is modified to make it possible to measure which of the two slits an individual photon passes through. Precisely this is what their experiment demonstrated. Aspect and his collaborators proved that *single* photons were in fact involved, not merely very low numbers greater than 1, as had applied to Taylor's attempts in 1909. Their average count rate of just two photons per second, with an average distance between photons corresponding to over 100,000 km, practically excluded any interaction among more than one photon. This experiment clearly did involve single-photon interference. Depending on the apparatus arrangement, it could demonstrate both particle properties and wave properties of light within the sub-Poissonian regime of single-photon states:

> Two triggered experiments have thus been performed, using the same source and the same triggering scheme by the detectors. They illustrate the wave-particle-duality of light. Indeed, if we want to use classical concepts, or pictures, to interpret these experiments, we must use a particle picture for the first one ('the photons are not split on a beam splitter') since we violate an inequality holding for any classical wave model. On the contrary, we are compelled to use a wave-picture ('the electromagnetic field is coherently split on a beam splitter') to interpret the second (interference) experiment. Of course, the two complementary descriptions correspond to mutually exclusive experimental set-ups.[25]

Further progress in experimental technology in the twenty-first century has facilitated experimentation with *single* photons. Nevertheless, due caution is warranted because true single-photon experiments remain finicky and costly.[26]

8.4   Alain Aspect on EPR Photon–Photon Correlations in the 1980s

Quantum mechanics developed in four versions by Heisenberg, Schrödinger, Dirac and Wiener, along with the subsequently named 'Copenhagen interpretation' proposed by Born, Bohr, Heisenberg, von Weizsäcker and others. Since their emergence, Einstein developed into one of the most strident critics of this novel theoretical framework. At the Solvay conference of 1927 and in many other contexts, he was indefatigable at coming up with new objections to the new paradigm, often elegantly dressed in clever thought experiments presented to the adherents. These in turn were equally intelligently analyzed and then rejected one by one, either as incompatible with their

---

[25]Grangier et al. (1986a) pp. 178–179; cf. Zeilinger (2005) pp. 230–232 and Gerry and Knight (2005) Chap. 6 on the implications of this experiment.

[26]Much of the material compiled by Roychoudhuri et al. (2003, 2006, 2008, 2009, 2015) involve criticism of purported single-photon experiments.

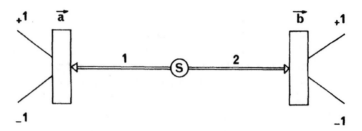

Fig. 8.4 The basic idea of the EPR thought experiment by Einstein et al. (1935). Two correlated elementary particles 1 and 2 (Einstein originally referred to electrons, here we shall use photons) are emitted from a light source S with a known spin. Traveling in opposite directions they hit detectors a and b, which each measure their spins by determining the direction of polarization of the associated luminous wave. *Source* Aspect et al. (1982a) p. 91. Reprinted by permission of the American Physical Society © 1982

premises or as unrealistic.[27] One of the most sophisticated and successful of these thought experiments was based originally on Einstein's idea for which Boris Podolsky (1896–1966) provided a fully elaborated formulation. They published it together with Nathan Rosen (1909–1995) in 1935: the Einstein-Podolsky-Rosen thought experiment (abbreviated as EPR, cf. Fig. 8.4).

Spin is the quantum-mechanical equivalent of intrinsic angular momentum (cf. Sect. 3.10 on the history of its discovery) which is a conserved quantity. Consequently, given that the initial and final states of the radiant source S are known, the spins of the two emitted elementary particles can be deduced. Einstein, Podolsky and Rosen were actually thinking of electrons in 1935, hence fermions with a spin of $\pm 1/2$, but this thought experiment works just as well with photons (bosons with a spin of ± 1). In the following we shall work this out for photons because then the later real experiments are more easily related with them.[28] If, for example, prior to the emission of the two photons the luminous source was in a state with total spin $S = 0$, then the spins of the two photons must have opposite spins. If detector a measures the spin of the first photon as $+1$, then it automatically follows (without any further measurement) that the spin of the second photon at detector b must be -1 (and vice versa). This prediction applies to all three directions in space that detectors a and b can measure (which is why they have vector arrows in Fig. 8.4). According to quantum mechanics, the spin of a system can be precisely determined only in one direction in space. If, for instance, the spin in the x direction is known, then the spins in the directions y and z are necessarily indefinite, etc. The thought experiment of 1935 by Einstein, Podolsky and Rosen concluded further that correlating photons 1 and 2 as arising out of a singlet state elegantly allows one to circumvent this quantum-mechanical

---

[27] See Born (1926), Heisenberg (1927, 1930, 1959), resp. Einstein (1949), Wheeler and Zurek (1983), Home and Whitaker (2007), Whitaker (2012) pp. 61 ff. and further literature cited there.

[28] David Bohm is the original author of this transposition from electrons to photons, which is why the EPR thought experiment is sometimes also called the Einstein-Podolsky-Rosen-Bohm experiment. See furthermore Näger and Stöckler in Friebe et al. (2015) Chap. 4 as well as Beller (1999) Chap. 7, Pais (1982) Chap. 25, and Brukner and Zeilinger (1997) on the limits of this transposition.

limitation: If detector a measures the spin of photon 1 in the x direction, then the spin of photon 2 on this axis can already be predicted reliably without measuring it. This means that while measuring at a, I can simultaneously determine the spin of photon 2 at detector b in another direction in space as well. Then, all in all, more would be known than is permissible according to orthodox quantum mechanics. Therefore, for Einstein and his collaborators, this orthodox quantum mechanics as Bohr, Heisenberg and others interpreted it[29] was at least "incomplete."[30] Thus the title of their paper: "Can quantum mechanical description be considered complete?" The answer to this question offered by these three authors—and soon other critics as well—was a vehement "no!"

For decades this argument by Einstein, Podolsky and Rosen was just a cunning thought experiment. However, advances in quantum optics and metrology since the 1960s produced the capability to perform this experiment on correlated pairs of elementary particles with mass. In 1967 Carl A. Kocher and Eugene D. Commins at the *University of California* in Berkeley, California, applied it to correlated photon pairs for the first time.[31] A more refined experiment in 1982 used detectors that alternated the direction in space of the spin components of photons 1 and 2 so rapidly and frequently that any mutual influence on the measurement processes at both ends was excluded. As has been shown by the above thought experiment, semiclassical theories that ascribe a separate, mutually independent local reality to each of photons 1 and 2 (cf. Sect. 9.3)—in other words, that presume that in principle their states can be determined as precisely as you want—lead to entirely different predictions than quantum mechanics. The difference is covered by Bell's theorem, derived by John Stewart Bell (1928–1990) for just these types of correlation experiments in 1964.[32] The experiments performed by Alain Aspect and his collaborators in Orsay agreed surprisingly well—within 1%—with the predictions of quantum mechanics; alternative theories led to a clear violation of Bell's theorem.

In 1982 the possibility remained to account the discrepancies to faulty detectors or elicit interactions between them relayed by "hidden variables." These could be whittled away very much in the years that followed by further improvements to the high-sensitivity detectors, by enlarging the distances between them and refining the measurement conditions (see Sect. 8.5 on delayed-choice experiments). By now, all the alternative interpretations count as eliminated by experiment.[33]

---

[29]Later referred to as the 'Copenhagen interpretation of quantum mechanics.' On its naming and the associated playing down of the differences between Bohr's, Heisenberg's and von Weizsäcker's interpretations, see Beller (1999) Chaps. 8 and 9, Howard (2004) as well as the literature cited in notes 27 and 105 of Chap. 3 above.

[30]This EPR argument also implicitly presupposes the locality: see here Sect. 9.3.

[31]They used a cascade transition in Ca gas. See Kocher and Commins (1967). Freedman and Clauser (1972) took improved measurements; cf. Freire (2006), Freire et al. (2013).

[32]I dispense here with deriving these inequations formally and reproducing the details of the experiments. See, e.g., Clauser and Shimony (1978), Paul (1985) pp. 164ff., Gerry and Knight (2005) Chap. 9, Whitaker (2012) pp. 87–282, Zeh (2012) pp. 14–17 and further citations there.

[33]Good review articles exist on the numerous confirmations of this experiment on correlated photons worldwide; see Freire (2006), Shadbolt et al. (2014) with primary citations.

8.5   Wheeler's 'Delayed Choice': Which-Way Experiments with Photons

One favorite way to side-step interpreting wave-particle duality and photon-correlation and anticorrelation experiments of the 1960s and 1970s was to point out that in principle it could be assumed that (metaphorically speaking) light would 'catch wind' of which of its properties are about to be measured inside the detector and still be able to 'prepare itself' ahead of time about whether to exhibit its wave characteristics or its particle characteristics. The physical mechanisms remained in the dark, however. How could the measurement instrumentation give such feedback about the object of analysis? Were obscure guiding fields behind this? Or was it a pernicious plot against guileless experimenters? These "conspiracy theories" were a thorn in the side for experimental physicists who wanted to close such loopholes in the argumentation.[34]

One brilliant suggestion about an at least conceivably realistic experimental test of these theories came from John Archibald Wheeler (1911–2008). Having laid the groundwork of 'geometrodynamics,' from the 1960s on he turned his attention increasingly to the fundamental debates about interpretations of quantum mechanics and made fundamental contributions there as well.[35] Referring back to earlier considerations by Carl Friedrich von Weizsäcker (1912–2007), Wheeler suggested in 1978 that the decision about whether to measure the characteristics due to the particulate or wave-like nature of photons be delayed until the actual interactive process inside the analyzer had already occurred, and only afterwards to commit the detectors to the type of values to measure.[36] The extreme speed of light condemned this suggestion to remain a pure thought experiment for a whole decade. In the second half of the 1980s, some research teams began to seriously consider attempting to perform it as a real experiment. The first two of these teams were German, then came an American one followed by a French team. This last attempt to carry out Wheeler's thought experiment, performed in 2007, was by far the most convincing. That is why I confine my account here to the latter. These French experimenters managed to avoid any causal interaction between the detectors, on one hand, which were engaged by a generator governed by chance, and the input of the photons into the apparatus, on the other.[37]

Philippe Grangier and Alain Aspect at the *Laboratoire Charles Fabry de l'Institut d'Optique* of the *École Polytechnique* in Palaiseau and a larger number of other experimental physicists at the *Laboratoire de Photonique Quantique et Moléculaire* of

---

[34]The expression "conspiracy theories" comes from Peter J. Lewis (2006); cf., e.g., Zajonc (1993), O'Brien (2010).

[35]See Wheeler (1978) and Wheeler and Zurek (1983). Wheeler's background is mentioned, for instance, in Wheeler and Ford (1998). On geometrodynamics as a visual geometrical reformulation of the general theory of relativity, see Hentschel (2014b) pp. 108ff., with further literature cited there.

[36]See von Weizsäcker (1931) p. 128 and (1941) as well as Wheeler (1978).

[37]See Hellmuth et al. (1987), Baldzuhn (1989), Jacques et al. (2007, 2008)

Fig. 8.5 Mach–Zehnder interferometer arrangement. *Source* Jacques et al. (2007) p. 967. Reprinted by permission of the American Association for the Advancement of Science

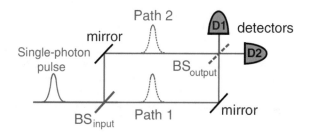

the *École Normale Supérieure* in Paris executed Wheeler's delayed-choice experiment in 2007 using a clockwork-triggered single-photon luminous source based on an isolated nitrogen vacancy center on a diamond nanocrystal. Its single-photon pulse was guided onto a Mach–Zehnder interferometer (see Fig. 8.5). A beam splitter set at 50% transmission led the entering light into two arms 48 m in length, which corresponds to a flight time in the interior of the instrument of 148 ns. Then the light from both arms reached a second beam splitter (analyzer) that was switched on or off by a random generator, behind which detectors $D1$ and $D2$ were positioned orthogonally to each other as extensions to each of the interferometer arms. Let us first look at this apparatus in the control-measurement mode, excluding the second beam splitter in front of the two detectors at the top right. The classical interpretation of light as a wave has a 50% portion of the incident light pass through the first semipermeable mirror BS into each of the two arms. This also implies an equal distribution of the light in both detectors. Einstein's theory of photons (a particle picture), however, predicts a perfect anticorrelation. For, the *single* photon entering into the apparatus by virtue of its design cannot cut itself in two and must consequently have already 'decided' at the first beam splitter which of the two interferometer arms it will enter. Accordingly, (if we leave out the analyzer) only one of the two detectors may respond and the correlation coefficient α should vanish. In the experiment α was not exactly zero, but it was low enough at 0.15 ± 0.01 to show that the experimenters were in fact operating very near the strict single-photon regime. Now we add the analyzer to the beam path. This electro-optically modulated beam splitter was randomly engaged by signals immediately prior to each measurement and the signals determined whether to measure a wave or a particle. Whenever the random generator switched the analyzer off, the detectors indicated the same rates of intensity and coincidence as the control measurement preceding it. Whenever the analyzer was switched on for partial reflection, the interference patterns anticipated by the wave theory appeared. The first case thus meant that the photon had taken just *one* of the two paths; the second case was only comprehensible as the photon having somehow been present in *both* of the interferometer arms in order to be able to generate the pattern. Both cases at once really ought not to be possible, but in some mysterious way the pattern that exactly suited the observation situation always appeared. The conspiracy theories that postulated feedback about the measurement result from the observation situation arranged by the experimenter had finally been disproved. This provided backing for Niels Bohr, who had correctly prognosticated in his contribution to a volume dedicated to Einstein in 1949:

> it obviously can make no difference as regards observable effects obtainable by a definite experimental arrangement, whether our plans of constructing or handling the instruments are fixed beforehand or whether we prefer to postpone the completion of our planning until a later moment when the particle is already on its way from one instrument to another.[38]

John Archibald Wheeler amplified the stark contrast between this experimental finding and our naive intuition that every object must always ever act like a particle or like a wave: Another level of the "schizophrenic quantum world" had been exposed by the delayed-choice experiment, "showing another quantum feature of the world that defies classical description: not only can a photon be in two places at once, but experimenters can choose, after the fact, whether the photon was in both places or just one. [...] In short, the experimental verdict is: the weirdness of the quantum world is real, whether we like it or not."[39]

8.6   Photon Bunching and the Hong-Ou-Mandel Dip 1987

With the first photon correlation experiments came the frequent observation that at very low intensities correlations between two photons from the same source were statistically not completely evenly distributed at all (see Sect. 8.3 on the HBT effect). For intervals lesser than or equal to the coherence time $\tau = \lambda/c$, the probability of counting two or more pulses at a time was higher than the statistical Poisson distribution would have led one to expect. In 1956 Purcell initially named this phenomenon 'clumping' in connection with the experiment by Hanbury Brown and Twiss, but later he mostly referred to "bunching of light" or "photon bunching."[40] This tendency of photons to appear in groups is derivable from Bose–Einstein statistics, which made it seem likely that a link existed.

The native German pioneer of quantum optics Leonard Mandel (1927–2001), who had fled the Nazis to England and was employed from 1964 at *Rochester University* in the USA, began to tackle this phenomenon specifically with his fellow researchers during the 1960s. They exposed a highly sensitive photomultiplier to the very weak radiation of a low-pressure mercury-vapor lamp at the 5461-Å line. Then they continued to diminish the emitted light intensity, which could be monitored well by the proportionately falling count rate. When the times of a series of recordings fell below two or three nanoseconds apart, which corresponded to the coherence time τ for this radiation to order of magnitude, then the average counts of registered photon coincidences suddenly rose noticeably. With a tungsten lamp, no such rise was visible because their coherence time was very much smaller (see Fig. 8.6).

---

[38]Bohr (1949) p. 230, also cited at the end of Jacques et al. (2008) p. 4.

[39]Wheeler in Tegmark and Wheeler (2001) p. 76.

[40]See Purcell (1956) p. 1450 and Silva and Freire (2013) pp. 474f. on the later success of the term 'bunching,' which was Twiss's preference.

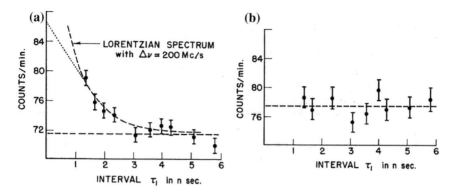

Fig. 8.6 Photon bunching: (left) a rise in the count rates of photon coincidences for light from a Hg[198]-vapor lamp under the coherence time τ of approx. 3 ns. By contrast, (right) from a tungsten lamp there was no rise for low τ. *Source* Morgan and Mandel (1966a) p. 1013. Reprinted by permission of the American Physical Society © 1966

Leonard Mandel conducted another experiment also based on the transmission of photons through a semitransparent mirror, with two of his graduate students, Chung-Ki Hong and Zhe-Yu Jeff Ou in 1987.[41] Among experts it is also known by the acronym HOM after these three quantum optics specialists in Rochester, New York. They used correlated photon pairs emitted from a potassium-dihydrogenphosphate crystal (KDP) irradiated by the weak but coherent beam of an argon-ion laser (see Fig. 8.7). These two photons hit the mirrors $M1$ and $M2$, which guided them onto a beam splitter BS (center left at the top of Fig. 8.7). Depending on whether the photons were reflected or transmitted, they continued along the upper or lower paths and after passing through an infrared filter IF finally each hit one of the detectors $D1$ and $D2$. Their output was then rapidly evaluated electronically. This output, a count rate of the measured photon–photon coincidences in detectors $D1$ and $D2$, was then analyzed in relation to the difference between the upper and lower paths. This path difference between the upper and lower paths was simply made by displacing the beam splitter by a small distance $\pm c\delta\tau$ within the coherence length of the generated pulse of light. Bafflingly, a strong decrease in this coincidence count rate appeared when the path difference was particularly small. The minimum count-rate decrease was from over 80 coincidences per minute to only about 10 per minute.

What was the explanation for this result? There was evidently a strong anticorrelation for a nearly vanishing difference in path. This meant that when more than one photon with exactly the same properties (wavelength, spin) impinged on the beam splitter virtually simultaneously from different sides, the probability that they

---

[41]On the following, see Hong et al. (1987, 1998) as well as Santori (2002) for a more sharply defined replication. Ou wrote his thesis in quantum optics in 1990 at the *University of Rochester*, where he currently has a professorship: see http://physics.iupui.edu/people/zhe-yu-jeff-ou-0. Hong earned his doctorate at Rochester in 1988 and is now a professor at the *Pohang University of Science and Technology* in South Korea. On Mandel see Bromberg (2010) pp. 6–8, (2016).

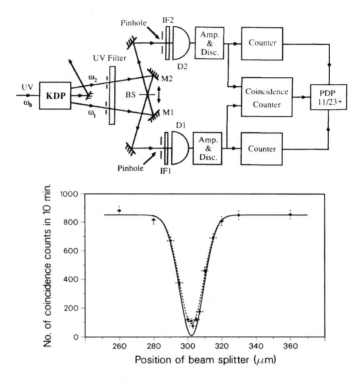

Fig. 8.7 Above: Setup by Hong et al. (1987). The semipermeable beam splitter BS could be moved slightly up or down, which led to a change in the coincidence count rate as a function of this mirror position. Below: The coincidence count rates producing the HOM dip (see the main text). *Source* Hong et al. (1987) pp. 2044 and 2046. Reprinted by permission of the American Physical Society © 1987

be reflected or transmitted was no longer statistically uncorrelated. Either they both continued to travel onward together in the lower direction or else both in the upper direction. Therefore, events in which a photon was measured in each of the two detectors were drastically reduced. "In the right circumstances, two photons can meet and 'coalesce' [...;] provided that the photons are in the same, single mode, [...] quantum mechanics predicts that a bunching, or coalescence effect occurs."[42] This is another example of the strange characteristic of photons to cluster together, which results quantum mechanically quite simply from the destructive interference of cases 2 and 3 (in Fig. 8.8), in which they both exit in different directions because the transmission of both (case 2) and the reflection of both (case 3) differ from each other exactly by half a phase.

A precision experiment in 2002 blocked the way out still open in 1987 to consider that the common origin of both photons from a common source could explain this strange correlation at a semipermeable mirror. Charles Santori and his collaborators

---

[42]Grangier (2002) p. 577; cf. further Grangier (2005), Sulcs (2003) pp. 380ff.

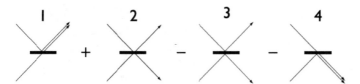

Fig. 8.8 Distinction between the four possible cases of reflection or transmission of two photons through a semitransparent mirror: Cases 2 and 3 destructively interfere with each other. Thus only cases 1 and 4 remain. In the former, both take the upper path, in the latter, both take the lower path. This leads to the strong anticorrelation observed by HOM between two photons with otherwise equal properties and a vanishing difference in path. *Source* Wikimedia, in the public domain

used a setup in which the two photons made to interfere at the mirror did not come from the same source but from many semiconductor quantum dots of the same kind.[43] It was thus shown irrefutably that photons in the same state become indistinguishable when they come close enough together. (Santori et al. recorded an overlap between the wave packets of up to 80%.) Thus we have a confirmation of Bose–Einstein statistics along with all its counterintuitive implications.

8.7   Photon Antibunching in Resonance Fluorescence

The research team headed by Leonard Mandel in Rochester also conducted complementary experiments to those just discussed. This time photon antibunching was their focus. Classical luminous sources contain many atoms that independently emit photons. Nonclassical luminous sources manifest temporal gaps that persist until an excitation is able to elicit the emission of another photon.

Using a tunable dye laser, Mandel's team managed to excite individual sodium atoms by directing laser light orthogonally onto an atomic beam in resonance fluorescence mode. The frequency of the laser light was set very close to the resonance frequency of the sodium atoms for transition into a defined excited state.[44] Setting the atomic beam parameters to a relatively broad bandwidth of $100\,\mu m$, which at a mean velocity of 10^4 m/s corresponds to an atomic flow density of approximately 10^{10} to 10^{11} atoms per square centimeter and second, they created the conditions by which, on average, at most only one atom per 100-ns time interval was raised into the excited state for individual measurement. Of course, after a very short period of time this atom returned from the excited state into the ground state again, thereby emitting a photon of a frequency that was very exactly calculable from the energy difference between the two states.

---

[43] See Santori (2002); cf. Grangier (2002, 2005), Zeilinger et al. (2005) p. 232 on the implications of this experiment. Santori is currently employed as a state scientist at the *Hewlett-Packard Laboratories* in Palo Alto, California: http://shiftleft.com/mirrors/www.hpl.hp.com/research/qsr/people/ Charles_Santori/index.html.

[44] Kimble, Dagenais and Mandel (1977), Dagenais and Mandel (1978), Paul (1985) pp. 92f., 159f.

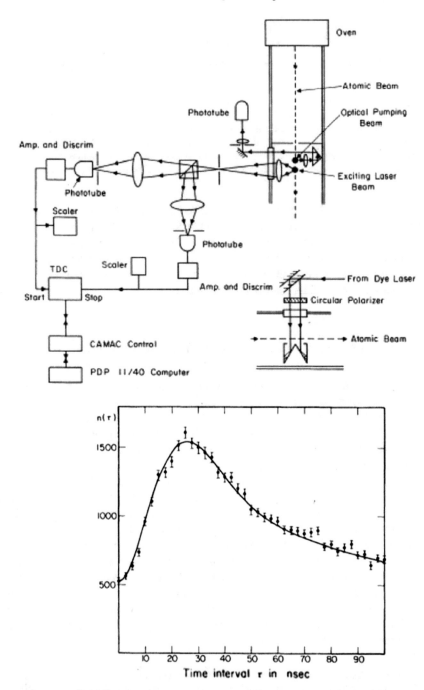

Fig. 8.9 The experimental setup (top) and measurement results with error bars (bottom), with an overlain comparison curve of the theoretical values calculated according to QED. *Source* Dagenais and Mandel (1978) pp. 2218 and 2222. Reprinted by permission of the American Physical Society © 1978

Using a semipermeable mirror (in the center of the upper image in Fig. 8.9), Leonard Mandel, Jeff Kimble and Mario Dagenais measured how frequently not one, but two photons were generated. These photons then hit the photodetectors (top figure, left) as well as below the central semipermeable mirror, and were electronically registered, increasing the number of coincidences n. As the graph (at the bottom of Fig. 8.9) shows, this number of coincidences n for a vanishing time difference τ was lower by a factor 3 than for time delays of 25 ns. This coincidence rate $\tau = 0$ s did not drop completely down to zero because of the noise generated by the thermal cooling of the detectors which could not be completely suppressed. With a quantum yield of 15%, they still had a dark count rate of about 100/s. Leonard Mandel and his collaborators interpreted this result as follows:

> We have demonstrated that the fluorescent photons exhibit antibunching in time in all cases, which may be regarded as a reflection of the fact that the atom makes a quantum jump to the ground state in the process of emission and is unable to radiate again immediately afterwards. No classical electromagnetic field is able to exhibit this behavior. [...] It was found that [...] fewer events were observed at $\tau = 0$ than were to be expected from random atomic emission alone, and that the emissions were anticorrelated [...] in agreement with [...] quantum electrodynamics, and in contradiction with any semiclassical emission theory. This provides good evidence that the atom makes a quantum jump to the ground state at the instant of photon emission.[45]

Thus we have here another case of "pure-culture photons" (Harri Paul), which semiclassical theories of light have failed to describe, making QED the only way to arrive at the correct predictions for these single quantum-like absorptive and emissive processes ('quantum jumps').

8.8   Recording the Birth and Death of a Photon in a Cavity

Observing a photon usually also spells its death. Whether the human retina or a photodetector is used to observe it, the photon has to be absorbed by such a light-sensitive surface, thereby destroying it. A clever measurement setup within the context of cavity quantum electrodynamics has more recently made it possible to record the birth and death of photons in the interior of a cavity.[46] The Nobel prize in physics for 2012 rewarded Serge Haroche (∗1944) for such 'quantum nondemolition experiments.'[47]

Haroche used a very attenuated atomic beam of rubidium which was transparent to light. The beam was put in a highly excited state, called the Rydberg state.

---

[45]Dagenais and Mandel (1978) p. 2225.

[46]On cavity QED, see Haroche and Kleppner (1989), Kuhn and Strnad (1995) Sect. 5.9, Gerry and Knight (2005), Haroche and Raimond (2006), Gleyzes, Haroche et al. (2007), Chiao and Garrison (2008) Chap. 12, Haroche (2012) and further literature cited there.

[47]S. Haroche, a native Moroccan, completed his studies at the *École Normale Supérieure* in Paris under Claude Cohen-Tannoudji, defended his thesis in 1967, and is a teacher and researcher in Paris, since 2001 as professor at the *Collège de France*. See the autobiography in Haroche (2012) written for his Nobel prize ceremony.

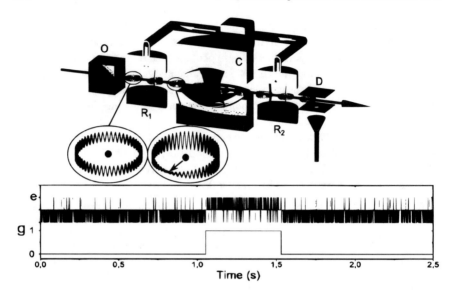

Fig. 8.10 The apparatus (above) and the results (below) by Haroche and his collaborators (see the main text). *Source* Haroche (2012) pp. 79 and 81

The electron shells of these rubidium atoms have a thousand times greater extension than for normal atoms and their half-life is in order of magnitude 30 ms. The atomic beam produced in oven *O* (see Fig. 8.10) was sent through a cavity *C* equipped with high-precision ultra-cooled niobium mirrors forming the two ends of the cavity 2.7 cm apart. Microwaves were reflected back and forth between them for one-tenth of a second. Between their generation and destruction, the microwaves thus traveled a distance roughly corresponding to the circumference of the Earth. Because the mirrors were spherical to a precision of almost one wavelength the electromagnetic radiation between the two mirrors was monochromatically quantized to very good approximation. The eigenfrequency of the microwaves of 51.1 GHz is such that these photons could not be absorbed by the Rydberg atoms which were sent perpendicularly through the photons' radiation field. Instead, their energy levels were raised very slightly. This is indicated by the alteration to the magnified eigenstate at the bottom left of the drawing (top of Fig. 8.10). These approximate changes of state, which only occur if a photon is present, were continuously measured by Fabry-Perot interferometers R_1 and R_2 at each end of the cavity. Because the Rydberg atoms travel at the slow rate of 500 m/s there are multiple interactions between each of these atoms and the microwaves reflected back and forth at the velocity of light. This considerably improved the statistics for this measurement of photon states.

In the lower part of Fig. 8.10 we see the raw signals obtained by this apparatus. Upward-drawn lines indicate when the probability of a photon being present is ≥ 0.5, and downward-drawn lines indicate when the probability is ≤ 0.5. A thought-out series of evaluations in multiple steps reduced this statistical spread of raw signals.[48] As a result we can clearly infer that a photon was present between the times 1.05 and 1.55 and that there was none before and afterwards. We can thus conclude that a photon was generated at the time 1.05 and was subsequently destroyed by natural absorption in one of the mirrors at the time 1.55. During its lifetime of approximately 0.5 s, it interacted with the atomic beam about one hundred times without this photon being destroyed by the measurement. That is why the term 'quantum nondemolition' experiments is appropriate.[49]

The hope is that in the future this experimental technique can be used to prepare nonclassical states and protect them from decoherence; it certainly justifies the award of the Nobel prize for these experiments.[50] The fate of a single photon could henceforth be observed from its birth up to its natural death, without the observer being immediately implicated in its extinction.

8.9   Quantum Entanglement and Quantum Teleportation

We have already become acquainted with some of the apparently paradoxical consequences of experiments on light quanta in correlated states, or 'entangled photons' (in Sects. 8.4 and 8.5). Einstein had dismissed them as "spooky actions at a distance." (*spukhafte Fernwirkung*).[51] Whole books could easily be devoted just to these ideas.[52] From among the numerous baffling entanglement experiments I chose just one other group of experiments, our third here, about 'quantum teleportation.' It holds the great promise of a future variety of new applications for surveillance-proof communications, as well as quantum computing. As early as 1982 William Kent Wootters (∗1961) and Wojciech Hubert Zurek (∗1951) showed that an individual quantum state cannot be copied or cloned without the initial state being altered in the process. Any monitoring of information involves such a copying process—thus the foundations are laid for quantum cryptography.[53] How can information about the existence of a quantum state be conveyed any distance from one place to another without monitoring? The answer is quantum teleportation. Its theoretical suggestion goes back to Wootters and his collaborators in 1993. The Viennese quantum physicist Anton Zeilinger (∗1945) and his collaborators first confirmed it by experiment in quantum optics with correlated photons in 1997.[54] In the following we shall consider only one

---

[48] See Gleyzes, Haroche et al. (2007) and Sayrin et al. (2012).

[49] See furthermore Grangier et al. (1998), Nogues et al. (1999) and Haroche (2012).

[50] Haroche (2012) Nobel lecture also touched on aspects in the history of science.

[51] Einstein's letter to Max Born from 3 Mar. 1947, Born (1969b) p. 158.

[52] Good examples include Espagnat (1971), Wheeler and Zurek (1983), Zeilinger (2003, 2005), Home and Whitaker (2007), Chiao and Garrison (2008) Chap. 6, pp. 19–20 and Whitaker (2012).

[53] See Wootters and Zurek (1982).

[54] See Bennett et al. (1993) resp. Zeilinger et al. (1997/98), further Boschi et al. (1998) and Zeilinger (2005b).

of the more recent real implementations of this thought experiment. In 2012 quantum information about a photon's spin was transmitted across a distance of 143 km, from La Palma in the Canary Islands to the neighboring island Teneriffe.[55] In June 2017 a satellite-supported exchange of signals in China breached over 1200 km.[56]

The basic precondition for quantum teleportation is the existence of *two* channels connecting the sender of the signal with the receiver: a classical information channel such as a telephone or telegraph line and a quantum channel, occasionally also called an 'EPR channel.' (Unlike in science fiction, information is what is transported, not matter.) At the sender's location (usually named Alice) an entangled photon pair in an antisymmetric total state is generated inside a barium borate crystal. That means the polarization state of one photon is orthogonal to the other (or they have opposite spins). It is not known which of the two photons is in which state. The first of these photons is conducted along the quantum channel at the speed of light to the receiver (let's call him Bob). The other one stays with Alice and is made to interfere with a third photon under controlled conditions. The quantum state of this third photon is the information that should be relayed from Alice to Bob. The interaction between the second and third photon allows Alice to calculate the quantum state of the second photon and from that, in turn, because of the entanglement of the first two photons, the state of the first photon. This state is then relayed to Bob along the second, classical line of communication. In this way, without having to conduct measurements on the second photon himself, Bob can learn from this relayed message the state of the arrived photon and reliably reproduce the initial state of the third photon which is still located with the sender by operations that do not destroy its state. The first photon at the receiver's location thereby becomes an exact copy of the information-bearing third photon.[57] If a spy tried to intercept the entangled light particles during transmission, the entanglement would get lost and the transfer of data would be disrupted. This is because in quantum teleportation the sender does not need to know either the transmitted information or the location of the recipient (as long as the two-channel connection remains open). Fixed conduits are not a prerequisite because such connections could also be established in open air.[58] The receiver, for his part, also does not need to read the information right away, but can theoretically store it indefinitely for later. Thus quantum teleportation seems to have a bright future with various secure encryption applications in information technology and computer science.

---

[55] On the following, see Zeilinger et al. (2012a, b) including the lengthy quantum-mechanical derivations and further related references.

[56] See Yin (2017) and Popkin (2017).

[57] Zeilinger and his collaborators on this pioneering experiment attained about 80% statistical reliability, far above the classical limit of 66% from information theory.

[58] In the case of Zeilinger (2012a), light signals were teleported between the *Jacobus Kapteyn-Teleskop* on La Palma and the *Optical Ground Station* of the *European Space Agency* on Teneriffe.

8.10 High-Energy Photon–Photon Scattering

From classical electrodynamics and James Clerk Maxwell's equations for the electromagnetic fields generated by electric charges and magnets it follows that two waves of light cross through each other without interacting. Many experiments in classical optics could confirm this finding. One expression of it is the strange behavior of light going through an aperture or a camera obscura. Ibn al-Haytham had repeatedly observed this inversion of images during the Middle Ages. Planck's and Einstein's quantum theory as well as the Bohr–Sommerfeld atomic model of early quantum mechanics from 1925 also incorporated this lack of interaction between two crossing beams. In 1936 Werner Heisenberg and his assistant Hans Euler (1909–1941) pointed out that in QED it was possible for photons to be scattered by other photons, even though the anticipated relative frequencies of such scattering (physicists refer to cross sections of activation σ) were very low, at $\sigma \sim \alpha^4 \simeq 3 \cdot 10^{-9}$—far too low to be measurable by experiment at that time. High-precision measurements of anomalous magnetic moments of an electron and a muon provided the first indirect evidence of such photon–photon scattering. Related Feynman diagrams (as in Fig. 8.11) describe photon–photon interactions as fourth-order and higher corrective processes (cf. also Fig. 3.14).

The activation cross sections for this photon–photon scattering increase with the electric charge Z of the participating collision partner as Z^2. That is why a heavy-ion accelerator such as the LHC in CERN near Geneva is an ideal instrument for experimental verification of this prediction from QED. The accelerator accelerates a positively and a negatively charged lead ion ($Z = 92$) almost to the speed of light and then guides them into an interaction zone surrounded by detectors to collide with each other head-on. This method makes it possible to prove the photon–photon interaction now directly. The sequence of events in the interaction between those two lead ions, depicted in the right-hand drawings (Fig. 8.11), occur within the tiniest fraction of a second. The ions are sent very close by each other in opposite directions and during this interaction they exchange energy in the form of photons. The two lead ions then disappear in the detector's in and out channel without leaving a

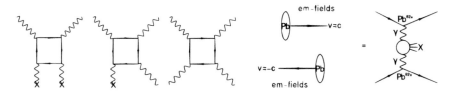

Fig. 8.11 Left: Three Feynman diagrams on the interaction between two photons. They are 'loop diagrams' with fermions, such as, electron-positron pairs, to convey this interaction. Four vertices, where the wavy lines for photons and the straight lines for fermions meet, indicate that these processes are at least to fourth order in the fine-structure constant $\alpha \simeq 1/137$ and consequently are very rare. Right: A schematic rendition of the high-energy scattering processes of lead ions through photon pairs observed at CERN. *Source* ATLAS collaboration (2017) p. 2

record but the detectors can determine the directions and energies of the participating photons and of other particles formed in this high-energy interaction. In February 2017, as my research for the present book was close to completion, a preprint by the ATLAS collaboration at CERN appeared reporting evidence of the effect. By a clever preliminary selection, the experimenters were able to pick out from enumerable observations conducted in 2015 thirteen collision processes between two lead ions, each with a total energy of 5.02 TeV. They interpreted these thirteen cases as good examples of photon–photon interaction because only two photons appeared in the detector and together carried practically the whole energy of the interaction but evidently must have interacted with each other prior to being recorded.[59] Once these processes will be calculated out in higher-order perturbation theory and measured to high precision, if these initial results are verified, this would be the first *direct* experimental proof of high-energy photon–photon scattering in addition to the many already available indirect indications of such processes in electron-photon interactions or massive boson-photon interactions. The statistical significance of these new findings is high and the ascertained interaction cross sections are compatible with the values expected from the standard model of elementary particle theory for strong, electromagnetic and weak interactions.[60]

---

[59] See ATLAS cooperation (2017)—the expectable statistical background induced by other processes amounted to only 2.6 ± 0.7 events. For a proposal of high-intensity laser photon-photon scattering, cf. Gies et al. (2018).

[60] The results are significant deviations from the standard by 4.4, and the measured $\sigma = 70 \pm 24(\text{stat.}) \pm 17(\text{syst.})$ nb is compatible with the prediction from the standard model at 49 ± 10 nb.

Chapter 9
What is Today's Mental Model of the Photon?

How do all these more recent surprising experiments add to our understanding of what a photon is? Have we come further along than Einstein with his sober resume that those many years of pondering had not brought him a single step closer to understanding light quanta?[1] The latest experiments *certainly have* brought us much further. Today we know that the predictions of quantum theory in their most radical reading must be taken seriously. All the salvaging attempts using semiclassical arguments by Planck, Louis de Broglie, Schrödinger and dozens of other physicists have failed in the face of these recent experiments. Anton Zeilinger, who together with his team in Vienna and Innsbruck had conducted many path-breaking experiments, and who had written many influential popular books about the new world of quanta, was categorical: "semiclassical radiation theories are a dead end. […] semiclassical ideas cannot account for all experimental observations."[2] His arguments included the lack of a minimum time required for electrons to be released from the metal surface in the photoelectric effect, which would otherwise have clearly suggested a corpuscular and practically instantaneous transmission of energy from radiation to matter. Classical local-realist theories *cannot* explain the many quantum correlation experiments in EPR-style, which have meanwhile also been done with correlated photon pairs and matter waves, either.[3]

In closing, let us go through each one of these points.

---

[1] See here the beginning of Sect. 9.5 below about Einstein's assessment in a letter to Michele Besso, 12 Dec. 1951.

[2] Zeilinger et al. (2005) p. 231.

[3] Ibid.; cf. further, e.g., Paul (1985, 1986), Scully and Zubairy (1997), Muthukrishnan et al. (2003), Gerry and Knight (2005) Chaps. 6–10, Chiao and Garrison (2008).

© Springer International Publishing AG, part of Springer Nature 2018
K. Hentschel, *Photons*, https://doi.org/10.1007/978-3-319-95252-9_9

9.1   Instrumentalistic Interpretation

Which adjustment screw in our current model of the photon needs to be turned? According to Zeilinger, we must give up the local-realist interpretation that is familiar to us from macroscopic experience and from particles since the days of the Greek atomists, Newton and, from 1900 on, the quantum pioneers who also could not resist the temptation to interpret light as a stream of 'atoms, 'corpuscles' or even 'molecules of light' or 'quanta.'

> The general conceptual problem is that we tend to reify—to take too realistically—concepts like wave and particle. Indeed if we consider the quantum state representing the wave simply as a calculational tool, problems do not arise. In this case, we should not talk about a wave propagating through the double-slit setup or the Mach–Zehnder interferometer; the quantum state is simply a tool to calculate probability. Probabilities of the photon being somewhere? No, we should be even more cautious and only talk about the probabilities of a photon detector firing if it is placed somewhere. One might be tempted, as was Einstein, to consider the photon as being localized at some place with us not knowing that place. But, whenever we talk about a particle, or more specifically a photon, we should only mean that which 'the click in the detector' refers to.[4]

The reading suggested here, known as the operationalistic or instrumentalistic interpretation of quantum mechanics, avoids implied 'reification.' That is, it avoids terms like 'light quantum,' which inherently turn a notion into something concrete. Its renaming into 'photon' from 1926 on (in linguistic analogy to electron, proton, neutron and other elementary particles)[5] only encouraged this misconception. The first step here obviously means dropping our habitual 'objectivizing' in everyday life, what is referred to as 'reification of entities' in natural philosophy. In a discussion about the meaning of the term 'photon' in the *American Journal of Physics*, which primarily targets science teachers and instructors in higher education, an experienced physics instructor revealed his almost startled reaction:

> Like most professors, I have long explained to my students that the basic observed facts about the photoelectric effect [...] unambiguously imply that the electromagnetic field is quantized, that its energy comes in 'lumps,' in short that photons exist. Although most textbooks still do it that way and many if not most of us still teach it that way, I have recently become belatedly aware that the situation is really much more murky.[6]

The special concern of this professional related only to the possibility of interpreting the photoelectric effect semiclassically, but the insight gained here applies much more generally. Obviously, the adjective "murky" is rather polemical and negative. Perhaps it would be better to formulate it more neutrally and to state that the situation is simply different from what our common sense and everyday habit purport. Photons are neither particles nor waves even though both of these concepts may help us to understand their behavior in certain experiments. These two intuitive concepts of wave and particle "are like a shark and a tiger, each supreme in its own element and

---

[4]Zeilinger et al. (2005) p. 233. Dieter Zeh's (1993, 2012) propositions are in the same vein.

[5]See Walker and Slack (1970) on the conscious linguistic analogy to these neologisms.

[6]Stanley (1996) p. 839; analogously also in Klassen (2011) p. 5.

helpless in that of the other".[7] So we have to be careful when to apply each of the two. The photon is merely a man-made concept created in an attempt to orientate ourselves in the physics laboratory and in the world we live in. Our intuition and our concepts must constantly be adapted to our environment and our knowledge about it in order for this orientation to be consistent and pragmatically appropriate. In this general debate about quantum-mechanical phenomena and about our specific understanding of photons, the following pointers should be borne in mind.

9.2 Avoid Naive Realism

Naive or exaggerated realism views photons as particles that really exist. It extends and embellishes our oldest semantic layer of quasi-Newtonian corpuscularity (Sect. 3.1 on pp. 40ff.) by distinguishing more or less sharply defined particles from their surroundings, which interact with matter exactly like billiard balls on a billiard table. One article for the *Rochester Meeting on Coherence and Quantum Optics* in 1972, that was actually intended as a defense of semiclassical interpretations of radiant phenomena, employed exaggerated polemics against the Copenhagen interpretation. Its author, Edwin Thompson Jaynes (1922–1998), thereby granted a glimpse into his own naively realistic mental model (which he shared with many other fellow opponents): "the Copenhagen theory slips here into mysticism. [...] it ends up having to deny the existence [...] of any 'objective reality' on the microscopic level. [...] I think that most physicists, even though they may profess faithful belief in the Copenhagen interpretation, still share with me a disreputable, materialistic prejudice that stones and trees cannot be either more—or less—real than the atoms of which they are composed."[8] Historically, this mental model was immensely influential and may well even still be haunting many a mind now, but it is nonetheless grossly misleading. Photons are not what is real; the associated events observed by us are. We model them as if photons existed because this mental model of quasi-particles works in some, but certainly not all experimental situations. Thus we have almost come full circle, back to Einstein's cautious formulation in 1905 of the light quantum as a "heuristical point of view." The Canadian physicist Harvey L. Armstrong (*1919) found a better solution: "it is plain what photons really are and are not. They are not particles like baseballs or shot; and photon theory is not a return to Newton's corpuscular theory of light. The photons are more like coefficients in a Fourier's series—or

---

[7] This quote from Heyl (1929), who in turn quoted Sir Oliver Lodge, is used by Stuewer (1975a) p. 331 and in the book title of Wheaton (1983) to capture the deep paradox of wave-particle duality.

[8] Jaynes (1973), pp. 48–50. On the career of this teacher at *Washington University* and adherent of semiclassical theories, see Clark et al. (2000). For the context of QED critics in USA, cf. Bromberg (2006) pp. 243–245, Bromberg (2016). The epistemic differences between classical particles and field quanta are described by Falkenburg in Esfeld (2012) pp. 158–184.

increments to the coefficients."[9] This was obviously going too far and objections to it were prompt. The Indian high-energy physicist Sardar Singh,[10] interjected that this attitude hit its limits where experiments (such as the ones discussed in Sects. 8.1–8.7) had found indications of individual photons, most particularly single-photon inter-ference experiments. The photons established there were more than mere formal mathematical coefficients of a series expansion. In a thought experiment he played out how physics could have developed, along with the emergence and interpretation of the photon concept:

> It might be informative to note that imposition of local gauge symmetry would lead to quan-tum electrodynamics even if Maxwell's equations and all information about electromagnetic fields were not known. Gauge invariance implies that a charged particle like the electron cannot exist by itself. There must be a massless vector particle with which it interacts. This massless spin 1 particle may, then, be identified with the *photon*. Maxwell's equations, too, follow from the local gauge invariance in quantum mechanics. I would say, if historically things had followed this path, the photons would not have been that strange.[11]

Seen from this perspective of symmetries, the photon is the massless spin-1 partner for electric charges with mass required for local gauge invariance, for reasons of symmetry. The wave characteristics related to the Maxwell equations and the wave theory of light derivable out of them would rank behind this, systematically as well as historically, just as in Singh's counterfactual thought experiment. A different historical chain of events would have decisively defused wave-particle duality. But even in this interpretational frame by an elementary particle physicist, photons in their role as exchange particles are not regarded as real particles but as virtual particles postulated to satisfy general principles of symmetry.

To summarize this subsection: 'Photons' are man-made theoretical constructs that do not describe real particles. Unlike classical, massive particles, photons have no size, shape or mass. The numbers of existing photons are also not subject to any conservation law.—Our intuition, schooled in such objects of the material world as peas or billiard balls, literally finds it too hard a nut to crack. I concede, no conservation law applies to massive particles in quantum field theory either (consider pair generation and annihilation). Nevertheless, there still are conservation values for mass-bearing electrons and positrons taken together: lepton numbers and electric charge, the analogous color charges of quarks, etc. Photons play a special role in this regard in that there is no conservation law for photon numbers.

---

[9] Armstrong (1983) p. 104. According to http://www.atomicheritage.org/profile/h-l-armstrong, H.L. Armstrong worked as a "Manhattan Project Veteran" for the *Tennessee Eastman Corporation* on the Y-12 plant in Oak Ridge, Tennessee.

[10] Professor emeritus of physics, at the *University of Rajasthan*, Jaipur, India; cf. https://scholar.google.co.in/citations?user=P4p2LbAAAAAJ&hl=en.

[11] Singh (1984) p. 11; for similar counterfactual variants, see Hund (1984), Pessoa (2000).

9.3 Avoid Illegitimate Attributions of Locality

Apart from discussing overly realistic assumptions, we must also treat the problem of the photon's supposed smallness or spatiotemporal locality. Of course, experiments were performed, particularly with high-energy gamma or x-rays, in which the energy and momentum of the radiation was transmitted over long distances without causing any noticeable broadening. This blatantly contradicts the classical model of a spreading spherically symmetrical wave and only encourages such metaphors as 'needle rays' or 'light bullets.' That was why experimenters like J.J. Thomson were motivated early on to consider corpuscular models, which Johannes Stark or A.H. Compton, for instance, correlated with Einstein's light-quantum hypothesis after 1905.[12] The scattering experiments by Compton and Simon 1922–1927, as well as the later ones by Bothe and Geiger (1924–1925) and by Raman in 1930, also seemed to suggest that after interacting with electrons electromagnetic radiation is emitted in very definite directions and not in the shape of spherical wave fronts.[13] However, in 1927 Arthur Jeffrey Dempster (1886–1950) and Harold F. Batho, two coworkers of Compton in the *Ryerson Physical Laboratory* at the *University of Chicago*, published diffraction experiments performed with an echelon grating at the helium line 4471 Å at very low light intensities corresponding to an average number of only 95 light quanta per second at this wavelength. A clearly recognizable interference pattern still appeared on a surface measuring 32 square millimeters set a distance of 34 cm away behind the echelon grating. The two experimenters concluded that even individual light quanta must be able to interfere with themselves over larger surfaces in space "when the quanta emitted from the volume of the source used were completely separated in time, showing that a single quantum obeys the classical laws of partial transmission and reflection at a half-silvered mirror and of subsequent combination with the phase difference required by the wave theory of light."[14] Subsequent papers criticized this pioneering experiment. Their argument was that the photon density achieved by them was still too high for such a far-reaching conclusion. But experiments with improved photon detectors and working increasingly clearly within the regime of lowest photon numbers and light intensities confirmed the hypothesis that in these cases photons occupy spatially very extended regions of space-time and definitely should not be imagined as almost point-shaped.[15] Thus the photon manifests the paradoxical consequences of wave-particle duality (Sect. 3.8) in its most extreme form: Depending on the experimental situation, it appears to us either as a wave or a particle, but prior to this measurement or the foregoing preparations associated with the quantum-mechanical state, a commitment to one or the other is impossible and

---

[12]These revealing metaphors and their context are discussed above in Sects. 2.5–2.6.

[13]On the Compton effect and its significance for the light-quantum hypothesis, see Sect. 5.2; on Bothe and Geiger (1924), see p. 128ff. above. On the Raman effect, a molecular scattering of electromagnetic radiation in UV, IF and visible light, which Chandrasekhar V. Raman (1888–1970) himself interpreted as the "optical analogue of the Compton effect," see Raman (1930) p. 270.

[14]Dempster and Batho (1927) p. 644.

[15]See the developments discussed in Sect. 8.1 above.

futile. Harry Paul resorted to anthropomorphic metaphor to describe this Janus-faced nature:

> The isolated photon thus proves to be an 'opportunist' of the first water, who can adapt himself effortlessly to different situations; there's no sign of 'independent values,' in other words, of real polarization properties [or already preset spatial probabilities]![16]

According to quantum electrodynamics, photons are quantized states of the electromagnetic field whose energy generally belongs to the whole region of space occupied by the radiation field. Localized states do not exist prior to a measurement, which depending on the interpretation leads either to a collapse of the wave function or to decoherence. The author of one of the first textbooks on QED, Walter Heitler (1904–1981), cautioned his readers:

> The existence of a discrete set of light quanta is only a result of the quantization [...] The particle properties of the light are comprised by the above-mentioned energy and momentum relations. But there is no indication that, for instance, the idea of the 'position of a light quantum' (or the probability of the position) has any simple physical meaning.[17]

Lev Landau (1908–1968) and his colleague Rudolf Peierls (1907–1995) already knew that extending Heisenberg's uncertainty principle to relativistic quantum mechanics would, at best, make it reasonable "to indicate the spatial probabilities of light quanta for regions that are large against the wavelength."[18] This brings us back within the classical range of geometrical optics.

Quantum field theorists regard virtual photons as "fluctuations of a random (zero-point) radiation filling the whole space"[19]; others emphasize the mathematical aspect and regard them as "propagating topological singularities"[20] or as "complex massless vector fields satisfying Maxwell's equations in vacuum".[21] Experimenters in quantum optics might rather see photons as something like standing waves in the interior of a lasing resonance cavity[22] —"in any case, the energy of a photon is distributed over the entire volume of the field and there is, in general, no use in attaching a coordinate to the photon."[23] Put more formally in the words of Newton and Wigner

---

[16]Paul (1985) p. 175, furthermore ibid., p. 179 on the photon as a "hybrid" (*Zwitter*), or on p. 11 about the "complicated structure, [...] a Janus-faced something that—depending on the kind of experimental conditions—'shows' itself one time as a corpuscle and another time as a wave." Han (2014) also stresses this.

[17]Heitler (1936b) pp. 63–64, also quoted by Armstrong (1983) pp. 103–104: "a photon is not a thing to which a position can be simply assigned." Earliest attempts in this direction include Heisenberg and Pauli (1929) and Landau and Peierls (1930). A thorough survey of the literature is offered by Keller (2005), a clear derivation is in Duncan (2012) pp. 159–164.

[18]Landau and Peierls (1931) p. 64; furthermore Bohr and Rosenfeld (1933), Newton and Wigner (1949), Mandel (1986) pp. 39f. and Keller (2005).

[19]Emilio Santos pleads in favor of this interpretation in Roychoudhuri et al. (2008), pp. 163–174.

[20]This is the view of the mathematician R.M. Kiehn in Roychoudhuri et al. (2008), pp. 251–270.

[21]Babaei and Mustafazadeh (2017) p. 1.

[22]See Muthukrishnan et al. (2003) S-24: "discrete excitation of the electromagnetic field in some cavity".

[23]Strnad (1986a) p. 650. The same point is raised by Gerry and Knight (2005) p. 18, Han (2014) pp. 47ff., Passon and Grebe-Ellis (2016) pp. 20ff., among many other quantum field theoreticians.

(1949): in general, "for higher but finite spin, beginning with $s = 1$ (i.e., Maxwell's equations), we found that no localized states [..., i.e., no position operator] exist."[24] In order to circumvent faulty assumptions about localizability, the Slovenian physics instructor Janez Strnad (1934–2015) recommends that physics teachers introduce photons exclusively as quanta of energy and momentum, and strictly avoid drawing any further analogy to electrons or other elementary particles with a rest mass. Otherwise the following would happen: "you think you make it simpler when you make it slightly wrong."[25]

9.4 Renounce Individualization

Not only does our intuition mislead us into thinking that particle-like quantum objects such as photons must always also be localizable down to the tiniest regions of space; there is another heritage of our evolutionary origins that we must abandon in the quantum world: the premise of their individualization. Although peas in a pot all look so similar, in principle it is possible to number them or attach different labels to each individual one of them. In the case of elementary particles or photons, this is no longer valid. They do not fall under classical statistics, for which the possibility of individualization (in Latin, *haecceitas*) is presumed.[26] They are subject to very different quantum statistics. Our indistinguishable light quanta (photons) are spin-1 particles, therefore bosons, which obey Bose-Einstein statistics (cf. Sect. 3.11). Electrons, muons, etc., are spin-1/2 particles, therefore fermions, which obey Fermi-Dirac statistics. Both kinds of statistics lead to effects that differ strongly from the behavior of classical particles: The Pauli exclusion principle applies to fermions, making two fermions with the same quantum numbers and spins 'get out of each other's way,' whereas bosons tend to cluster together, which leads among other things to the immense concentration of coherent light in a laser and to the famous Bose–Einstein condensates.[27]

We have seen (in Sects. 8.3–8.7) what enormous experimental investments are required to prepare genuine single-photon states.[28] Many experiments on photon interference, by Hanbury Brown and Twiss, for instance (Sect. 8.1), or on a pair of interfering laser beams, paradoxically encourage us to cling to the false classical picture of a particle and forget its nonlocality and the statistical fluctuations to which quantities like photon number are subject. The quantum optics specialist Harry

---

[24] Newton and Wigner (1949) p. 405; cf. Duncan (2012) pp. 159–164 on "local fields, non-localizable particles!" For a recent effort to construe localized states of the photon cf. Babaei and Mustafazadeh (2017).

[25] Strnad (1986a) p. 650. On pedagogically motivated oversimplifications, see Chap. 6 above.

[26] See, e.g., Redhead and Teller (1992), French (2015) as well as Lyre in Friebe et al. (2015) Chap. 3.

[27] See, e.g., Ketterle (1997, 2007) as well as further texts referenced here in Sects. 3.10, 8.1 and 8.6, and Fig. 8.2 on bunching and antibunching.

[28] Further literature on genuine single-photon experiments since 1996 are cited by Santori et al. (2002), Zeilinger (2005).

Paul (∗1931) at the *Zentralinstitut für Optik und Spektroskopie* of the East German *Akademie der Wissenschaften* in Berlin described what really happens when two beams of light interfere with each other:

> according to quantum mechanics, the photon number in a Glauber state is intrinsically indefinite, hence one is not justified in considering the number of photons in each beam (during one trial) to be a definite quantity, in the sense of classical reality. Formally, it is just this uncertainty in the photon number that brings into play the wave picture. The proper description of interference between independent photons will be as follows. What interferes with one another are waves, and when one wave is registered [...], one cannot say, on principle, from which laser it has come. What actually happens in that detection process is that an energy packet $h\nu$ is taken from the superposition field to which both lasers contribute equally, and hence it is only natural that this photon bears information on both laser fields that becomes manifest in the ultimate interference pattern.[29]

This inability to individualize, this impossibility to associate an individual photon with one of the two beams, is essential to the formation of the interference pattern. The which-way interference experiments have demonstrated this, with the interference pattern vanishing or not even appearing as soon as we have information about the path that an individual photon has taken.[30] Because—under normal conditions, at least— it is extremely difficult to examine photons separately, Edward M. Purcell and Emilio Panarella have proposed a "photon clump model" that places greater weight on the tendency of photons to cluster according to Bose–Einstein statistics. The question posed by Chandrasekhar Roychoudhuri and Negussie Tirfessa in the same collection of papers: "Do we count indivisible photons or discrete quantum events experienced by detectors?"[31] must therefore be answered in favor of the latter alternative in at least 99% of all the applied cases. This is also the reason why semiclassical theories work for so many experiments in which observed correlations are explained by fluctuations of electromagnetic fields that are classically expected.

But don't photographs already exist by now taken with ultrashort pulses of laser light, even down to attosecond bursts? What exactly do we see in Fig. 9.1?

These three images immediately demonstrate the immense technical progress made in the last few decades in the field of quantum optics. Around 1970 (left) we were still working within the range of picoseconds (10^{-12} s). In the late twentieth century we reached the femtosecond range (10^{-15} s), and in this twenty-first century we are already within the attosecond range (that is, 10^{-18} s). How are such brief laser pulses possible? As Felix Frank explains: "These pulses [were generated] by a process called high harmonic generation (HHG), using a high-power femtosecond laser system [...]. The near infrared femtosecond laser pulses are corralled through a waveguide and a series of specialized mirrors, causing them to be compressed in time. With their waveforms precisely controlled, these compressed pulses are

---

[29] Paul (1986) p. 221.

[30] See Sect. 8.5 above and additionally Paul (1985) pp. 98–123.

[31] See Purcell (1956) (cf. Sect. 8.1) as well as Panarella, resp., Roychoudhuri et al. (2008) pp. 111– 128 resp. 397–410.

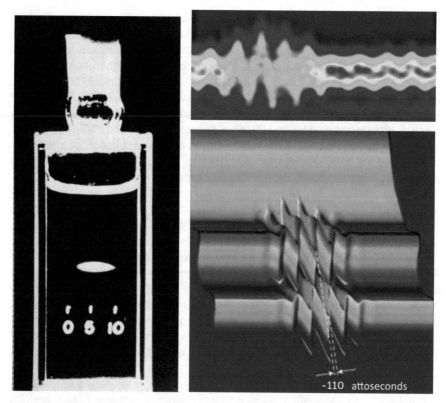

Fig. 9.1 Left: Ultrashort green laser pulse from a neodymium-doped gas laser within a Kerr cell that is triggered by the infrared pulse of the same laser. Exposure time: 10 ps. During this time the light travels the distance 2.2 mm, here from right to left. Photographed in *Bell Laboratories* around 1971. Source: Scully and Sargent (1972) p. 39. Reprinted by permission of Nokia Corporation. Top right: An attosecond burst produced in July 2012 by Felix Frank at *Imperial College* in London and elsewhere. Source: Frank (2012). Reprinted by permission of AIP Publishing LLC. Bottom right: Another spectrogram taken with an atomic transient recorder. Source: Kienberger and Krausz (2009) p. 39. Reproduced by the kind permission of Prof. Dr. Ferenc Krausz

then focussed into a gas target, creating an attosecond burst of extreme ultraviolet radiation."[32]

The two images on the right-hand side of Fig. 9.1 are not the ultrashort pulses of light themselves but the secondary radiation of photoelectrons set into very strong oscillation by these ultrashort pulses. These refined techniques of "attosecond science" convert the temporal sequence into a spatial sequence.[33]

How do we interpret such "fuzzy balls" as in the figure (left image of Fig. 9.1)? They are neither discrete photons nor states with a sharply defined photon number. As

---

[32]Felix Frank: The shortest artificial light burst in history, posted on 2 July 2012, on line at http://www.kurzweilai.net/the-shortest-artificial-light-bursts-in-history (accessed 19 Mar. 2016).

[33]For details, see Goulielmakis (2004), Frank et al. (2012), esp. pp. 19–20.

shown in Sect. 4.8, such Gaussian wave packets are produced by the superposition of a number of monochromatic coherent states that interfere with each other for a short time and lead to noteworthy constructive interference only within a relatively small region of space, whereas destructive interference causes the edge areas to disappear. 'Fuzzy' effects in the transition zone bring about the diffuse boundary between such axially or spherically symmetric structures. This is clearly visible in the other image of the figure (top right image in Fig. 9.1). The center of the wave packet is surrounded by a marginal zone of fainter but still clearly visible oscillations. The entire interference phase must be short enough, a few hundred attoseconds (one billionth of a billionth-second!) or at most a few femtoseconds (10^{-15} s) perhaps, but not shorter than that either. This is because due to the Heisenberg relation between the energy uncertainty ΔE and the time interval τ during which it holds, $\Delta E \cdot \tau \geq \hbar$, the energy uncertainty increases as the time period τ decreases, which destroys the coherence of the photon's state. In other words, the shorter the time is in which we measure a photon, the less precisely defined is its energy and frequency. And conversely: The more precisely we localize a photon in space (Δx), the less precise is its momentum $\Delta p = \Delta E/c = h/c\Delta\nu$, because $\Delta x \cdot \Delta p \geq \hbar$. An energetically exactly established quantized photon is completely undefined spatially; and conversely, complete fixation of a photon in space is at the expense of any information about its momentum and wavelength. Heisenberg's uncertainty relation is hence an expression of strange wave-particle duality. As Wolfgang Pauli put it: "you can look at the world with the p-eye or you can look at it with the q-eye, but if you want to open both eyes at the same time, you go crazy."[34]

According to the wave model of light, we regard light as energetically sharply defined waves. But then, of course, they have large spatial extension (which explains the considerable reach of interference phenomena over distances of 1 m or more). According to the particle model of light, we regard light with the p-eye, for example, in order to see collision processes or electromagnetic needle radiation with its virtually point-shaped zones of interaction. But then we lose the other half of the picture. In the Compton effect, for example, by observing a scattered electron we can calculate precisely the relativistic momentum balance between the incoming and outgoing radiation (cf. Sect. 5.2), but we cannot know exactly where this interaction occurred (nor do we need to know, for that matter!). Both mental models are legitimate and reasonable within each of their spheres of application. But if we try to transfer them into spheres of application that they are neither built for nor suited for, they become unreasonable. (This often happens unintentionally or from sloppy thinking, but occasionally also from a conscious reluctance or inability to shake off habitual interpretational approaches.) And this happened not only to Planck, Einstein, Bohr, Schrödinger and other great minds in the history of physics, but still continues to happen everywhere now. Concepts that have accreted slowly throughout history, such as the light quantum or the photon, carry their own semantic lives along

---

[34]Wolfgang Pauli in a letter to Werner Heisenberg, 9 Oct. 1926, in Meyenn, ed. (1979/85) vol. 1, doc. 143, p. 340; cf. Heisenberg (1927), Landau and Lifschitz (1947/73) 6th ed., pp. 44–47, 152–158, Paul (1985) pp. 33–37.

with themselves because they unite within themselves so many different semantic layers. Some of them, like the oldest layer of quasi-Newtonian corpuscularity, are compressed and folded virtually beyond recognition.[35]

9.5 The Photon: A "Mysterious Quantum Cheshire Cat" or "Elusive Beast"?

Recall Einstein's exclamation in 1951 that opened this book: "All those 50 years of careful pondering have not brought me closer to the answer to the question: 'What are light quanta?' Today any old scamp believes he knows, but he's deluding himself." What a skeptical resume this was—albeit, with some justification. One year after his death, the Hanbury Brown and Twiss experiment became known, ushering in the new series of optical precision experiments on photons and sounding the bell into a new era of quantum optics (see my brief summaries of some of the most crucial experiments in Chap. 8). Have we made significant progress during those decades since Einstein's death? Almost two decades later, in 1972, two quantum-optics experts, Marlan Ovil Scully (∗1939) and Murray Sargent (∗1941), exacted: "we need a logically consistent definition of the word 'photon'—a statement far more necessary than one might think, for so many contradictory uses exist of this elusive beast."[36] At around the same time, John Archibald Wheeler arrived at the same hardly encouraging assessment of the photon, calling it an ephemeral "smoky dragon" that escapes every attempt to pin it down.[37]

In 1984 the Canadian physical chemist Gordon R. Freeman (∗1930) was still forced to establish: "the nature of the photon is an unsolved problem." Almost twenty years later, the Nobel laureate Willis Lamb (1913–2008) pronounced in an article headed by the programmatic title 'Anti-photon': "there is no such thing as a photon. Only a comedy of errors and historical accidents led to its popularity among physicists and optical scientists. I admit that the word is short and convenient. Its use is also habit forming."[38]

Doesn't the history of such a term as the 'photon,' which appeared almost a century ago (in 1926), rather lead the modern user 'astray'? The physicist at the University of Wuppertal, Oliver Passon shares this opinion in common with the theoretical physicist Willis Lamb. In a message addressed to me, Passon compared the concept of an Einsteinian light quantum and the 'photon' with the 'phlogiston'

---

[35] See Sect. 1.3, especially on the concept of convolutions.

[36] Scully and Sargent (1972) p. 38. In 1997, Scully arrived at other conclusions in a textbook on quantum optics (see below). On Scully's vita and research context, cf. Bromberg (2006) pp. 245ff.

[37] Wheeler during the 72nd 'Enrico Fermi' summer school, published 1979, quoted here from Roychoudhuri and Roy (2003) S28. Another formulation in the same vein is also the title of the contribution by K. Michielsen, Th. Lippert and H. de Raedt in Roychoudhuri et al. (2015): "Mysterious quantum Cheshire cat: an illusion."

[38] See Freeman (1984) p. 11 resp. Lamb (1995) p. 77.

concept predating Lavoisier. Shouldn't the term 'photon' be supplanted by a completely new concept with a different name, the way the phlogiston concept vanished after Lavoisier's conception of combustion as oxidation led to Lavoisier's 'oxygène'?

> Aren't all these lines of development, which are reconstructible from 1905 down to today, grossly misleading? If one says: after 1923 (Compton), all physicists (almost) were convinced about Einstein's light-quantum hypothesis, isn't one insinuating (in the textbook literature) something to the effect of: "and legitimately so; and you, dear reader, should also be convinced!" But we got the QED photon in the end, which has almost no similarity with any of the various 'Einstein light quanta' between 1905 and 1916 (or thereabouts). I quite provocatively ask you as a historian: A line of development quite certainly exists from phlogiston to heat, energy and entropy in the modern meaning. But it makes so terribly much sense that we are not using the term phlogiston anymore because it has connotations that are false (according to the current understanding). Isn't Einstein's early light quantum such a 'phlogiston,' perhaps?[39]

In one respect Passon does have a point: In the later development after 1930, individual layers of meaning, in particular, localizability or individualizability, which are implied in naive classical Newtonian corpuscularity, are weakened or even wholly retracted again. That was precisely why some high-ranking physicists like Lamb occasionally pleaded that the photon term be dispensed with. Such proposals completely miss scientific practice, however, because science operates almost throughout with terms and concepts that come from earlier times and bring along notions of mental models that are partially obsolete. Consider, for instance, the terms 'force' or 'energy' with their anthropomorphic roots. Another example would be electromagnetic 'displacement density' with its roots in the aether notions from the nineteenth century. Radical replacements of scientific terminology are extremely seldom—Lavoisier's 'chemical revolution' of the late eighteenth century is one of the rare examples of this.[40] On the contrary, creeping conceptual changes and graded semantic shifts in the current—or just recently outdated—connotations in the definitions of scientific terms are entirely normal in the (natural) sciences. Scientists (and occasionally even historians and philosophers of science!) are not always sufficiently aware of this process, though. It would be wrong to assume that the expression 'photon' nowadays expresses exactly what it had meant to G.N. Lewis when he introduced it in 1926. But it would be just as wrong to contend that the photon of QED was an entirely different concept from Albert Einstein's light quantum of 1905. Both are interconnected in a long chain of small shifts of meaning and continual superposed semantic accretions. That is why I consider it important that a book like this be written. Terminological and conceptual developments follow such contorted paths that even neologisms (e.g., 'quarks,' introduced by Murray Gell-Mann (∗1929) in 1961 as a concept to cover subnuclear building blocks of elementary particles), are not left untouched.[41] Referential continuity of terms such as 'atom', 'electron' or

---

[39] Oliver Passon in an email to Klaus Hentschel, 2 Sep. 2016; cf. in a similar direction Simonsohn (1979), Lamb (1995) p. 80, and Sulcs (2003) p. 367.

[40] See, e.g., McEvoy (2010), Chang (2011) and further literature cited there.

[41] See, e.g., Gell-Mann and Ne'eman (1964), Walker and Slack (1970) and Johnson (1999) on semantic modifications of this concept with attributes like 'color,' 'flavor,' masses, etc.

'photon' does not imply epistemic and semantic stability, nor does a break in terminology (such as here from 'light-quantum' to 'photon') imply an epistemic shift or rupture. In our case, continuity in this period after 1926 was guaranteed by the many semantic layers that were taken over, with only a few of them being de-accentuated or withdrawn (in this instance, as localizability or individuality).

Wrapped inside my model of conceptual development, this means that the accretion of meaning or the enriching superposing of semantic layers does not proceed strictly cumulatively. Some layers may erode away again later or become much thinner—exactly as in the case of many other scientific concepts. The great majority of the twelve layers of meaning (Chap. 3) for the 'light quantum' or the 'photon' still continue to be valid, however, and were gradually made more precise. This, incidentally, is quite different from 'phlogiston,' which shortly before its demise even had to take on a negative mass in order to remain consistent with weighing experiments of chemical mass. This nonlinearity and noncumulativeness of concepts and their intertwining with mental models make the history of science even more exciting, but also terribly complex.

One of the editors of a collection of papers that appeared in 2008 on *The Nature of Light – What Is a Photon?* gave his own contribution a title that ventures within the world of Shakespearean tragedy: "Oh Photon, Photon, whither art Thou gone?"[42] This anthology was the first volume of what became a regular series of biannual conferences organized by the *International Society for Optics and Photonics* (SPIE), in which the eternal question: "What are photons?" reappears time and again[43]: These specialists in quantum optics thus ultimately refer us back to the level of cognition, of mental modeling, which should be performed without associated reification in order to avoid the errors of the past. Countless scientists and technologists still step into this so enticing trap nonetheless, in an effort to make things more intuitive. Wave aspects are covered at the same time as particle aspects, only to end up with far-fetched conceptual models, such as the "neutrino theory of light" or a "binary photon," in which the light quantum is interpreted as the link between two components, either "wavicles" (Arthur S. Eddington in 1928 and Charles Galton Darwin), or two neutrinos (thus Pascual Jordan 1935–1937, Jordan and Kronig (1936), Ralph de Kronig 1935–1936), matter and antimatter (thus American cell biologist Bruce Wayne 2009), or even two "semiphotons" with "bigness and sidedness as Newton 1730 would say."[44] Most of these models violate relativity theory or some other indispensable law of conservation, as sacrifices on the altar of supposed visualizability, which is why I shall leave them out here.

---

[42] Al F. Kracklauer in: Roychoudhuri et al. (2008) pp. 143–154.

[43] While I was writing these lines, six of these proceedings volumes were available to me, each with many dozens of contributions, although many of them range from the highly speculative to downright obscure toying around with ideas. See Roychoudhuri (2015) eds. or http://spie.org/Publications/Proceedings/Volume/9570 for the latest volume in this series.

[44] All foregoing quotes from Wayne (2009) p. 23, who offers a detailed survey.

Today the light quantum or photon is accepted as a convenient conceptual model, as a mental crutch or, to speak with Vaihinger, as a fiction.[45] Countless physicists, optics specialists and technicians rely on this very helpful conceptual crutch in their daily routine, true to Roy J. Glauber's motto: "I don't know anything about photons, but I know one when I see one."[46] Taking this shirt-sleeve attitude, we can very simply define the photon operationally: "A photon is what a photodetector detects."[47] But it remains controversial to this day whether these photons really exist, and if so, in what sense, in other words, by which underlying interpretation. Some (such as Marlan O. Scully or Anton Zeilinger) consider the photon a fascinating, intrinsically quantum-mechanical object that is indispensable for the purposes of quantum optics or quantum field theory.[48] Others consider it an anachronism, spare ballast or even a nuisance. Not long ago, the Russian physicist and member of the *Moscow Academy of Sciences*, Sergey Rashkovskiy, answered his own question, "Are there photons in fact?" with reference to the indisputably good performance of semiclassical theories without any photons, as follows: "we show that many quantum phenomena which are traditionally described by quantum electrodynamics can be described if light is considered within the limits of classical electrodynamics without quantization of radiation. These phenomena include the double-slit experiment, the photoelectric effect, the Compton effect, the Hanbury Brown and Twiss effect, the so-called multiphoton ionisation of atoms, etc. [...] We conclude that the concept of a 'photon' is superfluous in explanation of light-matter interactions."[49] But how seriously should we take this statement? Such heretical conclusions are the result of burgeoning usages of the word 'photon.' The ability to explain the outcomes of many semi-classical experiments without assuming photons does not prove that the concept is superfluous since there are, after all, many other experiments which do require such a concept. Just because slow particles seem to move according to Newton's prescriptions doesn't mean that relativity theory is superfluous, either: for particles with speeds close the velocity of light, it is indispensable. It is true: "a fully-quantized theory of matter-radiation interaction can lend a characteristic of spatial discreteness to the photon when it interacts with a finite-sized atom", but there are many other types of interaction, and in general "there is no particle-creation operation that creates a photon at an exact point in space".[50]

Let me close with a very apt paraphrasing of Stephen Shapin's witticism (originally conceived for the hotly debated 'scientific revolution' concept, playfully modified by me): "there are no photons, and this is the book about them."

---

[45] On the interpretation of relativity and quantum theory in a fictionalist perspective, cf. Zeh (2012) and Hentschel (2014a).

[46] Bonmot by Glauber at the summer school in Les Houches 1963, quoted as the motto to the paper by Holger Mack and Wolfgang Schleich in Roychoudhuri and Roy (2003) S28.

[47] This time quoting Glauber from the contribution by Scully et al. in Roychoudhuri and Roy, eds. (2003) S18.

[48] Ibid. and in Scully and Zubairy (1997) Chaps. 1 and 21, Grangier (2005) or Zeilinger (2005).

[49] Quoted from S.A. Rashkoskiy's abstract in Roychoudhuri et al. (2015).

[50] Quotes from Muthukrishnan, Scully and Zubairy (2003) S24–S25, thereby explicitly withdrawing earlier contrary claims in Scully and Sargent (1972).

Chapter 10
Summary

This book examines in detail the following six mental models forming the basis of the terms 'light quantum' or 'photon,' along with their associated conceptual thinking. Different from eloquent argumentation, most of these mental models were not explicitly enunciated. Nevertheless, they do merit specific attention if we want to gain a deeper understanding of the mental processes involved.

(1) The naive corpuscular model (Newton 1704, Newtonians, ...): Light is interpreted as a stream of very small, basically spherically shaped corpuscles. With Newton, and later around 1800, it is modified by additional hypotheses about their being dumbbell-shaped to account for polarization ("sidedness"). Later resumptions to this particulate model after the emergence of wave-particle duality interpret light quanta as particles with ultra-small volumes $V \sim (c/\nu)^3$, where the wavelength λ is related with the frequency ν and the velocity of light c as $\lambda = c/\nu$. This intuitive model offers a simple explanation for light reflection and other effects but has major problems interpreting interference, which suggests a wave-like character, hence having a spatially extended structure (see the debate between Einstein and Lorentz around 1909, discussed in Sect. 4.3).

(2) The singularity model (Einstein 1909a, Louis de Broglie 1923, Bohm and Hiley 1982, etc.): It is not clear whether Einstein ever did support the naive corpuscular model (1). It becomes evident latest since his discussions with Lorentz in 1909 that the mental model he was exploring was different (see Sect. 4.2 on his talk before the Scientists' Convention in 1909 with the subsequent debates recorded in the proceedings). At this time, light quanta rather seemed to Einstein denser centers in the field, physically strongly localized but owing to the surrounding field not point-shaped. (Newton's "globulus of light" somewhat resembles Einstein's later readoption of such considerations; cf. Fig. 4.1.) More recent versions of this mental model within the context of alternative interpretations of quantum mechanics identified this surrounding zone around the singularity with a hypothetical guiding field. Evidence of this guiding field has yet to be found.

© Springer International Publishing AG, part of Springer Nature 2018
K. Hentschel, *Photons*, https://doi.org/10.1007/978-3-319-95252-9_10

(3) The binary model of light quanta ("binary photons") [proposed in 1907 by William Henry Bragg, later also considered by Johannes Stark and still is being discussed from time to time by people with unconventional backgrounds and insufficient training in physics]: Light quanta are not elementary particles but extended structures made of two components whose oscillations and rotator degrees of freedom participate in the generation of wave characteristics, which could perhaps also explain polarization or spin. But then they must also generate dipolar and magnetic moments as well as other experimental effects which have thus far never been observed.

(4) The wave packet [Schrödinger 1926, Schrödinger's variant of quantum mechanics with Born's probability interpretation]: Light quanta, just as matter waves, are conceived here as a localized phenomenon confined within a relatively small region of space-time (see Sect. 4.8). The advantage of this interpretation is: it is closely connected with the Heisenberg uncertainty relation and is also amenable to wave-particle duality because this wave nature is adaptable to the given experimental conditions: the more like a wave the packet is, hence the broader it is in space-time, the more precise is its frequency/wavelength (resp. energy/momentum) and vice versa. Disadvantages: The coherence of the wave is destroyed as it spreads out over longer periods of time (dispersion of the wave function) or upon hitting against obstacles (diffraction), as well as from decoherence, that is, the collapse of the wave function allocated to the wave packet.

(5) The semiclassical model (Bohr et al. 1924; Jaynes 1965; Scully and Sargent 1972; Lamb 1995, etc.): A half-hearted attempt to combine concepts about classical electromagnetic waves with ones about quantized matter. Advantage: easier calculability of scattering processes and interactions. Disadvantage: asymmetry in the theoretical treatment of waves/fields, on one hand, and particles/matter, on the other. Non-applicability to genuine single-photon states.

(6) QED—Quantization of the electromagnetic field (Dirac 1927b, Heitler 1936, Feynman 1949, Tomonaga, Schwinger, Dyson, etc.): The photon as a massless, exchange 'particle' in electromagnetic interactions, which is modeled as if it bounced back and forth between the electrically charged partners in an interaction, either as a real interaction (e.g., as light) in time-like propagation along the light-cone, or virtually in a space-like connection outside the light-cone of Minkowski-diagrams. In this complex mental model, the photon conveys the electromagnetic interaction like a ping-pong ball that triggers small changes in momentum in the emitter and the absorber.[1]

This electromagnetic field quantum of QED (or as some would prefer to say: this excitation of modes of the electric field) substantially differs from semiclassical models in that it is generally nonlocal. In experimental situations of low field quantum numbers, it leads to highly nonclassical effects in quantum optics and to results that are statistically only describable by QED or quantum field theory: Examples include Bose-Einstein condensates, EPR correlations and quantum-optical effects of bunching and antibunching (Chap. 8).

---

[1]On this partial virtualization of the photon, and on the Feynman diagrams visualizing such interactions, see Sects. 3.12 and 9.2 and the texts cited there.

My picture of the emergence and development of concepts as layered semantic accretion seems to me the most suitable model for these very long-term, multilayered processes involving a gradual formation of signification and shifting over time. It is more stratified than Mark Turner's 'conceptual blending,' which only combines two levels of meaning at a time. In our history there were twelve clearly defined separate strata of meaning that came into play at different times, gradually gained significance and were partly pushed aside again (e.g., the corpuscularity and locality of light quanta, which receded again with the emergence of wave-particle duality, quantum-mechanical indeterminacy, and the virtualization of photons as exchange particles in QED). It has also been demonstrated that a combination of many such conceptual fusions or 'megablends' can also describe the historical events consistently and that both models of the formation of scientific concepts ought to be pursued and implemented. The unusually long process of development of a concept like the light quantum or photon is predestined for this kind of approach because here we find many developments stretched out in time as if in slow motion that in other cases are tightly packed together within a very short period. Both these models aim primarily at conceptual and cognitive changes, definitely not at the communicative processes within larger thought collectives (*Denkkollektive*), the focus of Ludwik Fleck's (1935/80) considerations about the gradual transition from diffuse preliminary ideas into scientific concepts. Or else Hans Blumenberg's (1957, 1960) thoughts about rampant metaphors that finally gain a life of their own and complement both of these modelings introduced here. Similar studies could be undertaken on many other concepts involving complex histories of ideas and subtle shifts in the attendant mental models (for instance, on the concepts of force or energy or on laws of nature). They may well profit from my methods proposed here. If this pioneering study has not only entertained and informed its readers but inspired them to take this up, the highest aim of this book will have been attained.

Some Interesting and Useful Web Sites

http://www.newtonproject.sussex.ac.uk/prism.php?id=1 Newton Project on line
(texts, mss., letters)

http://www.newtonproject.sussex.ac.uk/prism.php?id=47 Newton's optical papers

https://www.aip.org/history/exhibits/electron/jjinfo.htm# bibliography on the
history of the electron

http://www.mathpages.com/home/kmath677/kmath677.htm and

http://en.wikipedia.org/wiki/Poynting's_theorem

http://dbserv.ihep.su/hist/owa/hw.fulltextlist_txt downloads of many primary texts

http://www.aip.org/history/web-link.htm on-line exhibitions and links

http://alberteinstein.info/ Einstein Archives with on-line database

http://myweb.rz.uni-augsburg.de/~eckern/adp/history/Einstein-in-AdP.htm for
Einstein's articles in *Annalen der Physik*

http://press.princeton.edu/books/einstein11/c_biblio.pdf cumulative bibliography
of all primary and secondary sources cited in *Collected Papers of Albert Einstein*
vols. 1–10

http://www.pitt.edu/~jdnorton/Goodies/Einstein_stat_1905/ on Einstein's statistical
papers

https://www.youtube.com/watch?v=1RpLOKqTcSk Einstein Bose
CondensateColdest Place in the Universe

http://www.calcuttaweb.com/people/snbose.shtml on Bose

https://www.zwoje-scrolls.com/zwoje41/text10 and 11.htm on Natanson

http://nobelprize.org/physics/laureates/1927/compton-lecture.html Compton's
biography

http://eldorado.tu-dortmund.de/bitstream/2003/24257/1/006.pdf

http://www.ifpan.edu.pl/ON-1/Historia/natanson.htm

http://edition-open-access.de/studies/2/index.html quantum theory & quantum
mechanics textbooks

http://www.nobelprize.org/nobel_prizes/physics/laureates/1914 (von Laue), resp.
.../1915 (the Braggs), /1919 (Stark), /1921 (Einstein), /1922 (Bohr), /1923
(Millikan), /1927 (Compton), /2012 (Haroche), etc. for Nobel lectures

© Springer International Publishing AG, part of Springer Nature 2018
K. Hentschel, *Photons*, https://doi.org/10.1007/978-3-319-95252-9

http://www.europhysicsnews.org/articles/epn/abs/2000/03/epn00303/epn00303.
html on R.A. Millikan's struggle with theories

http://www.lorentz.leidenuniv.nl/history/spin/goudsmit.html

http://www.jeos.org/index.php/jeos_rp/article/view/10045s

https://www.nhn.ou.edu/~jeffery/astro/astlec/lec006.shtml

https://www.mpq.mpg.de/5020834/0508a_photon_statistics.pdf experiments on
photon statistics

https://www.sheffield.ac.uk/polopoly_rpfs/1.14183!/file/photon.pdf What is a
photon?

http://space.mit.edu/home/tegmark/PDF/quantum.pdf

http://mediathek.mpiwg-berlin.mpg.de/mediathekPublic/versionEins/Conferences-
Workshops/ HQ3/Quantum-Optics/Joan-Bromberg.html

https://www.researchgate.net/post/How_do_you_visualize_a_photon

https://physics.stackexchange.com/questions/273032/what-exactly-is-a-photon

http://www.nature.com/milestones/milephotons/index.html 23 Nature Milestones
Photons httpa://indico.cern.ch/event/423687 Proceedings of the Warsaw
Conference 2005 on "The Photon: Its First Hundred Years and the Future"

http://inspirehep.net/record/1623145/files/grangier.pdf on Grangier's experiments
with single photons 2005

References

This list of cited publications makes no distinction between primary and secondary sources. The entries are listed chronologically under author's name, followed by coauthored works. Coauthors are listed in the sequence given in the original publication. An author's full name is indicated only once, with first initials for coauthors after the first instance. Diacritical marks are generally ignored.

Achinstein, Peter. 2013. *Evidence and method: Scientific strategies of Isaac Newton and James Clerk Maxwell*. Oxford: Oxford University Press.

Ambroselli, Michael, and Chandrasekhar Roychoudhuri. 2015. Did Planck, Einstein, Bose count indivisible photons, or discrete emission/absorption processes in a black-body cavity? In *Proceedings SPIE* vol. 9570–9.

Andrade, Edward Neville da Costa. 1930/36. *The mechanism of nature*. London: G. Bell & Sons, 1st ed. 1930, 2nd ed. 1936.

Andrade, Edward Neville. da Costa. 1957. *An approach to modern physics*. London: G. Bell & Sons.

Anglin, James. 2010. Particles of light. *Nature* 468: 517–618 (cf. Klaers et al. 2010).

Arabatzis, Theodore. 2006. *Representing electrons*. A biographical approach to theoretical entities. Chicago: University of Chicago Press.

Arago, François. 1853. Mémoire sur la vitesse de la lumière. *Comptes Rendus hebdomadaires des Séances de l'Académie des Sciences, Paris* 36: 38–49.

Armstrong, Harvey L. 1983. No place for a photon?. *American Journal of Physics* 51 (2): 103–104 (comment on Berger 1981; cf. Singh 1984 and Freeman 1984).

Aspect, Alain, Philippe Grangier, and Gérard Roger. 1982. Experimental realization of Einstein-Poldolsky-Rosen-Bohm *Gedankenexperiment*. A new violation of Bell's inequalities, (a) *The Physical Review Letters, New Series* 49 (2): 91–94; (b) reprinted in Meystre & Walls, ed. 1991, 382–385.

Atlas Cooperation. 2017. Evidence for light-by-light scattering in heavy-ion collisions with the ATLAS detector at the LHC, CERN-EP-2016-316 (Feb 7, 2017), arXiv:1702.01625v1 [hep-ex], submitted to *Nature Physics*.

Babaei, Hassan, and Ali Mustafazadeh. 2017. Quantum mechanics of a photon, arxiv:1608.06479v3.

Bacciagaluppi, Guido, and Antony Valentini. 2009. *Quantum theory at the crossroads–Reconsidering the 1927 Solvay conference*. Cambridge: Cambridge University Press.

Badash, Lawrence. 1972. The completeness of 19th century science. *Isis* 63: 48–58.

© Springer International Publishing AG, part of Springer Nature 2018
K. Hentschel, *Photons*, https://doi.org/10.1007/978-3-319-95252-9

Badino, Massimiliano. 2009. The odd couple: Boltzmann, Planck and the application of statistics to physics 1900–1913. *Annals of Physics* 18: 81–101.

Badino, Massimiliano. 2015. *The Bumpy Road - Max Planck from radiation theory to the quantum 1896–1906*. Berlin: Springer.

Badino, M., and Jaume Navarro (eds.). 2013. *Research and pedagogy: A history of quantum physics through its textbooks*. Berlin: Edition Open Access. http://edition-open-access.de/studies/2/index.html.

Baldzuhn, J., E. Mohler, and W. Martienssen. 1989. A wave-particle delayed choice experiment with a single-photon state. *Zeitschrift für Physik* B77 (2): 347–352.

Band, W. 1927. Prof. Lewis' 'light corpuscles'. *Nature* 120: 405–406, = comm. on Lewis 1926c.

Barkla, Charles Glover. 1905. Polarized Röntgen radiation. *Philosophical Transactions of the Royal Society, London* 204 A: 467–479.

Barkla, Charles Glover. 1906. Polarization in secondary Röntgen radiation. *Philosophical Transactions of the Royal Society, London* A 77: 247–255.

Barkla, Charles Glover. 1907. The nature of x-rays. *Nature* 76: 761–662, see also Bragg 1908a.

Barkla, Charles Glover. 1908a. Homogeneous secondary Röntgen radiation. *Philosophical Magazine, London* 6 (16): 550–584.

Barkla, Charles Glover. 1908b. Der Stand der Forschung über die sekundäre Röntgenstrahlung. *Jahrbuch der Radioaktivität und Elektronik* 5: 246–324.

Barkla, Charles Glover. 1910. Erscheinungen beim Durchgange von Röntgenstrahlen. *Jahrbuch der Radioaktivität und Elektronik* 7: 1–15.

Batchelor, George K. 1996. *The life and legacy of G. I. Taylor*. Cambridge: Cambridge University Press.

Beaudouin, Charles. 2005. *Une histoire d'instruments scientifiques*. Paris: EDP sciences.

Bechler, Zed. 1973. Newton's search for a mechanistic model of color dispersion: A suggested interpretation. *Archive for History of the Exact Sciences, Berlin* 11: 1–37.

Bechler, Zed. 1974. Newton's law of forces which are inversely as the mass - a suggested interpretation of his later efforts to normalize a mechanistic model of optical dispersion. *Centaurus* 18: 184–222.

Beck, Guido. 1927. Zur Theorie des Photoeffekts. *Zeitschrift für Physik* 41: 443–452.

Beller, Mara. 1999. *Quantum dialogue: The making of a revolution*. Chicago: University of Chicago Press.

Bennet, Abraham. 1792. A new suspension of the magnetic needle invented for the discovery of minute quantities of magnetic attraction. *Philosophical Transactions of the Royal Society, London* 82: 81–98.

Bennett, Charles H., W.K. Wootters, et al. 1993. Teleporting an unknown quantum state via dual classical and Einstein-Podolsky-Rosen channels. *The Physical Review Letters, New Series* 70: 1895.

Berger, Steven B. 1981. Comment on the localization of the photon. *American Journal of Physics* 49 (2): 106 (comm. on Henderson 1980; cf. Armstrong 1983).

Bergia, Silvio. 1987. Who discovered the Bose-Einstein statistics? In *Symmetries in physics*, ed. G. Manuel, 223–280. Bellaterra: Doncel.

Bertolotti, Mario. 1999. *The history of the laser*. Bristol: IOP Publishing.

Beth, Richard. 1936. Mechanical detection and measurement of the angular momentum of light. *The Physical Review. A Journal of Experimental and Theoretical Physics, New Series* 2 (50): 115–125 & plates.

Bjerknes, Vilhelm. 1909. *Die Kraftfelder*. Braunschweig: Vieweg.

Bjorken, James D., and Sidney D. Drell. 1965/67. *Relativistic quantum fields*. New York: McGraw Hill.

Blair, Thomas. 1786. A proposal for ascertaining by experiments whether the velocity of light be affected by the motion of the body from which it is emitted or reflected and for applying instruments for deciding the question to several optical and astronomical enquiries, Archives of the Royal Society, Mss. L & P. VIII, 182, publ. in Eisenstaedt (2005).

Blondlot, René. 1903. Égalité des vitesses de propagation des rayons X et de la lumière dans l'air. *Archives des Sciences Physiques et Naturelles* 15: 5–29.

Blum, Alexander. 2014. From the necessary to the possible: The genesis of the spin-statistics theorem. *European Journal of Physics H* 39: 543–574.

Blum, A., and Christian Joas. 2016. The emergence of emergent entities in quantum field theory. *Studies in History and Philosophy of Modern Physics* 53: 1–8.

Blumenberg, Hans. 1957. Licht als Metapher der Wahrheit. *Im Vorfeld der philosophischen Begriffs-bildung, Studium Generale* 10: 432–447.

Blumenberg, Hans. 1960. Paradigmen zu einer Metapherologie. *Archiv für Begriffsgeschichte* 6: 68–88.

Bodenstein, Max. 1942. 100 Jahre Photochemie des Chlorknallgases. *Berichte der deutschen Chemischen Gesellschaft* 75: 119.

Bohm, David, and Basil Hiley. 1982. The de Broglie pilot wave theory and the further development and new insights arising out of it. *Foundations of Physics* 12 (10): 1001–1016.

Bohr, Niels. 1927/28. The quantum postulate and the recent development of quantum theory. *Nature Suppl.* 14 April 1928: 580–590 (Talk in Como on 16 Sep 1927).

Bohr, Niels. 1933. Licht und Leben. *Die Naturwissenschaften* 21 (13): 245–250.

Bohr, Niels. 1949. Discussion with Einstein on epistemological problems in physics, in: *Albert Einstein, Philosopher Scientist*, ed. P.A. Schilpp, Evanston, Ill., 199–242. Library of Living Philsophers, 1949 (reprint 1951 etc.).

Bohr, Niels. 1972. *Collected works*, 13 vols. Amsterdam: Elsevier, 1972–2008, reprinted 2008.

Bohr, N., Hendrik A. Kramers, and John C. Slater. 1924. The quantum theory of radiation, (a) *Philosophical Magazine, London* (6). 47: 785–802; (b) Über die Quantentheorie der Strahlung. *Zeitschrift für Physik* 24: 69–87.

Bohr, N., Ralph de L. Kronig, and J.C. Slater. 1925. Spinning electrons and the structure of spectra. *Nature* 117: 264–5, 550, 587.

Bohr, N., and Léon Rosenfeld. 1933. Zur Frage der Messbarkeit der elektromagnetischen Feld-grössen. *Kgl. Danske Videnskabernes Selskap, Mathem. Fys. Meddelser* 12, no. 8 (40 pp).

Bordoni, Stefano. 2009. Discrete models for electromagnetic radiation: J.J. Thomson and Einstein. In *Da Archimede a Majorana: La fisica nel suo divenire, (Atti del XXVI Congresso Nazionale SISFA - Roma 2006)*, ed. E. Giannetto, G. Giannini and M. Toscano, 247–260. Rimini: Guaraldi.

Bordoni, Stefano. 2011. Joseph John Thomson's models of matter and radiation in the early, 1890s. *Physis. Rivista Intern. di Storia della Scienza* 48: 197–240.

Born, Max. 1926. Zur Quantenmechanik der Stoßvorgänge. *Zeitschrift für Physik* 37 (12): 863–867.

Born, Max. 1969. *Albert Einstein–Max Born: Briefwechsel 1916–1955*, (a) Munich: Nymphen-burger; (b) transl. by Irene Born: *The Born–Einstein letters*. New York: Walker & Company, 1971.

Born, M., Werner Heisenberg, and Pascual Jordan. 1926. Zur Quantenmechanik II. *Zeitschrift für Physik* 35: 557–615; English translation in *Sources of quantum mechanics*, ed. B. L. van der Waerden, Dover, New York, 1968, paper 15.

Bortz, Fred. 2004. *The photon*. New York: Rosen Publ. Group (The Library of Subatomic Particles).

Boschi, D., et al. 1998. Experimental realization of teleporting an unknown pure quantum state, *Physical Review Letters* 80 (6): 1121–1125.

Bose, Satyendranath. 1924. Plancks Gesetz und Lichtquantenhypothese. *Zeitschrift für Physik* 26: 178–181.

Bothe, Walther. 1924. Die Emissionsrichtung durch Röntgenstrahlen ausgelöster Photoelektronen. *Zeitschrift für Physik* 26: 59–73.

Bothe, W., and Hans Geiger. 1924. Ein Weg zur experimentellen Nachprüfung der Theorie von Bohr, Kramers, und Slater. *Zeitschrift für Physik* 26: 44–58.

Bothe, W., and Hans Geiger. 1925. Über das Wesen des Comptoneffekts: Ein experimenteller Beitrag zur Theorie der Strahlung. *Zeitschrift für Physik* 32: 639–663 and summary in *Die Naturwis-senschaften. Wochenschrift für die Fortschritte der Naturwissenschaften, der Medizin und der Technik, Berlin* 13: 440–441.

Bradley, James. 1728. A Letter from the Reverend Mr. James Bradley Savilian Professor of Astronomy at Oxford and F.R.S. to Dr. Edmond Halley Astronom. Reg. &c. Giving an Account of a New Discovered Motion of the Fix'd Stars. *Philosophical Transactions of the Royal Society, London* 35: 637–661.

Bragg, William Henry. 1907. On the properties and natures of various electric radiations. *Philosophical Magazine, London* (6. 14: 429–449, cf. Barkla 1907).

Bragg, William Henry. 1908a. The nature of γ and X-rays. *Nature* 77: 270–271.

Bragg, William Henry. 1908b. The nature of the γ and X-rays. *Nature* 78: 271.

Bragg, William Henry. 1912. X-rays and crystals. *Nature* 90: 219, 360–361 (comment on Tutton 1912).

Bragg, William Henry. 1912/13. *Studies in Radioactivity*, (a) London: Macmillan 1912. (b) in German transl. by Max Iklé: *Durchgang der α-, β-, γ- und Röntgen-Strahlen durch Materie*, Leipzig: Barth, 1913.

Bragg, William Henry. 1913. Radiations old and new. *Nature* 90 (529–532): 557–560.

Bragg, William Henry. 1915. The diffraction of X-rays by crystals. *Nobel lectures, Physics*. Amsterdam: Elsevier, 1967: 370–382 (Nobel lecture on 6 Sep 1922, for the prize in physics for 1915).

Bragg, William Henry. 1921/22. *Electrons & Ether waves: Being the twenty-third Robert Boyle lecture, on 11th May 1921*, (a) Oxford: Oxford University Press, 1921; (b) *Scientific Monthly* 1921; 14: 153–160.

Bragg, W.H., and J.P.V. Marsden. 1908a. An experimental investigation of the nature of the γ-rays. *Philosophical Magazine, London* 6, 15: 663–675 & 16: 692–702, 918–939.

Bragg, W.H., and J.P.V. Marsden. 1908b. The quality of the secondary ionization due to β-rays. *Philosophical Magazine, London* 6 (16): 692–702.

Brandt, Siegmund, and Hans-Dieter Dahmen. 1985. *The picture book of quantum mechanics*. New York: Wiley.

Brannen, Eric, and Harry I.S. Ferguson. 1956. Question of correlation between photons in coherent light rays. *Nature* 178: 481–482 (cf. comment by Purcell 1956).

de Broglie, Louis. 1921/23. La relation $hv = \epsilon$ dans les phénomènes photo-électriques. In *Atomes et électrons – Rapports et Discussions du Conseil de Physique tenu à Bruxelles 1–6 Avril 1921*, 80–130. Paris: Gauthier-Villars, 1923 (with the following discussion).

de Broglie, Louis. 1922. Rayonnement noir et quanta de lumière. *Journal de Physique* 6 (3): 422–428.

de Broglie, Louis. 1923. Radiation: ondes et quanta. *Comptes Rendus hebdomadaires des Séances de l'Académie des Sciences, Paris* 177: 507–510.

de Broglie, Louis. 1924/25. *Recherches sur la théorie des quanta*. Paris: Masson, 1924. *Annales de Physique* 3: 22–138.

de Broglie, Louis. 1939. *Licht und Materie*. Hamburg: Ergebnisse der modernen Physik; reprinted as Fischer-Taschenbuch 1958.

de Broglie, Louis. 1949. L'Oeuvre d'Einstein et la dualité des ondes et des corpuscules. *Reviews of Modern Physics* 21: 345–347.

Bromberg, Joan Lisa. 1991. *The laser in America, 1950–70*. Cambridge, MA: MIT Press.

Bromberg, Joan Lisa. 2006. Divide physics vis-à-vis fundamental physics in Cold War America. *Isis* 97: 237–259.

Bromberg, Joan Lisa. 2010. *Modelling the Hanbury Brown-Twiss Effect; The Mid-Twentieth century revolution in optics; A talk for HQ3*; Berlin: MPI, http://mediathek.mpiwg-berlin. mpg.de/mediathekPublic/versionEins/Conferences-Workshops/HQ3/Quantum-Optics/Joan-Bromberg.html.

Bromberg, Joan Lisa. 2016. Explaining the laser's light: Classical versus quantum electrodynamics in the 1960s. *Archive for the History of Exact Sciences* 70: 243–266.

Brown, Laurie M. 2002. The Compton effect as one path to QED. *Studies in the History and Philosophy of Modern Physics* 33: 211–249.

Brukner, Časlav, and Anton Zeilinger. 1997. Nonequivalence between stationary matter wave optics and stationary light optics. *The Physical Review Letters, New Series* 79: 2599–2613.

Brunner, Otto, Werner Conze, and Reinhart Koselleck (eds.). 1979. *Geschichtliche Grundbegriffe: Historisches Wörterbuch zur politisch-sozialen Sprache in Deutschland.* Stuttgart: Klett-Cotta.

Brush, Stephen George. 1970. *Kinetische Theorie. Einführung und Originaltexte*, 2 vols. Berlin: Akademie Verlag.

Brush, Stephen George. 1974. Should the history of science be rated X? *Science* 183: 1164–1172.

Brush, Stephen George. 1976. *The kind of motion we call heat - A history of the kinetic theory of gases in the 19th Century, 2 vols.* Amsterdam: North Holland Publ.

Brush, Stephen George. 1987. *Die Temperatur der Geschichte. Wissenschaftliche und kulturelle Phasen im 19. Jahrhundert,* Braunschweig: Vieweg.

Brush, Stephen George. 2007. How ideas became knowledge: The light quantum hypothesis 1905–1935. *Historical Studies in the Physical (and Biological)/Natural Sciences, Berkeley* 37 (2): 205–246.

Bubb, Frank W. 1924. Direction of ejection of photo-electrons by polarized x-rays. *The Physical Review. A Journal of Experimental and Theoretical Physics, New Series* 2 (23): 137–143.

Buchwald, Jed Z., and Andrew Warwick (eds.). 2001. *Histories of the electron: The birth of microphysics.* Cambridge, MA: MIT Press.

Bunge, Mario. 1970. Virtual processes and virtual particles - real or fictitious. *International Journal of Theoretical Physics* 3: 507–508.

Büttner, Jochen, Olivier Darrigol, Dieter Hoffmann, and Jürgen Renn. (eds.). 2000. *Revisiting the quantum discontinuity.* Berlin: MPI-Preprint 150.

Cahan, David. 1989. *An Institute for an empire: The Physikalisch-Technische Reichsanstalt, 1871–1918.* Cambridge: Cambridge University Press.

Campbell, Norman Robert. 1909. Discontinuities in light emission. *Proceedings of the Cambridge Philosophical Society* 15: 310–328 & 513–527.

Cantor, Geoffrey N. 1983. *Optics after Newton. Theories of Light in Britain and Ireland, 1704–1840.* Manchester: Manchester University Press.

Cantor, G.N., and Michael J.S. Hodges (eds.). 1981. *Conceptions of Ether: Studies in the history of Ether theories 1740–1900.* Cambridge: Cambridge University Press.

Carazza, Bruno, and Helge Kragh. 1989. Bartoli and the problem of radiant heat. *Annals of Science* 46: 183–194.

Carey, Susan. 2009. *The origin of concepts.* Oxford: Oxford University Press.

Caso, Arthur Lewis. 1980. The production of new scientific terms. *American Speech* 55: 101–111.

Chang, Hasok. 2011. The persistence of epistemic objects through scientific change. *Erkenntnis* 75: 413–429.

Chiao, Raymond, and John Garrison. 2008. *Quantum Optics.* (a) Oxford: Oxford University Press, 1st ed.; (b) exp. 2nd ed. 2014.

Clark, John W., R.E. Norberg, and G.L. Bretthorst. 2000. Obituary of Edwin Thompson Jaynes. *Physics Today* 51 (1): 71–72.

Clauser, John Francis. 1974. Experimental distinction between the quantum and classical field-theoretic predictions for the photoelectric effect. *Physical Review D* 9: 853–860.

Clauser, J.F., and Abner Shimony. 1978. Bell's theorem: Experimental tests and implications. *Reports on Progress in Physics* 41: 1881–1927.

Cohen, I.Bernard (ed.). 1958. *Isaac Newton's papers and letters on natural philosophy and related documents.* Cambridge, MA: Harvard University Press.

Cohen-Tannoudji, Claude. 1983. Are photons essential? *Physikalische Blätter* 39 (7): 198–199.

Collins, Allan, and Dedre Gentner. 1987. How people construct mental models. In *Cultural models in language and thought*, ed. D. Holland and N. Quinn, 243–265. Cambridge.

Compton, Arthur Holly. 1921. The magnetic electron. *Journal of the Franklin Institute* 192 (2): 145–155.

Compton, Arthur Holly. 1922. Secondary radiation produced by x-rays and some of their applications to physical problems. *Bulletin of the National Research Council* 4, part 2 (20): 1–56.

Compton, Arthur Holly. 1923a. A quantum theory of the scattering of x-rays by light elements. *The Physical Review. A Journal of Experimental and Theoretical Physics, New Series* 2 (21): 483–502.

Compton, Arthur Holly. 1923b. The spectrum of scattered x-rays. *The Physical Review. A Journal of Experimental and Theoretical Physics, New Series* 2 (22): 409–413.

Compton, Arthur Holly. 1923c. The total reflection of x-rays. *Philosophical Magazine, London* 6 (45): 1121–1131.

Compton, Arthur Holly. 1924. A general quantum theory of the wavelengths of scattered x-rays. *The Physical Review. A Journal of Experimental and Theoretical Physics, New Series* 2 (24): 168–176.

Compton, Arthur Holly. 1925. Light waves or light bullets? *Scientific American, New York* 133 (Oct.): 246–247.

Compton, Arthur Holly. 1927. X-rays as a branch of optics, (a) *Nobel Lectures, Physics 1922–1941*. Amsterdam: Elsevier, 1965, 174–190 (lecture on 12 Dec 1927); (b) reprinted in *Journal of the Optical Society of America* 16. 1928): 71–87.

Compton, Arthur Holly. 1928. Discordances entre l'expérience et la théorie électromagnétique du rayonnement, (a) in: *Électrons et Photons. Rapports et Discussions de Cinquième Conseil de Physique*, ed. Institut International de Physique Solvay. Gauthier-Villars, Paris, 1928: 55–85; (b) in English transl.: Some experimental difficulties with the electromagnetic theory of radiation. *Journal of the Franklin Institute* 205: 155–178.

Compton, Arthur Holly. 1929a. What things are made of. *Scientific American, New York* 140: 110–113 (I), 234–236 (II).

Compton, Arthur Holly. 1929b. The corpuscular theory of light. *Physical Review Suppl.* 1: 74–89.

Compton, Arthur Holly. 1935/47. *Electrons (+ and -), Protons, Photons, Neutrons, Mesotrons, and Cosmic Rays*. Chicago: University of Chicago Press, (a) 1st ed. 1935; (b) 2nd ed. 1947.

Compton, Arthur Holly. 1961. The scattering of x rays as particles. *American Journal of Physics* 29: 817–820.

Compton, Arthur Holly. 1973. *Scientific papers of Arthur Holly Compton. X-Ray and other studies*, ed. Robert S. Shankland, Chicago: University of Chicago Press.

Compton, A.H., and Owen W. Richardson. 1913. The photoelectric effect. *Philosophical Magazine, London* (6). 51: 530.

Compton, A.H., and Alfred W. Simon. 1925. Directed quanta of scattered x-rays. *The Physical Review. A Journal of Experimental and Theoretical Physics, New Series* 2. 26, no. 3: 289–299 & pl.

Compton, Karl T. 1913. Note on the velocity of electrons liberated by photoelectric action. *The Physical Review. A Journal of Experimental and Theoretical Physics, New Series* 2 (1): 382–392.

Comstock, Daniel F., and Leonard T. Troland. 1917. *The nature of matter and electricity: An outline of modern views*. New York: Van Nostrand Co.

Cornelius, David W. 1913. The velocity of electrons in the photoelectric effect, as a function of the wave lengths of light. *The Physical Review. A Journal of Experimental and Theoretical Physics, New Series* 2 (1): 16–34.

CPAE, see: Stachel et al., (eds.). 1987. vol. 1, 1989. vol. 2; Klein et al., (eds.). 1993. vols. 3 and 5, 1995. vol. 4, 1996. vol. 6; Kormos Buchwald et al., ed. 2015. vol. 14.

Cramer, John G. 2015. *The quantum handshake: Entanglement*. Nonlocality and transactions. New York: Springer.

Crisp, Michael D., and Edwin T. Jaynes. 1969. Radiative effects in semiclassical theory. *The Physical Review. A Journal of Experimental and Theoretical Physics, New Series* 179 (5): 1253–1261.

Curie, Marie. 1904. *Untersuchungen über die radioaktive Substanzen*. Braunschweig: Vieweg, 1st ed.

Curie, Marie. 1912. *Die Radioaktivität*. Leipzig: Akademie-Verlagsgesellschaft.

Dagenais, Mario, and Leonard Mandel. 1978. Investigation of two-time correlations in photon emission from a single atom. *The Physical Review. A Journal of Experimental and Theoretical Physics, New Series* A18: 2217–2228 (cf. Kimble et al. 1977).

Dahl, Per F. 1997. *Flash of the cathode rays: A history of J.J. Thomson's electron*. London: Institute of Physics.

Darrigol, Olivier. 1986. The origin of quantized matter waves. *Historical Studies in the Physical (and Biological)/Natural Sciences, Berkeley* 16 (2): 197–253.

Darrigol, Olivier. 1988. Elements of a scientific biography of Tomonoga Sin-Itiro. *Historia Scientiarum* 35: 1–29.

Darrigol, Olivier. 1988/90. Statistics and combinatorics in early quantum theory. *Historical Studies in the Physical (and Biological)/Natural Sciences, Berkeley* 19: 17–80 & 21: 237–298.

Darrigol, Olivier. 1992. *From C-numbers to Q-numbers*. The classical analogy in the history of quantum theory. Berkeley: University of California Press.

Darrigol, Olivier. 1993. Strangeness and soundness in L. de Broglie's early works. *Physis* 30: 303–372.

Darrigol, Olivier. 2000. *Electrodynamics from Ampère to Einstein*. Oxford: Oxford University Press.

Darrigol, Olivier. 2000/01. The historians' disagreements over the meaning of Planck's quantum, (a) in Büttner et al., (eds.). 2000, 3–21; (b) republished in *Centaurus* 43: 219–239.

Darrigol, Olivier. 2012. *A history of optics from Greek antiquity to the nineteenth century*. Oxford: Oxford University Press, see Hentschel (2012/14).

Darrigol, Olivier. 2014. *The Quantum Enigma*. In eds. Janssen & Lehner, 117–142.

Davis, E.A., and Isabel Falconer. 1997. *J.J. Thompson and the Discovery of the Electron*. London: Taylor & Francis.

Davis, John, and Bernard Lovell. 2003. Robert Hanbury Brown. Biographical Memoirs of Fellows of the Royal Society, London 49: 83–106 and in Historical Records of Australian. *Science* 14: 459–483.

Davisson, Clinton J., and 1937. The discovery of electron waves, Nobel lecture, 13, Dec 1937. *In Nobel lectures, Physics 1922–1941, 387–394*, 1965. Amsterdam: Elsevier.

Davisson, C.J., and Lester H. Germer. 1927. Diffraction of electrons by a crystal of nickel. *The Physical Review A Journal of Experimental and Theoretical Physics, New Series* 30: 705–740.

Debye, Peter. 1909. Das Verhalten von Lichtwellen in der Nähe des Brennpunkts oder einer Brennlinie. *Annals of Physics* 4 (30): 755–776.

Debye, Peter. 1910. Der Wahrscheinlichkeitsbegriff in der Theorie der Strahlung. *Annals of Physics* 4 (33): 1427–1434.

Debye, Peter. 1911. Die Frage nach der atomistischen Struktur der Energie. *Vierteljahresschrift der Naturforschenden Gesellschaft in Zürich* 55: 156–167.

Debye, Peter. 1923. Zerstreuung von Röntgenstrahlen und Quantentheorie. *Physikalische Zeitschrift, Leipzig/Berlin* 24: 161–166.

Debye, P., and Arnold Sommerfeld. 1913. Theorie des lichtelektrischen Effektes vom Standpunkt des Wirkungsquantums. *Annalen der Physik, Leipzig* 4th ser. 41: 893–930.

Delbrück, Max. 1980. Was Bose-Einstein statistics arrived at by serendipity? *Journal of Chemical Education* 57: 467–470.

Dempster, A.J., and Harold F. Batho. 1927. Light quanta and interference. *The Physical Review. A Journal of Experimental and Theoretical Physics, New Series* 2 (30): 644–648.

Dickstein, S. 1913. *August Witkowski. Wiadomości Matematyczne* 17: 189–193.

Dirac, Paul Adrienne, and Maurice. 1927a. The Compton effect in wave mechanics. *Proceedings of the Cambridge Philosophical Society* 23: 500–507.

Dirac, Paul Adrienne, and Maurice. 1927b. The quantum theory of the emission and absorption of radiation. *Proceedings of the Royal Society, London A* 114: 243–265.

Dirac, Paul Adrienne Maurice. 1930. *The principles of quantum mechanics*. Oxford: Clarendon, (a) 1st ed., 1930; (b) 2nd ed., 1935; (c) 3rd ed., 1947; (d) 4th ed., 1958.

Dirac, Paul Adrienne Maurice. 1974/75. An historical perspective of spin, *Proceedings of the Summer Studies on High-Energy Physics with Polarized Beams, July 1974*, vol. 2, no. xxxii, 1–14. Ann Arbor: Argonne National Laboratory 1975.

Dodd, J.N. 1983. The Compton effect - a classical treatment. *European Journal of Physics* 4: 205–211.

Dörfel, Günter, and Falk Müller. 2003. Crookes' Radiometer und Geißlers Lichtmühle-Kooperation oder Konkurrenz? *NTM New Series* 11: 171–190.

Dorling, Jon. 1971. Einstein's introduction of photons - Argument by analogy or deduction from the phenomena? *British Journal for the Philosophy of Science, Cambridge* 22: 1–8.

Dresden, Max. 1987. *H.A. Kramers – Between tradition and revolution*. Heidelberg & New York: Springer.

Duncan, Anthony. 2012. *The conceptual framework of quantum field theory*. Oxford: Oxford University Press.

Eckert, Michael. 2014. How Sommerfeld extended Bohr's model of the atom. 1913–1916. *European Journal of Physics H* 39: 141–156.

Eckert, Michael. 2015. From aether impulse to QED: Sommerfeld and the Bremsstrahlen theory. *Studies in History and Philosophy of Modern Physics* 51: 9–22.

Ehrenfest, Paul. 1911. Welche Züge der Lichtquantenhypothese spielen in der Theorie der Wärmestrahlung eine wesentliche Rolle? *Annals of Physics* 4 (36): 91–118.

Ehrenfest, P., and Heike Kamerlingh Onnes. 1915. Vereinfachte Ableitung der kombinatorischen Formel, welche der Planckschen Strahlungstheorie zugrunde liegt, *Annalen der Physik, Leipzig* (4th ser. 46: 1021–1024).

Einstein, Albert. 1905. Über einen die Erzeugung und Umwandlung des Lichtes betreffenden heuristischen Standpunkt, dated Berne, 17 March 1905, (a) *Annalen der Physik, Leipzig* 9 June 1905 (4). 17: 132–148; (b) annotated reprint with commentary in *Collected Papers of Albert Einstein* (CPAE), vol. 2, J. Stachel et al. eds. 1989, doc. 14, 134–169 (transl. ed.: On a Heuristic Point of View Concerning the Production and Transformation of Light, 86–103).

Einstein, Albert. 1906a. Zur Theorie der Brownschen Bewegung, *Annalen der Physik, Leipzig* (4). 19: 371–382; reprinted in CPAE, vol. 2, 1989, doc. 32, 333–345 (transl. ed.: On the Theory of Brownian Motion, 180–190).

Einstein, Albert. 1906b. Zur Theorie der Lichterzeugung und Lichtabsorption, *Annalen der Physik, Leipzig* (4). 20: 199–206; reprinted in CPAE, vol. 2, 1989, doc. 34, 350–357 (trans. ed.: *On the theory of light production and light absorption*, 192–199).

Einstein, Albert. 1907a. Die Planck'sche Theorie der Strahlung und die Theorie der spezifischen Wärme, *Annalen der Physik, Leipzig* (4) 22: 180–190; reprinted in CPAE, vol. 2, 1989, doc. 38, 378–391 (trans. ed.: *Planck's theory of radiation and the theory of specific heat*, 214–224).

Einstein, Albert. 1907b. Über das Relativitätsprinzip und die aus demselben gezogenen Folgerungen, *Jahrbuch der Radioaktivität und Elektronik* 4: 411–462; reprinted in CPAE, vol. 2, 1989, doc. 47, 432–488 (trans. ed.: *On the relativity principle and the conclusions drawn from it*, 252–311).

Einstein, Albert. 1909. Über die Entwickelung unserer Anschauungen über das Wesen und die Konstitution der Strahlung [talk in Salzburg on 21 Sep. 1909] (a) *Physikalische Zeitschrift* 10: 817–825; (b) *Verhandlungen der Deutschen Physikalischen Gesellschaft* 7: 482–500; (c) reprinted in CPAE, vol. 2, 1989, doc. 60, 563–583 (trans. ed.: *On the development of our views concerning the nature and constitution of radiation*, 379–394).

Einstein, Albert. 1911. Über den Einfluss der Schwerkraft auf die Ausbreitung des Lichtes, *Annalen der Physik, Leipzig* (4) 35: 898–908; reprinted in CPAE, vol. 3, 1993, doc. 23, 485–497 (trans. ed.: *On the influence of gravitation on the propagation of light*, 379–387).

Einstein, Albert. 1911/12. Rapport sur l'état actual du problème des chaleur spécifiques, (a) in Langevin and de Broglie, ed. 1912, 407–450; (b) in German transl.: Zum gegenwärtigen Stande des Problems der spezifischen Wärmen, in: Arnold Eucken, ed. *Die Theorie der Strahlung und der Quanten*, Halle an der Saale, Knapp, 1914, 330–364; reprinted in CPAE, vol. 3, 1993, doc. 26, 520–548 (trans. ed.: *On the present state of the problem of specific heats*, 402–425).

Einstein, Albert. 1912a. Thermodynamische Begründung des photochemischen Äquivalentgesetzes, *Annalen der Physik, Leipzig* (4) 37: 832–838; reprinted in CPAE, vol. 4, 1995, doc. 2, 114–121 (trans. ed. *Thermodynamic proof of the law of photochemical equivalence*, 89–94).

Einstein, Albert. 1912b. Nachtrag zu meiner Arbeit: "Thermodynamische Begründung des photochemischen Äquivalentgesetzes" *Annalen der Physik, Leipzig* 4. 37: 881–884; reprinted in CPAE, vol. 4, 1995, doc. 5, 165–170 (trans. ed.: *Supplement to my paper: 'Thermodynamic proof of the law of photochemical equivalence*, 121–124). (See also Stark 1912 and Einstein's reply 1912c).

Einstein, Albert. 1912c. Antwort auf eine Bemerkung von J. Stark: "Über eine Anwendung des Planckschen Elementargesetzes ...", *Annalen der Physik, Leipzig* (4) 37: 888; reprinted in CPAE, vol. 4, 1995, doc. 5, 171–173 (trans. ed.: *Supplement to my paper: 'Thermodynamic proof of the law of photochemical equivalence,*' 121–124).

Einstein, Albert. 1913. Max Planck als Forscher, *Die Naturwissenschaften. Wochenschrift für die Fortschritte der Naturwissenschaften, der Medizin und der Technik, Berlin* 1: 1077–1079; reprinted in CPAE, vol. 4, 1995, doc. 23, 560–565 (trans. ed.: *Max Planck as scientist*, 271–275).

Einstein, Albert. 1916a. Strahlungs-Emission und -Absorption nach der Quantentheorie, *Verhandlungen der Deutschen Physikalischen Gesellschaft* 18: 318–323 (received on 17th of July 1916); reprinted in CPAE, vol. 6, 1996, doc. 34, 363–370 (trans. ed.: *Emission and absorption of radiation in quantum theory*, 212–217).

Einstein, Albert. 1916b. Zur Quantentheorie der Strahlung. *Mitteilungen der Physikalischen Gesellschaft Zürich* 18: 47–62; reprinted in CPAE vol. 6, 1996, doc. 38: 381–398 (trans. ed.: *On the quantum theory of radiation*, 220–233).

Einstein, Albert. 1917. Zur Quantentheorie der Strahlung. *Physikalische Zeitschrift* 18: 121–128 (a reprint of Einstein 1916b).

Einstein, Albert. 1924. Das Kompton'sche [sic] Experiment, *Berliner Tageblatt*, 20 April 1924, suppl. p. 1; reprinted CPAE vol. 14, 2015, doc. 236: 364–367 (trans. ed.: *The Compton experiment. Does science exist for its own sake?*, 231–234).

Einstein, Albert. 1924/25. Quantentheorie des einatomigen idealen Gases I-III, *Sitzungsberichte der Preußischen Akademie der Wissenschaften*, math.physik. Klasse (a) I, 1924: 261–267; (b) II & III, 1925: 3–14, 18–25 reprinted in CPAE vol. 14, 2015, docs. 283, 385 & 427, 433–441, 580–594, 648–657 (trans. ed.: *Quantum theory of the monatomic ideal gas*, 276–283, 371–383, 418–425).

Einstein, Albert. 1927. Theoretisches und Experimentelles zur Frage der Lichtentstehung. *Zeitschrift für angewandte Chemie* 40: 546.

Einstein, Albert. 1949. Autobiographical notes. In *Einstein: Philosopher-Scientist*, ed. P.A. Schilpp, 3–95. Evanston, Ill.: Library of Living Philosophers.

Einstein, A., and Paul Ehrenfest. 1923. Quantentheorie des Strahlungsgleichgewichts. *Zeitschrift für Physik* 19: 301–306.

Einstein, A., and Leopold Infeld. 1938. *The evolution of physics from early concepts to relativity and quanta*. New York: Simon & Schuster.

Einstein, A., Boris Podolsky, and Nathan Rosen. 1935. Can quantum-mechanical description of physical reality be considered complete? *The Physical Review. A Journal of Experimental and Theoretical Physics, New Series* 2 (47): 777–780.

Einstein, A., and Otto Stern. 1913. Einige Argumente für die Annahme einer molekularen Agitation beim absoluten Nullpunkt. *Annals of Physics* 4 (40): 551–560.

Eisenstaedt, Jean. 1991. De l'influence de la gravitation sur la propagation de la lumière en théorie newtonienne. L'archaeologie des trous noirs, *Archive for History of the Exact Sciences, Berlin* 42: 315–386.

Eisenstaedt, Jean. 1996. L'optique ballistique newtonienne à l'épreuve des satellites de Jupiter. *Archive for History of the Exact Sciences, Berlin* 50: 117–156.

Eisenstaedt, Jean. 2005. Light and relativity. A previously unknown 18th c. mss. by Robert Blair (1748–1828). *Annals of Science* 62: 347–376.

Eisenstaedt, Jean. 2007. From Newton to Einstein. A forgotten relativistic optics of moving bodies. *American Journal of Physics* 75: 74–79.

Eisenstaedt, Jean. 2012. The Newtonian Theory of Light Propagation, in: *Einstein and the Changing Worldview of Physics. Proceedings of the 7th Conference on the History of General Relativity*, Tenerife, 2005. Einstein Studies 12 (I), 23–37.

Eisler, Rudolf. 1909. *Wörterbuch der philosophischen Begriffe*. Berlin: Mittler.

Eisler, Rudolf. 1928. *Electrons et Photons/Electrons and Photons, rapports et discussions du Cinquième Conseil de Physique, tenu à Bruxelles du 24 au 29 octobre 1927 sous les auspices de l'Institut International de Physique Solvay*. Paris: Gauthier-Villars 1928.

Elliott, Paul. 1999. Abraham Bennet, FRS (1749–1799): A provincial Electrician in Eighteenth Century England. *Notes and Records of the Royal Society London* 53 (1): 59–78.

Ellis, Charles D. 1926. The light-quantum hypothesis. *Nature* 117: 895–897.

Ellis, John, and Daniele Amati (eds.). 2000. *Quantum reflections*. Cambridge: Cambridge University Press.

Elzinga, Aant. 2006. *Einstein's Nobel Prize*. A glimpse behind closed doors. Sagamore Beach: Science History Publ.

Esfeld, Michael (ed.). 2012. *Philosophie der Physik*. Frankfurt: Suhrkamp.

d'Espagnat, Bernard. 1971. *Conceptual foundations of quantum mechanics*, (a) New York, 1971; (b) 2nd ed., Reading, MA: Perseus Books 1999.

Eucken, Arnold, Otto Lummer, and E. Waetzmann (eds.). 1929. *Müller-Pouillets Lehrbuch der Physik*, 11th ed., vol. 2: *Lehre von der strahlenden Energie (Optik), Zweite Hälfte - Erster Teil*, Braunschweig: Vieweg.

Fano, Ugo. 1961. Quantum theory of interference effects in the mixing of light from phase independent sources. *American Journal of Physics* 29: 539–545.

Fauconnier, Gilles, and Mark Turner. 2002. *The way we think: Conceptual blending and the mind's hidden complexities*. New York: Basic Books.

Fellgett, Peter. 1957. The question of correlation between photons in coherent beams of light. *Nature* 179: 956–957.

Fellgett, P., R. Clark Jones, and Richard Q. Twiss. 1959. Fluctuations in photon streams. *Nature* 184: 967–969.

Fengler, Silke (ed.). 2014. *Kerne, Kooperation und Konkurrenz. Kernforschung in Österreich im Internationalen Kontext. 1900–1950*. Vienna: Böhlau.

Feynman, Richard P. 1949. Space-time approach to quantum electrodynamics, (a) *The Physical Review. A Journal of Experimental and Theoretical Physics, New Series*. 2. 76: 769–789; (b) reprint in Schwinger, ed. 1958, 236–258.

Feynman, Richard P. 1961. *Quantum electrodynamics*. New York: Benjamin.

Feynman, Richard P. 1985. *The strange theory of light and matter*. Princeton: Princeton University Press.

Fizeau, Armand-Hippolyte. 1851/53. Sur les hypothèses relatives à l'ether lumineux, et sur une expérience qui paraît démontrer que le mouvement des corps change la vitesse avec laquelle la lumière se propage dans leur intérieur, *Comptes Rendus hebdomadaires des Séances de l'Académie des Sciences, Paris* 33: 349–355.

Fizeau, Armand-Hippolyte. 1859/60. Sur les hypothèses relatives a l'ether lumineux, et sur une expérience qui paraît démontrer que le mouvement des corps change la vitesse avec laquelle la lumière se propage dans leur intérieur, (a) *Annales de Physique et Chimie* 57 (1859): 385–404; (b) in Engl. transl.: On the Effect of the Motion of a Body upon the Velocity with Which It Is Traversed by Light. *Philosophical Magazine, London* 19 (April 1860): 245–260.

Fleck, Ludwik. 1935/80. *Entstehung und Entwicklung einer wissenschaftlichen Tatsache: Einführung in die Lehre vom Denkstil und Denkkollektiv*, (a) Basel: Schwabe, 1935; (b) Frankfurt a.M.: Suhrkamp, 1980.

Fölsing, Albrecht. 1993. *Albert Einstein - Eine Biographie*. Frankfurt am Main: Suhrkamp.

Forman, Paul. 1968. The doublet riddle and atomic physics circa 1924. *Isis* 59: 156–174.

Forrester, A.Theodore, Richard A. Gudmundsen, and Philip O. Johnson. 1955. Photoelectric mixing of incoherent light. *The Physical Review. A Journal of Experimental and Theoretical Physics, New Series* 2 (99): 1691–1700.

Frank, Felix. 2012. The shortest artificial light burst in history, posted 2 July 2012, http://wwwurzweilai.net/the-shortest-artificial-light-bursts-in-history. (19. 3. 2016).

Frank, Felix, et al. 2012. Technology for attosecond science. *Review of Scientific Instruments* 83 (7): 071101.

Franklin, Allan. 2013. Millikan's measurement of Planck's constant. *European Physics Journal H* 38: 573–594.

Franklin, Allan. 2016. Physics textbooks don't always tell the truth. *Physics in Perspective* 18 (1): 3–57.

Fraser, Doreen. 2008. The fate of 'particles' in quantum field theories with interactions. *Studies in History and Philosophy of Modern Physics B* 39: 841–859.

Freedman, Stuart Jay, and John F. Clauser. 1972. Experimental test of local hidden-variable theories. *The Physical Review Letters, Lancaster* 28: 938–941.

Freeman, Gordon R. 1984. What are photons and electrons? *American Journal of Physics* 52, no. 1: 11 (comment on Armstong 1983).

Freire, Olival Jr. 2006. Philosophy enters the optics laboratory: Bell's theorem and its first experimental tests. 1965–1982). *Studies in the History and Philosophy of Science, Cambridge B* 37 (4): 577–616.

Freire, O. Jr., W.F. de Oliveira, and D.F.G. David. 2013. As contribuiçoes de John Clauser para o primeiro teste experimental do teorema de Bell: uma análise das técnicas e da cultura material. *Revista Brasileira de Ensino de Fisica* 35, 3, 3603: 1–7.

French, Steven. 2015. Identity and individuality in quantum theory, *Stanford encyclopedia of philosophy* at https://plato.stanford.edu/entries/qt-idind/ (3 Aug 2015).

Frercks., 2004. Das Verhältnis von Publikation zu Theorie und Experiment in Fizeaus Forschungsprogramm zur Äthermitführung. *NTM New Series* 12: 18–39.

Fresnel, Augustin. 1818. Lettre d'Augustin Fresnel à Franscois Arago sur l'influence du mouvement terrestre dans quelques phénomènes d'optique, (a) *Annales de chimie et de physique, Paris* IX (1818): 57–66; (b) reprinted in Fresnel 1866–70. vol. II, 627–636.

Fresnel, Augustin. 1866–70. *Oeuvres Complètes*. Paris: Imprimerie Impérial, 3 vols. 1866–70.

Friebe, Cord, Meinard Kuhlmann, Holger Lyre, Paul Näger, Oliver Passon, and Manfred Stöckler. 2015. *Philosophie der Quantenphysik*. Heidelberg: Springer Spektrum.

Friedel, E., and Fred Wolfers. 1924. Les variations de longueur d'onde des rayons X par diffusions et la loi de Bragg. *Comptes Rendus hebdomadaires des Séances de l'Académie des Sciences, Paris* 178: 199–200.

Friedman, Robert Marc. 2001. *The politics of excellence: Behind the nobel prize in science*. New York: Holt.

Fritzius, Robert S. 1990. The Ritz-Einstein Agreement to Disagree. *Physics Essays* 3: 371–374, also on-line at http://www.datasync.com/~rsf1/rtzein.htm (6 Mar 2016).

Galison, Peter. 1981. Kuhn and the quantum controversy. *British Journal for the Philosophy of Science, Cambridge* 32 (1): 71–85.

Galison, Peter. 1987. *How experiments end*. Chicago: University of Chicago Press.

Galison, Peter. 1991. *Image and logic: A material culture of microphysics*. Chicago: University of Chicago Press.

Gamow, George. 1966. *Thirty years that shook physics. the story of quantum theory*. New York: Dover.

Gearhart, Clayton A. 2002. Planck, the Quantum, and the Historians. *Physics in Perspective* 4: 170–215.

Gearhart, Clayton A. 2010. 'Astonishing successes' and 'bitter disappointment': The specific heat of hydrogen in quantum theory. *Archive for History of the Exact Sciences, Berlin* 64: 113–202.

Gell-Mann, Murray, and Yuval Ne'eman. 1964. *The eightfold way*. New York: Benjamin.

Gentner, Dedre, and Albert L. Stevens (eds.). 1983. *Mental models*. Hillsdale: Erlbaum.

Gerry, Christopher C., and Peter L. Knight. 2005. *Introductory quantum optics*. Cambridge: Cambridge University Press.

Ghosh, R., and Leonard Mandel. 1987. Observation of non-classical effects in the interference of two photons. *The Physical Review Letters, New Series* 59: 1903–1905.

Gies, Holger et al. 2018. Photon-photon scattering at the high-intensity frontier. arXiv: 1712.06450v2.

Giulini, Domenico. 2011. Max Planck und die Begründung der Quantentheorie. In *Rationalität und Irrationalität in den Wissenschaften*, ed. Ulrich Arnswald, and Hans-Peter Schütt, 111–137. Verlag für: Sozialwissenschaften.

Glauber, Roy Jay. 1963a. Photon correlations. *The Physical Review Letters, New Series* 10 (3): 84–86.

Glauber, Roy Jay. 1963b. The quantum theory of optical coherence. *The Physical Review. A Journal of Experimental and Theoretical Physics, New Series* 2 (130): 2529–2539.

Glauber, Roy Jay. 1963c. Coherent and incoherent states of the radiation field. *The Physical Review. A Journal of Experimental and Theoretical Physics, New Series* (2) 131: 2766–2788; (d) reprinted in Meystre & Walls eds. 1991, 2–24.

Glauber, Roy Jay, and 2005. One hundred years of light quanta, (Nobel lecture. Nobel Prize of, 2005. *78–98*. Nobel Prize Foundation: Stockholm.

Gleyzes, Sébastien, Stefan Kuhr, Christine Guerlin, Julien Bernu, Samuel Deléglise, Ulrich Busk Hoff, Michel Brune, Jean-Michel Raimond, and Serge Haroche. 2007. Quantum jumps of light recording the birth and death of a photon in a cavity. *Nature* 446: 297–301.

Goudsmit, Samuel. 1965. Die Entdeckung des Elektronenspins. *Physikalische Blätter* 21: 445–453.

Goudsmit, Samuel. 1971. The discovery of the electron spin, http://www.lorentz.leidenuniv.nl/history/spin/goudsmit.html (last accessed May 28, 2007).

Goudsmit, S., and George E. Uhlenbeck. 1925. Ersetzung der Hypothese vom unmechanischen Zwang durch eine Forderung bezüglich des innern Verhaltens jedes einzelnen Elektrons. *Die Naturwissenschaften. Wochenschrift für die Fortschritte der Naturwissenschaften, der Medizin und der Technik, Berlin* 13: 953–954.

Goudsmit, S., and George E. Uhlenbeck. 1976. It might as well be spin. *Physics Today* 29 (6): 40–48.

Goulielmakis, Eleftherios, et al. 2004. Direct measurement of light waves. *Science* 305: 1267–1269.

Grafton, Anthony. 2006. The history of ideas: Precept and practice, 1950–2000 and beyond. *Journal of the History of Ideas* 67 (1): 1–32.

Grangier, Philippe. 2002. Single photons stick together. *Nature* 419: 577, see also Santori (2002).

Grangier, Philippe. 2005. Experiments with single photons. *Séminaire Poincaré* 2: 1–26.

Grangier, P., Juan Ariel Levenson, and Jean-Philippe Poizat. 1998. Quantum non-demolition measurements in optics. *Nature* 396: 537–542.

Grangier, P., Gérard Roger, and Alain Aspect. 1986. Experimental evidence for a photon anticorrelation effect on a beam splitter: A new light on single-photon interferences, (a) *Europhysics Letters* 1: 173–179; (b) reprinted in Meystre & Walls, ed. 1991, 336–342.

Grattan-Guiness, Ivor. 1990. *Convolutions in French Mathematics, 1800–1840*. Basel: Birkhäuser (Science Networks: Historical Studies, no. 2).

Greenberger, Daniel, Klaus Hentschel, and Friedel Weinert (eds.). 2009. *Compendium of quantum physics*. Concepts, Experiments, History and Philosophy, New York: Springer.

Greene, John C. 1957. Objectives and methods in intellectual history. *The Mississippi Valley Historical Review* 44 (1): 58–74.

Hagmann, Johannes Geert, and Wilhelm Füssl. 2012. *Konstruierte Wirklichkeit. Philipp Lenard 1862–1947. Biografie-Physik-Ideologie*. Munich: Deutsches Museum.

Hall, Rupert A. 1993. *All Was Light*. Clarendon: An introduction to Newton's opticks. Oxford.

Hallwachs, Wilhelm, and Ludwig Franz. 1888a. Über den Einfluß des Lichtes auf electrostatisch geladene Körper. *Annals of Physics* 3 (33): 301–312.

Hallwachs, Wilhelm, and Ludwig Franz. 1888b. Ueber die Electrisierung von Metallplatten durch Bestrahlung mit electrischem Licht. *Annals of Physics* 3 (34): 731–734.

Hallwachs, Wilhelm, and Ludwig Franz. 1889. Ueber den Zusammenhang des Electrizitätsverlustes durch Beleuchtung mit der Lichtabsorption. *Annals of Physics* 37: 666–675.

Hallwachs, Wilhelm, and Ludwig Franz. 1916. Die Lichtelektrizität. In *Handbuch der Radiologie*. Leipzig: Akad. Verlagsgesellschaft, vol. 3, 245–564, see also Marx 1916.

Halpern, Otto. 1924. Zur Theorie der Röntgenstrahlstreuung. *Zeitschrift für Physik* 30: 152–172.

Halpern, O., and Hans Thirring. 1928/29. Die Grundgedanken der neueren Quantentheorie, *Ergebnisse der exakten Naturwissenschaften* (a) 7: 384ff., (b) 8: 367–508.

Han, Moo-Young. 2014. *From Photons to Higgs*. A story of light. Singapur: World Scientific.

Hanbury, Brown Robert. 1991. *Boffin: A personal story of the early days of radar*. Radio astronomy and quantum optics. London: Hilger.

Hanbury Brown, R., and Richard Q. Twiss. 1956a. Correlation between photons in two coherent beams of light. *Nature* 177: 27–29.

Hanbury Brown, R., and Richard Q. Twiss. 1956b. A test of a new type of stellar interferometer on Sirius. *Nature* 178: 1046–1048.

Hanbury Brown, R., and Richard Q. Twiss. 1957. The question of correlation between photons in coherent beams of light. *Nature* 179: 1128–1129.

Hanson, Norwood Russel. 1963. *The concept of the positron*. Cambridge: Cambridge University Press.

Haroche, Serge. 2012. Controlling photons in a box and exploring the quantum to classical boundary (Nobel Lecture, Dec. 8, 2012), in *Nobel lectures in physics* 2012, Stockholm: Nobel Foundation, 63–107 (also freely available online).

Haroche, S., and Daniel Kleppner. 1989. Cavity quantum electrodynamics. *Physics Today* 42 (1): 24–30.

Haroche, S., and Jean-Michel Raimond. 2006. *Exploring the quantum. Atoms, cavities, and photons*. Oxford: Oxford University Press.

Harper, William L. 2011. *Isaac Newton's Scientific Method. Turning Data into Evidence about Gravity and Cosmology*, Oxford: Oxford University Press.

Harré, Rom. 1988. Parsing the amplitudes. In *Philosophical foundations of quantum field theory*, ed. Harvey R. Brown, and Rom Harré, 59–71. Oxford: Oxford University Press.

Hecht, Jeff. 2005. *Beam: The race to make the laser*. Oxford: Oxford University Press.

Heilbron, John Lewis. 1983. The origin of the exclusion principle. *Historical Studies in the Physical (and Biological)/Natural Sciences. Berkeley* 13: 261–310.

Heilbron, John Lewis. 1986. *The dilemmas of an upright man: Max Planck and the Fortunes of German Science*, (a) Berkeley: University of California Press, 1986; (b) 2nd, exp ed, 2000. Cambridge, MA: Harvard University Press.

Heisenberg, Werner. 1922. Zur Quantentheorie der Linienstruktur und der anomalen Zeemaneffekte. *Zeitschrift für Physik* 8: 273–297.

Heisenberg, Werner. 1925. Über quantentheoretische Umdeutung kinematischer und mechanischer Beziehungen. *Zeitschrift für Physik* 33: 879–893.

Heisenberg, Werner. 1927. Über den anschaulichen Inhalt der quantenmechanischen Kinematik und Mechanik. *Zeitschrift für Physik* 43: 172–198.

Heisenberg, Werner. 1930. *Physikalische Prinzipien der Quantentheorie* (lectures at the University of Chicago 1930. (a) Mannheim: BI Hochschultaschenbuch 1930 etc.; (b) revised ed., Heidelberg: Spektrum, 1991.

Heisenberg, Werner. 1959. *Physik und Philosophie*. Stuttgart: Hirzel, (a) 1st ed., 1959; (b) 5th ed., 1990.

Heisenberg, Werner. 1969. *Der Teil und das Ganze. Gespräche im Umkreis der Atomphysik*, Munich: Piper.

Heisenberg, W., and Hans Euler. 1936. Folgerungen aus der Diracschen Theorie des Positrons. *Zeitschrift für Physik* 98: 714–732.

Heisenberg, W., and Wolfgang Pauli. 1929. Zur Quantentheorie der Wellenfelder. *Zeitschrift für Physik* 56: 1–61 & 168–190.

Heisenberg, W., and Arnold Sommerfeld. 1922. Die Intensität der Mehrfachlinien und ihrer Zeemankomponenten. *Zeitschrift für Physik* 11: 131–154.

Heitler, Walter. 1936. *The Quantum Theory of Radiation*. Oxford: Oxford University Press, (a) 1st ed., 1936; (b) 2nd ed., 1944; reprinted 1947; (c) 3rd ed., 1957.

Hellmuth, Thomas, Herbert Walther, Arthur Zajonc, and Wolfgang Schleich. 1987. Delayed-choice experiments in quantum interference. *Physical Review A* 35: 2532–2541.

Henderson, Giles. 1980. Quantum dynamics and a semiclassical description of the photon. *American Journal of Physics* 48: 604–611, see also the comment by Berger 1981.

Hendrick, R.E., and Anthony Murphy. 1981. Atomism and the illusion of crises: The danger of applying Kuhnian categories to current particle physics. *Philosophy of Science* 48: 454–468.

Hendry, John. 1980. The development of attitudes to the wave-particle duality of light and quantum theory, 1900–1920. *Annals of Science* 37: 59–79.

Henny, M., S. Oberholzer, C. Strunk, T. Heizel, K. Ensslin, M. Holland, and C. Schönenberger. 1999. The fermionic Hanbury Brown and Twiss experiment. *Science* 284: 296–298.

Henri, Victor, and René Wurmser. 1913. Action des rayons ultraviolets sur l'eau oxygénée. *Comptes Rendus hebdomadaires des Séances de l'Académie des Sciences, Paris* 157: 126–128.

Hentschel, Ann M. 2005. *Albert Einstein: "Those Happy Bernese Years*. Bern: Stämpfli.

Hentschel, Klaus. 1990. *Interpretationen und Fehlinterpretationen der speziellen und der allgemeinen Relativitätstheorie durch Zeitgenossen Albert Einsteins*, vol. 6 (Diss. Univ. Hamburg, 1989), Basel: Birkhäuser Science Networks.

Hentschel, Klaus (ed.). 1996. *Physics and national socialism. An anthology of primary sources*. Basel: Birkhäuser; 2nd printing 2010.

Hentschel, Klaus. 1997. *The Einstein Tower–An intertexture of dynamic construction. Relativity Theory and Astronomy*, Stanford: Stanford University Press.

Hentschel, Klaus. 1998a. The interplay of instrumentation, experiment and theory: Patterns emerging from case studies on solar redshift 1890–1960. *Philosophy of Science* 64: S53–S64.

Hentschel, Klaus, and 1998b. *Zum Zusammenspiel von Instrument, Experiment und Theorie. Rotverschiebung im Sonnenspektrum und verwandten spektrale Verschiebungseffekte von 1880 bis 1960.* 2 vols. Hamburg: Verlag Dr Kovač.

Hentschel, Klaus. 2001. Das Brechungsgesetz in der Fassung von Snellius Rekonstruktion seines Entdeckungspfades und eine Übersetzung seines lateinischen Manuskriptes sowie ergänzender Dokumente. *Archive for History of the Exact Sciences, Berlin* 55 (4): 297–344.

Hentschel, Klaus. 2005. Einstein und die Lichtquantenhypothese. Zur stufenweisen Anreicherung der Bedeutungsschichten von 'Lichtquantum'. *Naturwissenschaftliche Rundschau* 58, no. 6: 311–319 & 7 (July): 363–371.

Hentschel, Klaus. 2005/07. The route to light quanta: From Newton to Einstein, in Anwar Hossain, ed., *Albert Einstein and S.N. Bose* [Proceedings of an intern. conference on Einstein and Bose 2005, ed. by the Bangladesh Academy of Sciences], Dhaka: Duck Int., 2007, 45–69.

Hentschel, Klaus. 2007a. *Unsichtbares Licht? Dunkle Wärme? Chemische Strahlen? Eine wissenschaftshistorische und -theoretische Analyse von Argumenten für das Klassifizieren von Strahlungssorten 1650–1925*. Diepholz: GNT-Verlag.

Hentschel, Klaus. 2007b. Light quanta: The maturing of a concept by the stepwise accretion of meaning, *Physics and Philosophy* 1(2): 1–20, https://eldorado.tu-dortmund.de/handle/2003/24257.

Hentschel, Klaus. 2008. Kultur und Technik in Engführung: Visuelle Analogien und Mustererkennung am Beispiel der Findung der Balmerformel, *Themenheft Forschung* [Universität Stuttgart] no. 4: 100–109.

Hentschel, Klaus. 2009a. Light quantum. In Greenberger et al. (eds.), 339–346.

Hentschel, Klaus. 2009b. Quantum jumps. In Greenberger et al. (eds.), 599–601.

Hentschel, Klaus. 2009c. Quantum theory, crisis period. In Greenberger et al. (eds.), 613–617.

Hentschel, Klaus. 2009d. Spin. In Greenberger et al. (eds.), 726–730.

Hentschel, Klaus. 2009e. 'jj-coupling', 'Landé g-factor', 'quantum number', 'Russell-Saunders coupling', 'Selection rules', 'Spectroscopy', 'spin', 'Stark effect', 'Zeeman effect' for *Compendium of Quantum Physics. Concepts, Experiments, History and Philosophy*, Dordrecht, Heidelberg, New York: Springer.

Hentschel, Klaus, and 2012/14. Review of Olivier Darrigol: A History of Optics from Greek Antiquity to the 19th Century, from 2012. Annals of Science. *London* 71 (4) (2014): 586–588.

Hentschel, Klaus. 2014a. Zur Rezeption von Vaihingers Philosophie des Als-Ob in der Physik. In *Fiktion und Fiktionalismus*, ed. Matthias Neuber, 161–186. Königshausen & Neumann: Beiträge zu Hans Vaihingers Philosophie des Als-Ob, Würzburg.

Hentschel, Klaus. 2014b. *Visual cultures in science and technology.* A comparative study. Oxford: Oxford University Press.

Hentschel, Klaus. 2015. Die allmähliche Herausbildung des Konzepts 'Lichtquanten'. *Berichte zur Wissenschaftsgeschichte* 38: 1–19.

Hentschel, K., and Renate Tobies. 1996. Interview mit Friedrich Hund (durchgeführt am 15. Dezember 1994 in Göttingen). *NTM* New Series 4: 1–18.

Hentschel, K., and R. Tobies (eds.). 1999. *Brieftagebuch zwischen Max Planck, Carl Runge, Bernhard Karsten und Adolf Leopold.* Berlin: ERS-Verlag, 1999; 2nd exp. ed., 2003.

Hentschel, K., and Magda Waniek. 2011. Nicht zu unterscheiden. *Physik-Journal* 10 (5): 39–43.

Hentschel, K., and Ning Yan Zhu (eds.). 2017. *Gustav Robert Kirchhoff's Treatise "On the Theory of Light Rays" (1882), English Translation, Analysis and Commentary.* Singapore: World Scientific.

Herivel, John. 1965. *The Background to Newton's Principia: A Study of Newton's Dynamical Researches in the Years 1664–84.* Oxford: Clarendon Press.

Hermann, Armin. 1968. Die frühe Diskussion zwischen Stark und Sommerfeld über die Quantenhypothese. *Centaurus* 12: 38–40.

Hermann, Armin. 1969/71. *Frühgeschichte der Quantentheorie (1899–1913),* (a) Baden: Verlag Physik, 1969; (b) in Engl. transl. by Claude W. Nash: *The Genesis of Quantum Theory (1899–1913),* Cambridge, MA: MIT Press, 1971.

Hertz, Heinrich. 1887. Ueber den Einfluss des ultravioletten Lichtes auf die electrische Entladung. *Annals of Physics* 267: 983–1000.

Heyl, Paul R. 1929. The history and present of the physicist's concept of light. *Journal of the Optical Society of America* 18: 189.

Hildebrand, Joel H. 1947. Gilbert Newton Lewis, (a) *Obituary Notices of Fellows of the Royal Society* 5 (March 1947); (b) reprinted in *Biographical Memoirs of the National Academy of Sciences* 1958: 210–235.

von Hirsch, Rudolf, and Robert Döpel. 1928. Die "Axialität" der Lichtemission und verwandte Fragen. *Physikalische Zeitschrift* 29: 394–398 (comment on Stark 1927).

Hoffmann, Dieter. 1982. Johannes Stark - eine Persönlichkeit im Spannungsfeld von wissenschaftlicher Forschung und faschistischer Ideologie. *Philosophie und Naturwissenschaften in Vergangenheit und Gegenwart* 22: 90–101.

Hoffmann, Dieter (ed.). 2010. *Max Planck und die moderne Physik.* Heidelberg: Springer.

Hoffmann, D., and Jost Lemmerich. 2000. *Quantum theory centenary.* Bad Honnef: Deutsche Physikalische Gesellschaft.

Hoffmann, D., and Mark Walker (eds.). 2007. *The German Physical Society in the Third Reich: Physicists between Autonomy and Accommodation.* Cambridge: Cambridge University Press.

Hofstadter, Douglas, & Fluid Analogies Research Group. 1995. *Fluid concepts and creative analogies.* Computer models of the fundamental mechanisms of thought. New York: Basic Books.

Holton, Gerald. 1973. *Thematic origins of scientific thought - Kepler to Einstein.* Cambridge, MA: Harvard University Press.

Holton, Gerald. 1975. On the role of themata in scientific thought. *Science* New Series 188 (4186): 328–334.

Holton, Gerald. 1984. *Themata.* Zur Ideengeschichte der Physik, Braunschweig: Vieweg.

Holton, Gerald. 2000. Millikan's struggle with theory. *Europhysics News* 31, no. 3 (2000), also on-line at http://www.europhysicsnews.org/articles/epn/abs/2000/03/epn00303/epn00303.html

Holweck, Fernand. 1927. *De la lumière au rayons X.* Paris: Presses Univ. de France.

Homberg, Wilhelm. 1708. Sur la force de presser & de poussez des rayons du Soleil, in *Histoire de l'Academie Royale des Sciences,* Paris, section on 'Histoire', p. 21 (just Fontenelle's abstract of Homberg's talk).

Home, Dipankar, and Andrew Whitaker. 2007. *Einstein's struggles with quantum theory.* A reappraisal. New York: Springer.

Hong, Chong-Ki, Zhe-Yu Ou, and L. Mandel. 1987. Measurement of subpicosecond time intervals between two photons by interference. *The Physical Review Letters, New Series* 59: 2044–2046; see also Ou & Mandel 1988.

Horst, Christoph auf der. 1998. Intellectual History in der Medizinhistoriographie am Beispiel des Naturbegriffs. In *Medizingeschichte*, ed. Thomas Schlich and Norbert Paul, 186–215. Campus: Aufgaben - Probleme - Perspektiven, Frankfurt.

Howard, Don. 2004. Who invented the Copenhagen interpretation? A study in mythology. *Philosophy of Science* 71: 669–682.

Howard, D., and John Stachel (eds.). 2000. *Einstein, The formative years 1879–1909*. Basel: Birkhäuser.

Huggett, Nick. 2000. Philosophical foundations of quantum field theory. *British Journal for the Philosophy of Science, Cambridge* 51: 617–637.

Hughes, Arthur Llewelyn. 1912. On the emission velocities of photo-electrons. *Philosophical Transactions of the Royal Society, London A* 212: 205–226.

Hughes, Arthur Llewelyn. 1914a. On the long-wave limits of the normal photoelectroc effect. *Philosophical Magazine, London* 6th ser., 27: 473–475.

Hughes, Arthur Llewelyn. 1914b. *Photo-electricity*. Cambridge: Cambridge University Press.

Hughes, A.L., and Lee Alvin DuBridge. 1932. *Photoelectric phenomena*. New York & London: McGraw Hill.

Hund, Friedrich. 1984. *Geschichte der Quantentheorie*, Mannheim: Bibliographisches Institut, (a) 1st ed., 1984; (b) expanded 4th ed., 1996.

Huygens, Christian. 1678/90. *Traité de la lumière*, written 1678, (a) publ. in Leiden: Pierre van der Aa, 1690, (b) reprinted in *Oeuvres Complètes* vol. XIX: *Mécanique théorique et physique 1666–1695*, La Hague: Nijjhoff, 1937, 463–466.

Irons, F.E. 2004. Reappraising Einstein's 1909 application of fluctuation theory to Planck's radiation. *American Journal of Physics* 72: 1059.

Ives, Herbert E. 1952. Derivation of the mass-energy-relation. *Journal of the Optical Society of America* 42 (8): 540–543.

Jacques, Vincent, E. Wu, Frédéric Grosshans, François Treussart, Philippe Grangier, Alain Aspect, and Jean-François Roch. 2007. Experimental realization of Wheeler's delayed-choice gedanken experiment. *Science* 315: 966–968.

Jacques, Vincent, E. Wu, Frédéric Grosshans, François Treussart, Philippe Grangier, Alain Aspect, and Jean-François Roch. 2008. Delayed-choice test of quantum complementarity with interfering single photons. *The Physical Review Letters, New Series* 100 (22): 220402.

Jammer, Max. 1961/74. *Concepts of mass in classical and modern physics*, (a) Cambridge, MA: Harvard University Press, 1961; (b) German trans.: *Der Begriff der Masse in der Physik*, Darmstadt: Wissenschaftliche Buchgesellschaft, 1974.

Jammer, Max. 1966. *The conceptual development of quantum mechanics*. New York: McGraw-Hill.

Jammer, Max. 1974. *The philosophy of quantum mechanics: The interpretations of quantum mechanics in historical perspective*. New York: Wiley.

Jánossy, Lajos, and Zs. Náray. 1957. The interference phenomena of light at very low intensities. *Acta Physica Academiae Scientiarum Humgaricae* 7: 493–495.

Janssen, Michel, and Christoph Lehner (eds.). 2014. *The Cambridge companion to Einstein*. Cambridge: Cambridge University Press.

Jauch, Josef M., and Fritz Rohrlich. 1955. *The theory of photons and electrons*. New York, Heidelberg & Berlin: Springer, (a) 1st ed., 1955; (b) 2nd ed., 1976, reprinted 1980.

Javan, Ali, E.A. Ballik, and W.L. Bond. 1962. Frequency characteristics of a continuous H-Ne optical maser. *Journal of the Optical Society of America* 52: 96–98.

Jaynes, Edwin Thompson. 1965. Is QED necessary? In *Proceedings of the Second Rochester Conference on Coherence and Quantum Optics*, ed. by Leonard Mandel & Emil Wolf, New York: Plenum, 21–23.

Jaynes, Edwin Thompson. 1973. Survey of the present status of neoclassical radiation theory. In *Coherence and quantum optics*, ed. Leonard Mandel, and Emil Wolf, 35–81. New York: Plenum.

Jenkin, John. 2004. William Henry Bragg in Adelaide. *Isis* 95: 58–90.

Jenkin, John. 2007. *William and Lawrence Bragg, Father and Son*. The Most Extraordinary Collaboration in Science, Oxford: Oxford University Press.

Joas, Christian, Christoph Lehner, and Jürgen Renn (eds.). 2008. *HQ1-Conference on the History of Quantum Physics, 350.* Berlin: Max Planck Institute preprint.

Johnson, George. 1999. *Strange beauty.* Murray Gell-Mann and the Revolution in Twentieth-Century physics, New York: Alfred Knopf.

Joly, John. 1921a. A quantum theory of vision. *Philosophical Magazine, London* 6 (41): 289–304.

Joly, John. 1921b. A quantum theory of colour vision. *Proceedings of the Royal Society, London B* 92: 219–232.

Jones, D.G.C. 1991. Teaching modern physics - misconceptions of the photon that can damage understanding. *Physics Education* 26 (2): 93–98.

Jones, D.G.C. 1994. Two slit interference - classical and quantum pictures. *European Journal of Physics* 15: 170–178.

Jordan, Pascual. 1941. *Physik und das Geheimnis des organischen Lebens.* Braunschweig: Vieweg.

Jordan, P., and Ralph de Kronig. 1936. Lichtquant und Neutrino. *Zeitschrift für Physik* 100: 569–583.

Jost, Res. 1995. *Das Märchen vom Elfenbeinernen Turm.* Reden und Aufsätze, Berlin: Springer.

Kadesch, William Henry. 1914. The energy of photo-electrons from sodium and potassium as a function of the frequency of the incident light. *The Physical Review. A Journal of Experimental and Theoretical Physics, New Series* 2 (3): 367–374.

Kahn, F.D. 1958. On photon coincidences and Hanbury Brown's interferometer. *Optica Acta* 5: 93–100.

Kaiser, David. 2004. *Drawing theories apart: The dispersion of Feynman diagrams in postwar physics.* Chicago: University of Chicago Press.

Kaiser, David (ed.). 2005. *Pedagogy and the practice of science: Historical and contemporary perspectives.* Cambridge. Mass.: MIT Press.

Kangro, Hans. 1970. *Vorgeschichte des Planckschen Strahlungsgesetzes.* Wiesbaden: Steiner.

Kangro, Hans. 1970/71. Ultrarotstrahlung bis zur Grenze elektrisch erzeugter Wellen: Das Lebenswerk von Heinrich Rubens *Annals of Science, London* 26: 235–259 & 28: 165–170.

Karlik, Berta, and Johannes Seidl. 2005. Schweidler Egon Ritter von. *Österreichisches Biographisches Lexikon 1815–1950* (12): 39–40.

Karlik, B., and Erich Schmid. 1982. *Franz Serafin Exner und sein Kreis.* Vienna: Verlag der Österreichischen Akademie der Wissenschaft.

Kastler, Alfred. 1983. On the historical development of the indistinguishability concept for microparticles. In *Old and New Questions in Physics, Cosmology and Theoretical Biology,* ed. A. van der Merwe, 607–623. New York & London: Plenum.

Kastner, Ruth E. 2012. *The transactional interpretation of quantum mechanics: The reality of possibility.* Cambridge: Cambridge University Press.

Katzir, Shaul. 2006. Thermodynamic deduction versus quantum revolution: The failure of Richardson's theory of the photoelectric effect. *Annals of Science* 63 (4): 447–469.

Kayser, Heinrich. 1936. In *Erinnerungen aus meinem Leben* [original typescript], ed. by Matthias Dörries & K. Hentschel, Munich: Deutsches Museum, 1996.

Keck, Jonas. 2016. *Die Einsteine des Einsteinjahres 2005 – zur vielstimmigen Darstellungen Albert Einsteins in populären Zeitschriften,* Bachelor's thesis at the GNT section, University of Stuttgart.

Keller, Ole. 2005. On the theory of spatial localization of photons. *Physics Reports* 411: 1–232.

Keller, Ole. 2007. Historical papers on the particle concept of light. *Progress in Optics* 50: 51–95.

Kelley, Donald. 1990. What is happening to the history of ideas? *Journal of the History of Ideas* 51 (1): 3–25.

Kelley, Donald. 2002. *The descent of ideas: The history of intellectual history.* Aldershot: Ashgate.

Kepler, Johannes. 1608. *Ausführlicher Bericht von dem .. 1607 erschienenen Haarstern oder Cometen vnd seinen Bedeutungen,* (a) Halle: Hynitzsch; (b) reprinted in *Opera Omnia,* vol. 7, ed. by Ch. Frisch, Frankfurt: Heyder & Zimmer, 1868, 23–41; (c) in *Gesammelte Werke,* vol. IV. 1941, ed. by Max Caspar & Franz Hammer, 58–76, 426–429.

Ketterle, Wolfgang. 1997. Bose-Einstein-Kondensate: eine neue Form von Quantenmaterie. *Physikalische Blätter* 53: 677.

Ketterle, Wolfgang. 2007. Bose Einstein condensation: Identity crisis for indistinguishable particles. In *Quantum Mechanics at the Crossroads*, ed. J. Evans, et al., 159–182. Berlin: Springer.

Kidd, Richard, James Ardini, and Anatoi Anton. 1989. Evolution of the modern photon, *American Journal of Physics* 57: 27–35; see also Raymer 1990.

Kienberger, Reinhard, and Ferenc Krausz. 2009. Elektronenjagd mit Attosekundenblitzen. *Spektrum der Wissenschaft* Feb: 32–40.

Kiesel, Harald, Andreas Renz, and Franz Hasselbach. 2002. Observation of Hanbury Brown-Twiss anticorrelation for free electrons. *Nature* 418: 392–394.

Kimble, H. Jeffrey, Mario Dagenais and Leonard Mandel. 1977. Photon antibunching in resonance fluorescence, (a) *The Physical Review Letters, New Series* 39: 691–695; (b) reprinted in Meystre and Walls, ed. 1991, 100–104; see also Dagenais and Mandel 1978.

Kirchhoff, Gustav Robert. 1860. Ueber das Verhältnis zwischen dem Emissionsvermögen und dem Absorptionsvermögen der Körper für Wärme und Licht, (a) *Annalen der Physik, Leipzig* (2) 109: 275–301 & pls. II-III; (b) Engl. transl.: On the relation between the radiating and absorbing powers of different bodies for light and heat, *Philosophical Magazine, London* (4) 20: 1–21; (c) reprinted in Ostwalds Klassiker series, no. 100, ed. by Max Planck.

Kirsten, Christa, and Hans-Jürgen Treder (eds.). 1979. *Albert Einstein in Berlin 1913–1933*, 2 vols.: *Teil I: Darstellung und Dokumente. Teil II: Spezialinventar.* Berlin: Akademie-Verlag.

Klaers, Jan, Julian Schmitt, Frank Vewinger, and Martin Weitz. 2010. Bose-Einstein condensation of photons in an optical microcavity, *Nature* 468: 545–547, see also Anglin (2010).

Klassen, Stephen. 2008. The photoelectric effect: Rehabilitating the story for the physics classroom. Science & Education, *Proceedings of the 2nd International Conference on Story in Science Teaching*, Munich, July 2008, 1–16.

Klassen, Stephen. 2011. The photoelectric effect: Reconstructing the story for the physics classroom, *Science and Education* 20: 719–731, see also Niaz, Klassen & Metz (2010).

Klassen, Stephen, Mansoor Niaz, Don Metz, Barbara McMillan, and Sarah Dietrich. 2012. Portrayal of the history of the photoelectric effect in laboratory instruction. *Science and Education* 21 (5): 729–743.

Kleeman, R.D. 1907. On the different kinds of γ-rays of radium and the secondary rays they produce. *Philosophical Magazine, London* 6 (15): 638–663.

Klein, Martin J. 1959. Ehrenfest's contribution to the development of quantum mechanics. *Proceedings of the Academy of Sciences Amsterdam B* 62: 41–62.

Klein, Martin J. 1964. Einstein and the wave-particle duality. *The Natural Philosopher* 3: 3–49.

Klein, Martin J. (ed.). 1970. *The Making of a Theoretical Physicist*. Paul Ehrenfest, Amsterdam: North Holland.

Klein, Martin J., et al. (eds.). 1993a. *Collected Papers of Albert Einstein*, vol. 3: *The Swiss Years: Writings, 1909–1911*, ed. by M.J. Klein, A.J. Kox, J. Renn, R. Schulmann, J. Buchwald, J. Eisenstaedt, D. Howard, J. Norton and T. Sauer, Princeton: Princeton University Press; suppl. English ed. trans. by Anna Beck, Don Howard, consultant.

Klein, Martin J., et al. (eds.). 1993b. *Collected Papers of Albert Einstein*, vol. 5: *The Swiss Years: Correspondence, 1902–1914*, ed. by M.J. Klein, A.J. Kox, R. Schulmann, P. Brenni, K. Hentschel and J. Renn, Princeton: Princeton University Press; suppl. English ed. trans. by Anna Beck, Don Howard, consultant.

Klein, Martin J., et al. (eds.). 1995. *Collected Papers of Albert Einstein*, vol. 4: *The Swiss Years: Writings, 1912–1914*, ed. by M.J. Klein, A.J. Kox, J. Renn, R. Schulmann, S. Bergia, J. Illy, M. Janssen, J.D. Norton, and T. Sauer, Princeton: Princeton University Press; suppl. English ed. trans. by Anna Beck, Don Howard, consultant.

Klein, Martin J., et al. (eds.). 1996. *Collected Papers of Albert Einstein*, vol. 6: *The Berlin Years: Writings 1914–1917*, ed. by A.J. Kox, M.J. Klein, R. Schulmann, J. Illy and J. Eisenstaedt, Princeton: Princeton University Press; suppl. English ed. trans. by Alfred Engel, Engelbert Schucking, consultant.

Klein, Oscar, and Yoshio Nishina. 1928. The scattering of light by free electrons according to Dirac's new relativistic dynamics. *Nature* 122: 398–399.

Klein, Oscar, and Yoshio Nishina. 1929. Über die Streuung von Strahlung durch freie Elektronen nach der neuen relativistischen Quantenmechanik von Dirac. *Zeitschrift für Physik, Braunschweig and Berlin* 52: 853–868.

Kleinert, Andreas. 1983. Das Spruchkammerverfahren gegen Johannes Stark. *Sudhoffs Archiv* 67: 13–24.

Kleinert, Andreas. 2002. "Die Axialität der Lichtemission und Atomstruktur". Johannes Starks Gegenentwurf zur Quantentheorie, in: *Chemie – Kultur – Geschichte. Festschrift für Hans-Werner Schütt anlässlich seines 65. Geburtstages*, Berlin/Diepholz, 213–222.

Kocher, Carl A., and Eugene D. Commins. 1967. Polarization correlation of photons emitted in an atomic cascade. *The Physical Review Letters, New Series* 18: 575–577.

Kojevnikov, Alexei. 2002. Einstein's fluctuation formula and the wave-particle duality. In *Einstein Studies in Russia*, ed. Yuri Balashov, and Vladimir Vizgin, 181–228. Boston: Birkhäuser.

Kormos Buchwald, Diana, et al. (eds.). 2015. *Collected Papers of Albert Einstein*, vol. 14: *The Berlin Years: Writings & Correspondence, April 1923–May 1925*, ed. by D. Kormos Buchwald, J. Illy, Z. Rosenkranz, T. Sauer, O. Moses, A.J. Kox, I. Unna & D. Lehmkuhl, Princeton: Princeton University Press; selected suppl. English ed. trans. by Ann M. Hentschel & Jennifer Nollar James, Klaus Hentschel, consultant.

Koselleck, Reinhart. 2000. *Zeitschichten. Studien zur Historik*, Frankfurt am Main: Suhrkamp.

Koselleck, Reinhart. 2010. *Begriffsgeschichten, Studien zur Semantik und Pragmatik der politischen und sozialen Sprache*, Frankfurt am Main: Suhrkamp.

Kox, A.J. (ed.). 2008. *The Scientific Correspondence of H.A. Lorentz*, New York: Springer.

Kragh, Helge S. 1985. The Fine Structure of Hydrogen and the Gross Structure of the Physics Community, 1916–26. *Historical Studies in the Physical (and Biological)/Natural Sciences, Berkeley* 15 (2): 67–125.

Kragh, Helge S. 1990. *Dirac: A scientific biography*. Cambridge: Cambridge University Press.

Kragh, Helge S. 1992. A sense of history: History of science and the teaching of introductory quantum theory. *Science and Education* 1: 349–363.

Kragh, Helge S. 2009. Bohr-Kramers-Slater theory. In Greenberger et al. (eds.), 62–64.

Kragh, Helge S. 2012. *Niels Bohr and the quantum atom: The Bohr Model of atomic structure 1913–1923*. Oxford: Oxford University Press.

Kragh, Helge S. 2014a. Photon – New light on an old name (just available online at http://arxiv.org/ftp/arxiv/papers/1401/1401.0293.pdf).

Kragh, Helge S. 2014b. The names of physics: Plasma, fission, photon. *European Physical Journal H* 39: 263–281.

Kragh, Helge S. 2014c. Planck's second quantum theory. In H.S. Kragh and James M. Overduin: *The weight of the vacuum*, 13–18. New York: Springer.

Krutkow, Yuri. 1914. Aus der Annahme unabhängiger Lichtquanten folgt die Wiensche Strahlungsformel, *Physikalische Zeitschrift* 15, 133–136, 363–364 (reply to Wolfke 1913/14).

Kubli, Fritz. 1971. Louis de Broglie und die Entdeckung der Materiewellen. *Archive for History of the Exact Sciences, Berlin* 7: 26–68.

Kuhn, Thomas S. 1962. *The Structure of Scientific Revolutions*, (a) 1st ed., Chicago: University of Chicago Press 1962; (b) 2nd ed. with a postscript, 1970.

Kuhn, Thomas S. 1978. *Black-body theory and the quantum discontinuity 1894–1912*. Oxford: Oxford University Press.

Kuhn, Wilfried, and Janez Strnad. 1995. *Quantenfeldtheorie*. Photonen und ihre Deutung, Braunschweig: Vieweg.

Kunz, Jakob. 1909. On the photoelectric effect of sodium-potassium alloy and its bearing on the structure of the ether. *The Physical Review. A Journal of Experimental and Theoretical Physics, New Series* New Series 29: 212–228.

Kunz, Jakob. 1911. On the positive potential of metals in the photoelectric effect and the determination of wave-length equivalent of Roentgen rays. The Physical Review. *A Journal of Experimental and Theoretical Physics, New Series* 33: 208–214.

Kurlbaum, Ferdinand, and Otto Lummer. 1898. Über eine Methode zur Bestimmung der Strahlung in absolutem Maß und der Strahlung des schwarzen Körpers zwischen 0 und 100 Grad, *Annalen der Physik, Leipzig* (3rd ser. 65: 746–760).

Kurlbaum, Ferdinand, and Otto Lummer. 1901. Der elektrisch geglühte 'schwarze' Körper. *Annalen der Physik, Leipzig* 310 (4th ser.) 5: 829–836.

Ladenburg, Rudolf. 1907. Ueber Anfangsgeschwindigkeit und Menge der photoelektrischen Elektronen in ihrem Zusammenhange mit der Wellenlänge des auslösenden Lichtes. *Physikalische Zeitschrift* 8: 590–594.

Ladenburg, Rudolf. 1909. Die neueren Forschungen zur Emission negativer Elektronen. *Jahrbuch der Radioaktivität und Elektronik* 6: 425–484.

Lamb, Willis E. 1995. Anti-photon. *Applied Physics B* 60: 77–84.

Lamb, W.E., and Marlan O. Scully. 1969. *The photoelectric effect without photons*, (a) CTS-QED-68-1 (preprint); (b) in *Polarisation, matière et Rayonnement*, 363–369. Paris: Presses Univ. de France.

Landau, Lev, and Evgeni Mikhailovich Lifschitz. 1947/73. *Lehrbuch der theoretischen Physik*, vol. III: *Quantenmechanik*. Berlin: Akademie-Verlag, 6th ed., 1979; 1st ed., Moscow, 1947; 3rd Russian ed., 1973.

Landau, L., and Rudolf Peierls. 1930. Quantenelektrodynamik im Konfigurationsraum. *Zeitschrift für Physik* 62: 188–200.

Landau, L., and Rudolf Peierls. 1931. Erweiterung des Unbestimmtheitsprinzips für die relativistische Quantentheorie. *Zeitschrift für Physik* 69: 56–69.

Langevin, Paul, and Maurice de Broglie (eds.). 1911/12. *La théorie du rayonnement et les quanta: Rapports et discussions de la réunion tenue à Bruxelles, du 30 octobre au 3 novembre 1911, sous les auspices de M.E. Solvay*, Paris: Gauthier-Villars, 1912.

Laub, Jacob. 1907/08. Zur Optik der bewegten Körper, *Annalen der Physik, Leipzig* 4th ser. 23, 1907: 738–744 & 25, 1908: 175–176.

Laue, Max von. 1907. Die Mitführung des Lichtes durch bewegte Körper nach dem Relativitätsprinzip, *Annalen der Physik, Leipzig* (4th ser. 23: 989–990).

von Laue, Max. 1911. *Das Relativitätsprinzip*. Braunschweig: Vieweg.

Laue, Max von. 1913. Das Relativitätsprinzip, *Jahrbücher der Philosophie* 1: 99–128, reprinted in Laue 1961, 264–293.

von Laue, Max. 1914. Die Freiheitsgrade von Strahlenbündeln. *Annals of Physics* 4 (44): 1197–1232.

Laue, Max von. 1920. *Über die Auffindung der Röntgenstrahlinterferenzen*, Karlsruhe: Müller, 1922 (Nobel lecture on 3 June 1920 Nobel prize for physics 1914).

Laue, Max von. 1923. Book review of Stark. 1922. (a) *Die Naturwissenschaften. Wochenschrift für die Fortschritte der Naturwissenschaften, der Medizin und der Technik, Berlin*, 11, no. 2, 12 Jan. 1923: 29–30; (b) English transl. in Hentschel, ed. 1996. doc. 2, pp. 6–7.

von Laue, Max. 1961. *Gesammelte Schriften und Vorträge*. Berlin: Friedr. Vieweg & Sohn.

Lawrence, Ernest O., and Jesse W. Beams. 1928. The element of time in the photoelectric effect. *The Physical Review. A Journal of Experimental and Theoretical Physics, Lancaster* 2 (32): 478–485.

Lebedew, Pjotr Nikolajewitsch. 1901. Untersuchungen über die Druckkräfte des Lichts, *Annalen der Physik, Leipzig*. (4th ser. 6: 433–458).

Lemmerich, Jost. 1987. *Zur Geschichte der Entwicklung des Lasers*. Berlin: ERS-Verlag.

Lenard, Philipp. 1894. Über Kathodenstrahlen in Gasen von atmosphärischem Druck und im äussersten Vakuum. *Annals of Physics* 3 (51): 225–267.

Lenard, Philipp. 1895. Ueber die Absorption der Kathodenstrahlen. *Annals of Physics* 3 (56): 255–275.

Lenard, Philipp. 1898. Ueber die electrostatischen Eigenschaften der Kathodenstrahlen. *Annals of Physics* 3 (64): 279–289.

Lenard, Philipp. 1900. Erzeugung von Kathodenstrahlen durch ultraviolettes Licht. *Annals of Physics* 4 (2): 359–375.

Lenard, Philipp. 1902. Ueber die lichtelektrische Wirkung, *Annalen der Physik, Leipzig* (4) 8: 149–198 & pl. I.

Lenard, Philipp. 1906. *Über Kathodenstrahlen*, Nobel lecture on 28 May 1906: (a) Heidelberg: C. Winters; (b) Engl. transl.: On cathode rays. In *Nobel Lectures. Physics 1901–1921*, 105–134. Amsterdam: Elsevier.

Lenard, Philipp. 1918. *Quantitatives über Kathodenstrahlen*. Heidelberg: C. Winters.

Lenard, Philipp. 1944. *Wissenschaftliche Abhandlungen*. Vol. 3: *Kathodenstrahlen, Elektronen, Wirkungen ultravioletten Lichtes*. Leipzig: Hirzel.

Lewis, Edward S. 1998. *A biography of distinguished Scientist Gilbert Newton Lewis*. New York: Edwin Mellen Press.

Lewis, Gilbert Newton. 1926a. The nature of light. *Proceedings of the National Academy of Sciences* 12: 22–29, see also Tolman & Smith 1926.

Lewis, Gilbert Newton. 1926b. Light waves and light corpuscles. *Nature* 117: 236–238.

Lewis, Gilbert Newton. 1926c. The conservation of photons. *Nature* 118: 874–875.

Lewis, Peter J. 2006. Conspiracy theories of quantum mechanics. *British Journal for the Philosophy of Science, Cambridge* 57 (2): 359–381.

Ley, Lothar. 2005. Einstein und die Natur des Lichts, speech held in Erlangen, http://www.didaktik. physik.uni-erlangen.de/fortbildung/fortbildung-archiv/Einstein-Ley.pdf.

Lorentz, Hendrik Antoon. 1892. La théorie électromagnétique de Maxwell et son application aux corps Mouvants, (a) *Archives néerlandaises des Sciences exactes et naturelles* 25: 363–552; (b) shortened in *Archives néerlandaises des Sciences exactes et naturelles* 35: 1–190; (c) reprinted in Lorentz's *Collected Papers*, 9 vols. 1935–39, vol. 2: 162–343 The Hague: Nijhoff.

Lorentz, Hendrik Antoon. 1895. *Versuch einer Theorie der elektrischen und optischenen Erscheinungen in bewegten Körpern*. Leiden: Brill.

Lorentz, Hendrik Antoon. 1909. *The Theory of Electrons and Its Applications to the Phenomena of Light and Radiant Heat. A Course of Lectures delivered in Columbia University, New York, in March and April 1906*, (a) New York: Columbia University Press, 1909; (b) 2nd ed., 1915; (c) Leipzig: Teubner, 1916.

Lorentz, Hendrik Antoon. 1910a. Die Hypothese der Lichtquanten. *Physikalische Zeitschrift* 11: 349–354.

Lorentz, Hendrik Antoon. 1910b. Alte und neue Fragen der Physik. *Physikalische Zeitschrift* 11: 1234–1257.

Loudon, Rodney. 1973. *Quantum theory of light*. Oxford: Clarendon Press, (a) 1st ed.; (b) 2nd ed., 1983; (c) Oxford University Press, 3rd ed., 2000.

Louradour, F., F. Reynaud, B. Colombeau, and C. Froehly. 1993. Interference fringes between two separate lasers, *American Journal of Physics* 61: 242–245 (see also Wallace 1994).

Lovejoy, Arthur O. 1936. *The Great Chain of Being*, Cambridge, MA, 1st ed.

Lovell, Bernard. 2002. Robert Hanbury Brown (1916–2002). *Nature* 416: 34.

Lukishova, Svetlana G. 2010. Valentin A. Fabrikant – Negative absorption, his 1951 patent application for amplification of electromagnetic radiation (ultraviolet, visible, infrared and radio spectral regions. and his experiments, *Journal of the European Optical Society* 5, http://www.jeos.org/index.php/jeos_rp/article/view/10045s.

Lummer, Otto. 1918. *Grundlagen, Ziele und Grenzen der Leuchttechnik (Auge und Lichterzeugung)*, Munich.

Lummer, O., and Ernst Pringsheim. 1897. Die Strahlung eines 'schwarzen' Körpers zwischen 100 und 1300°C. *Annals of Physics* 3 (63): 395–410.

Lummer, O., and Ernst Pringsheim. 1899a. Die Verteilung der Energie im Spektrum des schwarzen Körpers, *Verhandlungen der Deutschen Physikalischen Gesellschaft* 1: 23–41 bzw. (b) summary in *Verhandlungen der Gesellschaft Deutscher Naturforscher und Ärzte* 71: 58.

Lummer, O., and Ernst Pringsheim. 1899c. Die Verteilung der im Spektrum des schwarzen Körpers und des blanken Platins. *Verhandlungen der Deutschen Physikalischen Gesellschaft* 1899: 215–235.

Lummer, O., and Ernst Pringsheim. 1900. Über die Strahlung des schwarzen Körpers für lange Wellen. *Verhandlungen der Deutschen Physikalischen Gesellschaft* 2: 163–180.

Mach, Ernst. 1921. *Die Prinzipien der physikalischen Optik. Historisch und erkenntnispsychologisch entwickelt.* Leipzig: Barth.

Magyar, George, and Leonard Mandel. 1963. Interference fringes produced by superposition of two independent maser light beams. *Nature* 198: 255–256.

de Mairan, Jean Dartons. 1747. Éclairissement sur l'impulsion des rayons solaires, *Histoire de l'Académie Royale des Sciences* 1747: 423–435.

Mandel, Leonard. 1986. Non-classical states of the electromagnetic field. *Physica Scripta T* 12: 34–42.

Mandel, L., and Emil Wolf. 1961. Correlation in the fluctuating outputs from two square-law detectors illuminated by light of any state of coherence and polarization, *The Physical Review. A Journal of Experimental and Theoretical Physics, New Series* (2) 124,6: 1696–1702.

Mandelbaum, Maurice. 1965. The history of ideas, intellectual history and the history of philosophy. *History and Theory, Suppl* 5: 33–66.

Martinez, Alberto A. 2004. Ritz, Einstein and the emission hypothesis, *Physics in Perspective* 6: 4–28.

Marx, Erich. 1905. Die Geschwindigkeit der Röntgenstrahlen, (a) *Physikalische Zeitschrift* 6: 768–777; (b) *Annalen der Physik, Leipzig* (4) 20: 677–722.

Marx, Erich. 1916. Entwicklung der Lichtelektrizität von Januar 1914 bis Oktober 1915, *Handbuch der Radiologie*, vol. 3, Leipzig: Akad. Verlagsgesellschaft, 565–588 (suppl. to Hallwachs 1916).

Mascart, Eleuthère Eli Nicolas. 1889–94. *Traité d'Optique*. Paris: Gauthier-Villars, 3 text vols. 1889–1893 & 1 plate vol. 1894.

Maxwell, James Clerk. 1873. *Treatise on Electricity and Magnetism*, (a) Oxford: Clarendon, 1st ed., 1873; (b) rev. ed. prep. by Maxwell until 1879, finished by William Davidson Niven, publ. 1881; (c) 3rd ed. arranged by J.J. Thomson 1892, also reprinted New York: Dover, 1954 etc.

McCormmach, Russell. 1967. J.J. Thomson and the structure of light. *British Journal for the History of Science, London* 3: 362–387.

McCormmach, Russell. 1968/69. John Michell and Henry Cavendish: weighing the stars. *British Journal for the History of Science, London* 4: 126–155.

McCormmach, Russell. 1982. *Night thoughts of a classical physicist*. Cambridge, MA: Harvard University Press.

McCormmach, Russell. 2012. *Weighing the world. The Reverend John Michell of Thornhill*. Dordrecht, Heidelberg, London & New York: Springer.

McEvoy, John G. 2010. *The historiography of the chemical revolution: Patterns of interpretation in the history of science*. London: Routledge.

Mehra, Jagdish. 1994. *The beat of a different drum - The life and science of Richard Feynman*. Oxford: Clarendon Press.

Mehra, J., and Helmut Rechenberg. 1982–2001. *The Historical Development of Quantum Theory*. New York: Springer, 6 vols., esp. vol. 1: *The Quantum Theory of Einstein, Bohr, Planck and Sommerfeld 1900–1925 — Its Foundation and the Rise of Its Difficulties*, 1982; vol. 5: *Erwin Schrödinger and the Rise of Wave Mechanics*, 1987; vol. 6: *The Completion of Quantum Mechanics 1926–1941*, 2001.

Meidinger, Walter. 1934. Photochemie. *Chemiker-Zeitung* 58: 629.

Meier, Helmut Günter. 1971. Begriffsgeschichte, in *Historisches Wörterbuch der Philosophie* 1, 788–808.

Meitner, Lise. 1922. Über die Wellenlänge der γ-Strahlen. *Die Naturwissenschaften. Wochenschrift für die Fortschritte der Naturwissenschaften, der Medizin und der Technik, Berlin* 10: 381–384.

Melvill, Thomas. 1754. Discourse concerning the cause of the different refrangibility of the rays of light, *Philosophical Transactions of the Royal Society, London* 48: 261–270.

Melvill, Thomas. 1756. Observations on light and colours, *Essays and Observations Physical and Literary, read before the Philosophical Society in Edinburgh* 2: 12–90.

Melvill, Thomas. 1784. On the means of discovering the distance, magnitude etc. of the fixed stars, in consequence of the diminution of the velocity of their light..., *Philosophical Transactions of the Royal Society, London* 74: 35–57.

Meyenn, Karl von (ed.). 1979/85. *Wolfgang Pauli: Wissenschaftlicher Briefwechsel mit Bohr, Einstein, Heisenberg u. a.*. Berlin: Springer (vol. 1) 1919–1929, 1979; (vol. 2) 1930–1939.

Meyenn, Karl von (ed.). 1980/81. Paulis Weg zum Ausschließungsprinzip, *Physikalische Blätter* 36: 293–298, 37: 13–19.

Meystre, Pierre, and Daniel F. Walls (eds.). 1991. *Nonclassical effects in quantum optics: A collection of reprints*. New York: American Institute of Physics.

Michell, John. 1784. On the Means of Discovering the Distance, Magnitude, etc. of the Fixed Stars, in Consequence of the Diminution of the Velocity of Their Light, in Case Such a Diminution Should be Found to Take Place in any of Them, and Such Other Data Should be Procured from Observations, as Would be Farther Necessary for That Purpose. In a Letter to Henry Cavendish, Esq. F.R.S. and A.S., *Philosophical Transactions of the Royal Society, London* 74: 35–57.

Michelson, Albert Abraham. 1903. *Light waves and their uses*. Chicago: Chicago University Press.

Miller, Arthur I. (ed.). 1994. *Early quantum electrodynamics. A sourcebook*. Cambridge: Cambridge University Press.

Millikan, Robert A. 1913. Atomic theories of radiation. *Science*, New Series 37: 119–133.

Millikan, Robert A. 1914. A direct determination of h. *The Physical Review. A Journal of Experimental and Theoretical Physics*, New Series 2 (4): 73–75.

Millikan, Robert A. 1916a. Einstein's photoelectric equation and contact electromotive force. *The Physical Review. A Journal of Experimental and Theoretical Physics, New Series* 2 (7): 18–32.

Millikan, Robert A. 1916b. A direct photoelectric determination of Planck's h. *The Physical Review. A Journal of Experimental and Theoretical Physics, New Series* 2 (7): 355–388.

Millikan, Robert A. 1917. *The electron, its isolation and measurement and the determination of some of its properties*. Chicago: University of Chicago Press.

Millikan, Robert A. 1924. The electron and the light-quant from the experimental point of view, (Nobel lecture on 23 May 1924), in: *Nobel Prizes in Physics 1922–41, incl. presentation speeches and the laureates' biographies*, Stockholm, Nobel Foundation, 1964, 54–66.

Millikan, Robert A. 1935. *Electrons (+ and -), protons, photons, neutrons, and cosmic rays*, (a) Chicago: University of Chicago Press, 1st ed.; (b) 2nd exp. ed., 1947.

Millikan, Robert A. 1949. Albert Einstein on his seventieth birthday. *Reviews in Modern Physics* 21: 343–345.

Millikan, Robert A. 1950. *The autobiography*. New York: Prentice Hall.

Millikan, R.A., et al. 1931. Professor Einstein at the California Institute of Technology, addresses at the dinner in his honor. *Science* 73: 375–379.

Milner, Peter. 2013. A short history of spin. In *Proceedings of Science* 3, http://pos.sissa.it/archive/conferences/182/003/PSTP2013_003.pdf.

Milonni, Peter. 1984. Wave-particle dualism. In *The wave-particle dualism. A tribute to Louis de Broglie on his 90th birthday*, 27–67. Dordrecht: Reidel, S.

Milonni, Peter. 2007. Field quantization in optics. *Progress in Optics* 50: 97–135.

Mittelstraß, Jürgen, et al. (eds.). 1980–96. *Enzyklopädie Philosophie und Wissenschaftstheorie*, 4 vols. Mannheim: Bibliographisches Institut.

Monaldi, Daniela. 2009. A note on the prehistory of indistinguishable particles. *Studies in History and Philosophy of Modern Physics* 49: 383–394.

Morgan, Brian Lealan, and Leonard Mandel. 1966. Measurement of photon bunching in a thermal light beam, (a) *The Physical Review Letters, New Series* 16 (12): 1012–1015; (b) reprinted in Meystre & Walls, ed. 1991, 33–37.

Morse, Philip M. 1982. John Clarke Slater 1900–1976. *Biographical Memoirs of the National Academy of Sciences*: 296–321.

Müller, Ernst, and Falko Schmieder. 2008. *Begriffsgeschichte der Naturwissenschaften: Zur historischen und kulturellen Dimension naturwissenschaftlicher Konzepte*. Berlin: Walter de Gruyter.

Muthukrishnan, Ashok, Marlan O. Scully, and M. Suhail Zubairy. 2003. The concept of the photon - revisited. *Optics and Photonics. New Trends* 3 (1): S18–S27.

Natanson, Ladislas. 1911. Statistische Theorie der Strahlung, *Physikalische Zeitschrift* 12: 659–666 (in Polish in *Bulletin de l'Académie des Sciences de Cracovie* A 1911: 134–148).

Natarajan, Vasant, V. Balakrishnan, and N. Mukunda. 2005. Einstein's miraculous year. *Resonance* 1 (March): 35–56.

Navarro, Jaome. 2005. J.J. Thomson on the nature of matter: Corpuscles and the continuum. *Centaurus* 47: 259–282.

Navarro, Luiz, and Enric Pérez. 2004. Paul Ehrenfest on the necessity of quanta 1911. *Archive for History of the Exact Sciences, Berlin* 58 (2): 97–141.

Needell, Alan A. 1980. *Irreversibility and the failure of classical dynamics: Max Planck's work on the quantum theory 1900–1915.* Ann Arbor: Michigan, University Microfilms.

Newton, Isaac. 1671/72. New theory about light and colors, (a) *Philosophical Transactions of the Royal Society, London* 6, no. 80: 3075–3087; (b) reprinted in I.B. Cohen, ed. 1958, 47–59.

Newton, Isaac. 1672. Answer upon some considerations upon the Doctrine of light and colours. *Philosophical Transactions of the Royal Society, London* 6 (88): 5083–5103.

Newton, Isaac. 1675. An hypothesis explaining the properties of light, submitted to the *Royal Society* in 1675, (a) publ. in Thomas Birch: *The History of the Royal Society*, vol. 3, London, 1757, 247–305, esp, 248–269; (b) reprinted in I.B. Cohen, ed. 1958, 177–199.

Newton, Isaac. 1687. *Philosophiae Naturalis Principia Mathematica*, (a) London, 1687; (b) in Engl. transl. by Andrew Motte: *The mathematical principles of natural philosophy*, London, 1729; (c) transl. by I. Bernard Cohen & Anne Whitman: *Mathematical principles of natural philosophy*. Berkeley: University of California Press, 1999.

Newton, Isaac. 1704. *Opticks*. New York: Dover reprinting of the 1704-edition, with a new preface by Albert Einstein, ed. by I.B. Cohen 1954.

Newton, Théodore Dudell, and Eugene Paul Wigner. 1949. Localized states for elementary systems. *Reviews of Modern Physics* 21: 400–406.

Niaz, Mansoor, Stephen Klassen, and Don Metz. 2010. A historical reconstruction of the photoelectric effect and its implication for general physics textbooks, *Science and Education* 94: 903–931, see also Klassen et al. (2012) and Klassen (2008), (2011).

Nichols, Ernest Fox, and Gordon Ferrie Hull. 1901. A preliminary communication on the pressure of heat and light radiation. *The Physical Review. A Journal of Experimental and Theoretical Physics, New Series* 2 (13): 307–320.

Nichols, Ernest Fox, and Gordon Ferrie Hull. 1903. The pressure due to radiation. *The Astrophysical Journal* 17: 315–351.

Nickelsen, Kärin. 2013. *Explaining photosynthesis. Modelling biochemical mechanisms 1840–1960*. Dordrecht: Springer.

Nickelsen, Kärin. 2016. Otto Warburg, die Quanten und die Photosynthese. *Acta Historica Leopoldina* 65: 37–63.

Niedderer, Hans. 1982. Unterschiedliche Interpretationen des Fotoeffekts. Eine historisch-wissenschaftstheoretische Fallstudie. *Der Physikunterricht* 16: 39–46.

Nisio, Sigeko. 1973. The formation of the Sommerfeld quantum theory of 1916. *Japanese Journal in the History of Science* 12: 703–713.

Nogues, G., A. Rauschenbeutel, S. Osnaghi, M. Brune, J.-M. Raimond, and S. Haroche. 1999. Seeing a photon without destroying it. *Nature* 400: 239–242.

Norton, John. 2004. Einstein's investigations of Galilean covariant electrodynamics prior to 1905. *Archive for History of the Exact Sciences, Berlin* 59: 45–105.

Norton, John. 2008. *Einstein's miraculous argument of 1905: The thermodynamic grounding of light quanta, HQ1*, Max-Planck-Institute Preprint, 63–78.

Norton, John. 2016. How Einstein did not discover. *Physics in Perspective* 18 (3): 249–282.

Nudds, John R. 1986. The life and work of John Joly (1857–1933). *Irish Journal of Earth Sciences* 8: 81–94.

O'Brien, Allyson. 2010. *Wheeler's delayed choice experiment*, only available online at http://einstein.drexel.edu/~bob/TermPapers/WheelerDelayed.pdf.

Ogborn, Jon, Gunther Kress, Isabel Martins, and Kieran McGillicuddy. 1996. *Explaining science in the classroom*. Buckingham: Open University Press.

Okun, Lev Borisovich. 2008. Photon: history, mass, charge, online since 2008, https://ia601009.us.archive.org/23/items/arxiv-hep-ph0602036/hep-ph0602036.pdf.

Olesko, Kathryn M. 1991. *Physics as a Calling: Discipline and Practice in the Koenigsberg Seminar for Physics*. Ithaca, NY: Cornell University Press.

Ornstein, Leonard Salomon, and Frits Zernike. 1920. Energiewisselingen der zweite Straling en light-atomen?, *Verslagen van de gewone vergaderingen der Afdeling Natuurkunde. Koniklijke Nederlandse Akademie van Wetenschappen te Amsterdam* 28: 281–292.

Ou, Zhe-Yu, and Leonard Mandel. 1988. Violation of Bell's inequality, and classical probability in a two-photon correlation experiment, (a) *The Physical Review Letters, New Series* 61: 50–53; (b) reprinted in Meystre & Walls, ed. 1991, 386–389; see also Hong, Ou, and Mandel 1987. *Oxford English Dictionary*. 1989), Oxford: Oxford University Press, 2nd ed.

Pais, Abraham. 1948. *Developments in the theory of electrons*. Institute for Advanced Study: Princeton.

Pais, Abraham. 1979. Einstein, and the quantum theory. *Reviews of Modern Physics* 51: 863–914.

Pais, Abraham. 1982. *Subtle is the Lord*. Oxford: Oxford University Press.

Pais, Abraham. 1991. *Niels Bohr's times in physics, philosophy, and polity*. Oxford: Oxford University Press.

Park, David. 1997. *The fire within the eye: A historical essay on the nature and meaning of light*. Princeton: Princeton University Press.

Passon, Oliver. 1997. Photonen in der Schulphysik, *Physik in der Schule*, 54, nos. 7/8: 257–258.

Passon, Oliver. 2014. Was sind Elementarteilchen und gibt es virtuelle Photonen? *Physik in der Schule* 63 (5): 44.

Passon, O., amd Johannes Grebe-Ellis. 2016. Misguided quasi-history in the teaching of the photon concept, unpubl. Mss., (received by the present author on 28 Sep. 2016, after having already finished this book).

Passon, O., and Johannes Grebe-Ellis. 2017. Planck's radiation law, the light quantum, and the prehistory of indistinguishability in the teaching of quantum mechanics. *European Journal of Physics* 38 (3):

Paul, Harry. 1985. *Photonen. Experimente und ihre Deutung*, (a) Berlin: Akademie-Verlag, 1st ed.; (b) exp. ed. with new subtitle: *Photonen. Eine Einführung in die Quantenoptik*. Wiesbaden: Vieweg, 1995.

Paul, Harry. 1986. Interference between independent photons. *Reviews of Modern Physics* 58: 209–223.

Pauli, Wolfgang. 1924/25. Über den Einfluß der Geschwindigkeitsabhängigkeit der Elektronenmasse auf den Zeemaneffekt, *Zeitschrift für Physik* 31: 373–385 (received 2 Dec. 1924, publ. 19 Feb. 1925).

Pauli, Wolfgang. 1925. Über den Zusammenhang des Abschlusses der Elektronengruppen im Atom mit der Komplexstruktur der Spektren, *Zeitschrift für Physik* 31: 765–783 (received 16 Jan., publ. March 21).

Pérez, Enric, and Tilman Sauer. 2010. Einstein's quantum theory of the monatomic ideal gas: non-statistical arguments for a new statistics. *Archive for History of the Exact Sciences* 64: 561–612.

Pessoa, Osvaldo Jr. 2000. Historias contrafactuais: o surgimento da fisica quantica. *Estudos Avançados* 14 (39): 175–204.

Pfleegor, Robert L., and Leonard Mandel. 1967. Interference of independent photon beams. *Physical Review* 2 (159): 1084–1088.

Pfleegor, Robert L., and Leonard Mandel. 1968. Further experiments on interference of independent photon beams at low light levels. *Journal of the Optical Society of America* 58 (7): 946–950.

Pipkin, Francis M. 1978. Atomic physics tests of the basic concepts in quantum mechanics. *Advances in Atomic and Molecular Physics* 4: 281–340.

Planck, Max. 1897/99. Über irreversible Strahlungsvorgänge, erste bis fünfte Mitteilung. *Sitzungs-berichte der Königlich-Preußischen Akademie der Wissenschaften* (Berlin), (I): 57–68 (1897); (II): 715–717 (1897); (III): 1122–1145 (1897); (IV): 449–476 (1898); (V): 440–480 (1899).

Planck, Max. 1900a. Über eine Verbesserung der Wienschen Spektralgleichung. *Verhandlungen der Deutschen Physikalischen Gesellschaft* 2: 202–204.

Planck, Max. 1900b. Zur Theorie des Gesetzes der Energieverteilung im Normalspektrum. *Verhandlungen der Deutschen Physikalischen Gesellschaft* 2: 237–245.

Planck, Max. 1901. Über das Gesetz der Energieverteilung im Normalspektrum. *Annals of Physics* 4 (4): 553–563.

Planck, Max. 1907. Zur Dynamik bewegter Systeme, (a) *Situngsberichte der Preußischen Akademie der Wissenschaften*, math.-physik. Klasse 1907: 542–570; (b) *Annalen der Physik, Leipzig* (4. 26. 1908): 1–34.

Planck, Max. 1910. Zur Theorie der Wärmestrahlung, (a) *Annalen der Physik, Leipzig* (4) 31: 758–768; (b) reprinted in Planck 1958. vol. II, 237–247.

Planck, Max. 1911. Eine neue Strahlungshypothese, *Verhandlungen der Deutschen Physikalischen Gesellschaft* 13: 138–148; reprinted in Planck 1958: 249–259.

Planck, Max. 1912. Über die Begründung das Gesetzes des schwarzen Strahlung, *Annalen der Physik, Leipzig* (4) 37: 642–656; reprinted in Planck 1958, 287–301.

Planck, Max. 1913. *Vorlesungen über die Theorie der Wärmestrahlung*. Leipzig: Barth.

Planck, Max. 1919. Das Wesen des Lichts (Talk held at the plenary session of the Kaiser-Wilhelm-Gesellschaft on 28 Oct. 1919). In Physikalische Rundblicke. *Gesammelte Reden und Aufsätze von Max Planck, Leipzig, Hirzel* 1922: 129–147.

Planck, Max. 1923. Die Bohrsche Atomtheorie. *Die Naturwissenschaften. Wochenschrift für die Fortschritte der Naturwissenschaften, der Medizin und der Technik, Berlin* 11: 535–537.

Planck, Max. 1924. Über die Natur der Wärmestrahlung. *Annals of Physics* 4 (73): 272–288.

Planck, Max. 1927. Die physikalische Realität der Lichtquanten. *Die Naturwissenschaften. Wochenschrift für die Fortschritte der Naturwissenschaften, der Medizin und der Technik, Berlin* 15: 529–531.

Planck, Max. 1943. Zur Geschichte der Auffindung des physikalischen Wirkungsquantums. *Die Naturwissenschaften. Wochenschrift für die Fortschritte der Naturwissenschaften, der Medizin und der Technik, Berlin* 31: 53–84.

Planck, Max. 1948. *Wissenschaftliche Selbstbiographie*. Leipzig: Johann Ambrosius Barth.

Planck, Max. 1958. *Physikalische Abhandlungen und Vorträge*, Braunschweig: Vieweg, 3 vols.

Plotnikow, Johannes. 1920. *Allgemeine Photochemie*. Berlin & Leipzig: De Gruyter; 2nd ed. 1936.

Pohl, Robert Wichard. 1912. *Die Physik der Röntgenstrahlen*. Braunschweig: Vieweg.

Pohl, R.W., and Peter Pringsheim. 1913. On the long-wave limit of the normal photoelectric effect. *Philosophical Magazine, London* 6 (26): 1017–1024.

Poincaré, Henri. 1900. La théorie de Lorentz et le principe de réaction, (a) *Archives Néerlandaises des Sciences Exactes et Naturelles* (2) 5: 252–278; (b) in Engl. transl.: The Theory of Lorentz and The Principle of Reaction, on-line at http://physicsinsights.org/poincare-1900.pdf.

Polkinghorne, John. 2002. *Quantum theory: A very short introduction*. Oxford: Oxford University Press.

Pont, Jean-Claude (ed.). 2012. *Le Destin Douloureux de Walther Ritz, physicien théoricien de génie*. Sion: Archives de l'Etat de Valais.

Popkin, Gabriel. 2017. China's quantum satellite achieves 'spooky action' at record distance, *Science news:* 06.

Popper, Karl Raymund. 1934. Zur Kritik der Ungenauigkeitsrelationen, *Die Naturwissenschaften. Wochenschrift für die Fortschritte der Naturwissenschaften, der Medizin und der Technik* 22, no. 48: 807–808; reprinted. *Frühe Schriften*, Tübingen: Mohr 2006: 393–397.

Popper, Karl Raymund. 1935. *Logik der Forschung*. Vienna: Springer (= Schriften zur wiss. Weltauffassung, 9).

Poynting, John Henry. 1884. On the transfer of energy in the electromagnetic field. *Philosophical Transactions of the Royal Society, London* 175: 343–361.

Principe, Lawrence M. 2008. Wilhelm Homberg et la chimie de la lumière, *Methodos* 8, online at http://methodos.revues.org/1223.

Przibram, Karl (ed.). 1963. *Schrödinger, Planck, Einstein, Lorentz - Briefe zur Wellenmechanik*, Vienna: Springer.

Purcell, Edward M. 1956. Question of correlation between photons in coherent light rays, *Nature* 178: 1148–1150 (comment on Brannen & Ferguson 1956).

Pyenson, Lewis. 1985. *The Young Einstein. The Advent of Relativity*, Bristol: Hilger.

Rahhou, Adnane, Fatiha Kaddari, Abdelrhani Elachquar, and Mohammed Oudrhiri. 2015. The role of the history of science in the understanding of the concept of light, Procedia. *Social and Behavioral Sciences* 191: 2593–2597.

Raman, Chandrasekhar Venkata. 1930. The molecular scattering of light, Nobel Lecture, 11 Dec. 1930. In *Nobel Lectures Physics, 1922–41*, Amsterdam: North Holland, 1965, 267–275.

Ramsauer, Carl. 1914. Über die lichtelektrische Geschwindigkeitsverteilung und ihre Abhängigkeit von der Wellenlänge. *Annals of Physics* 4 (45): 1121–1159.

Raymer, Michael G. 1990. Observation of the modern photon, *American Journal of Physics* 58: 11 (comment on Kidd et al. (1989)).

Redhead, Michael, and Paul Teller. 1992. Particle labels and the theory of indistinguishable particles in quantum mechanics. *British Journal for the Philosophy of Science, Cambridge* 43: 201–218.

Regener, Erich. 1915. *Über Kathoden-, Röntgen- und Radiumstrahlen*. Berlin & Vienna: Urban & Schwarzenberg.

Richardson, Owen W. 1914. Note on the direct determination of h. *The Physical Review. A Journal of Experimental and Theoretical Physics, New Series*. 2 (4): 522–523.

Richardson, O.W., and Karl T. Compton. 1912. The photoelectric effect, (a) *Philosophical Magazine*, London (6) 24: 575–594; (b) *The Physical Review. A Journal of Experimental and Theoretical Physics, New Series* 34: 393–396.

Richardson, O.W., and F.J. Rogers. 1915. The photoelectric effect III. *Philosophical Magazine, London* 6 (29): 618–623.

Richter, Melvin. 1987. Begriffsgeschichte and the history of ideas. *Journal of the History of Ideas* 48: 247–263.

Rigden, John. 2005. Einstein's revolutionary paper. *Physics World* 18 (April): 18–19.

Ringbauer, Martin. 2017. *Exploring quantum foundations with single photons*. Heidelberg: Springer.

Ritter, Joachim, Karlfried Gründer, and Gottfried Gabriel (eds.). 1971–2007. *Historisches Wörterbuch der Philosophie*. Basel: Schwabe Verlag, 13 vols.

Ritz, Walther. 1908a. Du rôle de l'éther en physique. *Rivista di Scienza* 4: 260–274.

Ritz, Walther. 1908b. Recherches critiques sur l'Électrodynamique Générale, *Annales de chimie et de physique. Paris* 13: 145.

Ritz, Walther. 1911. *Gesammelte Werke/Oeuvres*, ed. by the *Société Suisse de Physique*. Paris: Gauthier-Villars.

Ritz, W., and Albert Einstein. 1909. Zum gegenwärtigen Stand des Strahlungsproblems, *Physikalische Zeitschrift* 10: 323–324 (see also *Physikalische Zeitschrift* 10: 185, 224 and 9. 1908: 903).

Robison, John. 1790. On the motion of light, as affected by refracting and reflecting substances which are also in motion, *Transactions of the Royal Society of Edinburgh* 2: 83–111.

Robison, John. 1797. Optics, *Encyclopedia Britannica*, 3rd ed., vol. 13, 231–364.

Robotti, Nadia. 1990. Quantum numbers and electron spin. *Archives Internationales d'Histoire des Sciences* 40: 305–331.

Roditschew, Wladimir Ivanovič, and U.I. Frankfurt (eds.). 1977. *Die Schöpfer der physikalischen Optik*. Berlin: Akademie-Verlag.

Römer, Ole. 1676. Demonstration touchant le mouvement de la lumière..., *Journal des Savants*, 7 Dec: 233–236.

Roychoudhuri, Chandrasekhar. 2006. Do we count indivisible photons or discrete quantum events experienced by detectors? In *Advanced Photon Counting Techniques* (SPIE vol. 6372, no. 29).

Roychoudhuri, Chandrasekhar. 2008. What is a photon? In Roychoudhuri et al. (eds.), 129–141.

Roychoudhuri, Chandrasekhar. 2009. Indivisibility of the photon In Roychoudhuri et al. (eds.) 2009, 1–9.

Roychoudhuri, Chandrasekhar. 2015. Replacing the paradigm shift model in physics with continuous evolution of theory by frequent iterations. In *Sixty Years after Albert Einstein (1879–1955)*, ed. Charles Tandy, 157–180. Ria Univ. Press. Ann Arbor.

Roychoudhuri, C., and Rajarshi Roy. 2003. The Nature of Light. What is a Photon?, *Optics and Photonics News Supplement* OPN-Trends, Oct.: S1–S35.

Roychoudhuri, C., Al F. Kracklauer, and Katherine Creath (eds.). 2008. *The Nature of light. What is a photon?*. Boca Raton: CRC Press.

Roychoudhuri, C., A.F. Kracklauer, and Andrei Yu. Khrennikov (eds.). 2009. *The nature of light. What are photons? III*. Bellingham, WA: SPIE, http://spie.org/Publications/Proceedings/Volume/7421.

Roychoudhuri, C., A.F. Kracklauer, and Hans de Raedt (eds.). 2015. *The Nature of Light. What are Photons? VI*. Bellingham, WA: SPIE, http://spie.org/Publications/Proceedings/Volume/9570.

Rubens, Heinrich. 1917. Das ultrarote Spektrum und seine Bedeutung für die Bestätigung der elektromagnetischen Lichttheorie, *Sitzungsberichte der Preußischen Akademie der Wissenschaften*: 47–63.

Rubens, H., and Ernest Fox Nichols. 1896. Über Wärmestrahlen von großer Wellenlänge. *Naturwissenschaftliche Rundschau* 11: 545–549.

Rubens, H., and Ferdinand Kurlbaum. 1900. Über die Emission langwelliger Wärmestrahlen durch den schwarzen Körper bei verschiedenen Temperaturen. *Sitzungsberichte der Preußischen Akademie der Wissenschaften* 1900: 929–941.

Rubens, H., and Ferdinand Kurlbaum. 1901. Anwendung der Methode der Reststrahlen zur Prüfung des Strahlungsgesetzes. *Annalen der Physik, Leipzig* 309 (4th ser. 4: 649–666).

Russo, Arturo. 1981. Fundamental research at Bell laboratories: The discovery of electron diffraction. *Historical Studies in the Physical (and Biological)/Natural Sciences, Berkeley* 12 (1): 117–160.

Rutherford, Ernest. 1904. Nature of the γ-rays from Radium. *Nature* 69: 436–437.

Rutherford, E., and C. Andrade. 1914a. The wave-length of the soft γ rays from Radium B'. *Philosophical Magazine, London* 6 (27): 854–868.

Rutherford, E., and C. Andrade. 1914b. The spectrum of the penetrating γ rays from Radium B and Radium C. *Philosophical Magazine, London* 6 (28): 263–273.

Rynasiewicz, Robert, and Jürgen Renn. 2006. The turning point for Einstein's annus mirabilis. *Studies in the History and Philosophy of Modern Physics* 37: 5–35.

Santori, Charles, D. Fattal, J Vučković, G.S. Solomon, and Y. Yamamoto. 2002. Indistinguishable photons from a single-photon device, *Nature* 419: 594–597, see also Grangier (2002).

Sayrin, C., I. Dotsenko, S. Gleyzes, M. Brune, J.M. Raimond, and S. Haroche. 2012. Optimal time-resolved photon number distribution reconstruction of a cavity field by maximum likelihood. *New Journal of Physics* 14: 115007.

Schaffer, Simon. 1979. John Michell and black holes. *Journal for History of Astronomy* 10: 42–43.

Schaffer, Simon. 1989. Glass works. Newton's prism and the uses of experiment, in David Gooding, Trevor Pinch and S. Schaffer, ed. *The Uses of Experiment*. Cambridge: Cambridge University Press, 67–104; arguing against this: Shapiro 1996.

Schilpp, Paul (ed.). 1959. *Albert Einstein - Philosopher-Scientist*. New York: Harper.

Schönbeck, Charlotte. 2000. Albert Einstein und Philipp Lenard: Antipoden im Spannungsfeld von Physik und Zeitgeschichte, *Schriftenreihe der mathematisch-naturwissenschaftlichen Klasse der Heidelberger Akademie der Wissenschaften*, no. 8: 1–42.

Schrödinger, Erwin. 1926a. Der stetige Übergang von der Mikro- zur Makromechanik. *Die Naturwissenschaften. Wochenschrift für die Fortschritte der Naturwissenschaften, der Medizin und der Technik, Berlin* 14: 664–666.

Schrödinger, Erwin. 1926b. Quantisierung als Eigenwertproblem (Zweite Mitteilung). *Annals of Physics* 4 (79): 489–527.

Schrödinger, Erwin. 1927a. Über den Comptoneffekt, *Annalen der Physik, Leipzig* (4) 82 (=387, no. 2): 257–264.

Schrödinger, Erwin. 1927b. *Abhandlungen zur Wellenmechanik*. Leipzig: Barth.

Schweber, Silvan S. 1994. *QED and the men who made it: Dyson, Feynman, Schwinger and Tomonaga*. Princeton: Princeton University Press.

Schneider, Martin. 2015. Die Deutsche Physik - Wissenschaft im Dienst von Ideologie und Macht. *Naturwissenschaftliche Rundschau* 69 (3): 125–134.

von Schweidler, Egon. 1904. Die lichtelektrischen Erscheinungen (Die Emission negativer Elektronen von belichteten Oberflächen). *Jahrbuch der Radioaktivität und Elektronik* 1: 358–400.

von Schweidler, Egon. 1910. Zur experimentellen Entscheidung der Frage nach der Natur der γ-Strahlen. *Physikalische Zeitschrift* 11: 614–619.

von Schweidler, Egon. 1915. Photoelektrizität, in Leo Graetz, ed. Handbuch der Elektrizität, III, 131–192.

Schwinger, Julian (ed.). 1958. *Selected papers on quantum electrodynamics*. New York: Dover.

Scully, Marlan Ovil, and Murray Sargent. 1972. The concept of the photon. *Physics Today* 25 (3): 38–47.

Scully, M.O., and Muhammad Suhail Zubairy. 1997. *Quantum optics*. Cambridge: Cambridge University Press 2nd ed., 2003.

Seidl, Johannes. 2010. Schweidler, Egon Ritter von. *Neue Deutsche Biographie (NDB)* 24: 40–41.

Sepper, Dennis L. 1994. *Newton's optical writings: A guide*. New Brunswick: Rutgers.

Serwer, Dan. 1977. Unmechanischer Zwang: Pauli, Heisenberg and the rejection of the mechanical atom, 1923–1925. *Historical Studies in the Physical (and Biological)/Natural Sciences*. 8: 189–256.

Shadbolt, Peter, Jonathan C.F. Mathews, Anthony Laing, and Jeremy L. O'Brien. 2014. Testing foundations of quantum mechanics with photons. *Nature Physics* 10: 278–286.

Shankland, Robert. 1963. Conversations with Albert Einstein I/II. *American Journal of Physics,* 31: 47–57.

Shapiro, Alan. 1993. *Fits, passions and paroxysms. Physics, method and chemistry and Newton's theories of colored bodies and fits of easy reflection*. Cambridge: Cambridge University Press.

Shapiro, Alan. 1996. The gradual acceptance of Newton's theory of light and color 1672–1727. *Perspectives in Science* 4: 59–140 (critique of Schaffer 1989).

Shapiro, Alan. 2009. Kinematic optics. *British Journal for the Philosophy of Science, Cambridge* 60: 253–269.

Shrader-Frechette, Kristin. 1977. Atomism in crisis: An analysis of the current high-energy paradigm. *Philosophy of Science* 44 (3): 409–440.

Sibum, Otto. 2008. Science and the changing sense of reality circa 1900. *Studies in the History and Philosophy of Science, Cambridge* 39: 295–297.

Sillitto, Richard M. 1957. Correlation between events in photon detectors. *Nature* 179: 1127–1128.

Sillitto, Richard M. 1960. Light waves, radio waves and photons. *The Institute of Physics Bulletin* 11 (5): 129–134.

Silva, Indiarana, and Olival Freire Jr. 2011. O modelo do grande elétron: o background classico do efeito Compton, *Revista Brasileira de Ensino de Fisica* 33, no. 4: 4061 1–7.

Silva, Indiarana, and Olival Freire Jr. 2013. The concept of the photon in question - the controversy surrounding the HBT effect 1957–58. *Historical Studies in the Natural Sciences* 43 (4): 453–491.

Simon, Josep (ed.). 2012. Cross-National Education and the Making of Science, Technology and Medicine. Special issue no. 168 of *History of Science* 50, no. 3: 251–374.

Simon, Josep (ed.). 2013. Cross-national and comparative history of science education. Special issue of *Science and Education* 22 (4): 763–866.

Simonsohn, Gerhard. 1979. Der fotoelektrische Effekt. Geschichte - Verständnis - Mißverständnis. In *Vorträge der Frühjahrstagung 1979*, ed. A. Scharmann and W. Kuhn, 45–54. Deutsche Physikalische Gesellschaft, Fachausschuss Didaktik der Physik: Giessen.

Simonsohn, Gerhard. 1981. Probleme mit dem Photon im Physikunterricht. *Praxis der Naturwissenschaften* 9 (30): 257–266.

Singh, Sardar. 1984. More on photons. *American Journal of Physics* 52, no. 1: 11 (comment on Armstrong 1983).

Slater, John C. 1924. Radiation and atoms. *Nature* 113: 307–308.

Small, Henry. 1981. *Physics citation index 1920–1929, Vol. 1 citation index, Vol. 2: Corporate index & source index*. Philadelphia: Institute for Scientific Information.

Small, Henry. 1986. Recapturing physics in the 1920s through citation analysis. *Czechoslovak Journal of Physics* 36: 142–147.

Smith, A.Mark. 2014. *From sight to light: The passage from ancient to modern optics*. Chicago: University of Chicago Press.

Solovine, Maurice (ed.). 1956. *Albert Einstein - Lettres à Maurice Solovine*. Paris: Gauthier-Villars.

Sommerfeld, Arnold. 1909. Über die Verteilung der Intensität bei der Emission von Röntgenstrahlen, *Physikalische Zeitschrift* 10: 969–976 (comment on Stark 1909b).

Sommerfeld, Arnold. 1911a. Das Plancksche Wirkungsquantum und seine allgemeine Bedeutung für die Molekularphysik. *Physikalische Zeitschrift* 12: 1057–1069.

Sommerfeld, Arnold. 1911b. Über die Struktur der γ-Strahlen, *Sitzungsberichte der Bayerischen Akademie der Wissenschaften, math.-physik. Klasse* 1911: 1–60.

Sommerfeld, Arnold. 1919. *Atombau und Spektrallinien*, Braunschweig: Vieweg, (a) 1st ed., 1919; (b) 2nd ed., 1921. (c) 4th rev. ed., 1924.

Sommerfeld, Arnold. 1920. Zur Kritik der Bohrschen Theorie der Lichtemission, *Jahrbuch der Radioaktivität und Elektronik* 17: 417–421 (reply to Stark 1920).

Sommerfeld, Arnold. 1930. *Wellenmechanischer Ergänzungsband*. Braunschweig: Vieweg.

Spence, John C.H. 2002. Quantum physics: Spaced-out electrons. *Nature* 418: 377–379.

Speziali, Pierre (ed.). 1972. *Albert Einstein - Michele Besso: Correspondance 1903–1955*. Paris: Hermann.

Średniawa, Bronisław. 1997. Wladyslaw Natanson (1864–1937), fizyk, ktory wyprzedzal swoj epoke [Wladyslaw Natanson (1864–1937), a physicist far ahead of his times]. *Kwartalnik Historii Nauki i Techniki* 42: 3–22.

Średniawa, Bronisław. 2001. Historia filozifii przyrody i fizyki w Uniwesytecie Jagiellońskim [History of natural philosophy and physics at the Jagiellonen Universität]. (*Rotprawy z dziejów nauki i techniki*, vol. 12), Warsaw.

Średniawa, Bronisław. 2007. Wladyslaw Natanson (1864–1937). *Concepts of Physics* 4: 705–723.

Stachel, John. 1989. Editorial headnote: Einstein's Early Work on the Quantum Hypothesis, in Stachel et al. 1989, 134–148.

Stachel, John. 2000. Einstein's light quantum hypothesis, or why didn't Einstein propose a quantum gas a decade-and-a-half earlier? In *Einstein: The Formative Years 1879–1909*, ed. Don Howard, and John Stachel, 231–251. Boston: Birkhäuser.

Stachel, John. 2002. *Einstein from B to Z*. Basel: Birkhäuser.

Stachel, John. 2002. Einstein and the quantum. *Fifty years of struggle, in Stachel* 2002: 367–402.

Stachel, John. 2002. Einstein's light quantum hypothesis. *Stachel* 2002: 427–444.

Stachel, John et al. (eds.). 1987. *Collected Papers of Albert Einstein*, vol. 1, *The early years, 1879–1902* ed. by J. Stachel, David C. Cassidy, Robert Schulmann & Jürgen Renn, Princeton: Princeton University Press.

Stachel, John et al. (eds.). 1989. *Collected Papers of Albert Einstein*, vol. 2, *The Swiss Years: Writings, 1900–1909* ed. by J. Stachel, David C. Cassidy, Jürgen Renn, Robert Schulmann, Don Howard & A.J. Kox, Princeton: Princeton University Press; suppl. English ed. transl. by Anna Beck, Peter Havas consultant.

Stanley, Robert Q. 1996. What (if anything) does the photoelectric effect teach us? *American Journal of Physics* 64 (7): 839.

Stark, Johannes. 1908a. Neue Beobachtungen zu Kanalstrahlen in Beziehung zur Lichtquantenhypothese *Physikalische Zeitschrift* 9: 767–773.

Stark, Johannes. 1908b. Weitere Bemerkungen über die thermische und chemische Absorption im Bandenspektrum. *Physikalische Zeitschrift* 9: 889–894.

Stark, Johannes. 1908c. Über die zerstäubende Wirkung des Lichtes und die optische Sensibilisation. *Physikalische Zeitschrift* 9: 894–900.

Stark, Johannes. 1909a. Über Röntgenstrahlen und die atomistische Konstitution der Strahlung. *Physikalische Zeitschrift* 10: 579–586.

Stark, Johannes. 1909b. Zur experimentellen Entscheidung zwischen Ätherwellen- und Lichtquantenhypothese. I. Röntgenstrahlung, *Physikalische Zeitschrift* 10: 902–913, see also Sommerfeld 1909 and Stark 1910.

Stark, Johannes. 1910a. Zur experimentellen Entscheidung zwischen der Lichtquantenhypothese und der Ätherimpulstheorie der Röntgenstrahlen. *Physikalische Zeitschrift* 11: 24–31.

Stark, Johannes. 1910b. Weitere Beobachtungen über die disymmetrische Emission von Röntgenstrahlen. *Physikalische Zeitschrift* 11: 107–112.

Stark, Johannes. 1912a. Über eine Anwendung des Planckschen Elementargesetzes auf photochemische Prozesse. Bemerkung zu einer Mitteilung des Hrn. Einstein, *Annalen der Physik, Leipzig* (4) 38: 467–469 (attack on Einstein 1912a); see also the reply by Einstein 1912c, *ibid.*, p. 888 and Stark 1912b.

Stark, Johannes. 1912b. Antwort an Hrn. A. Einstein, *Annalen der Physik, Leipzig* (4) 38: 496.

Stark, Johannes. 1912c. Zur Diskussion über die Struktur der γ-Strahlen. *Physikalische Zeitschrift* 13: 161–162.

Stark, Johannes. 1920. Zur Kritik der Bohrschen Theorie der Lichtemission, *Jahrbuch der Radioaktivität und Elektronik* 17: 161–173, see also the reply by Sommerfeld. 1920.

Stark, Johannes. 1922. *Die gegenwärtige Krisis in der deutschen Physik*. Leipzig: Barth.

Stark, Johannes. 1927. *Die Axialität der Lichtemission und Atomstruktur*. Berlin: Polytechn. Buchhandlung A. Seydel; continued in 9 parts in *Annalen der Physik, Leipzig* 1928–1930.

Stark, Johannes. 1930. Die Axialität der Lichtemission und Atomstruktur – VI. Folgerungen über den elementaren Vorgang der Lichtemission, *Annalen der Physik, Leipzig* (5) 4, no. 6: 686–709 (cf. Stark 1927).

Stark, Johannes. 1950. *Erfahrungen und Theorie über Licht und Elektron*. Traunstein, Obb.: Stifel.

Stokes, George Gabriel. 1845. On the aberration of light. *Philosophical Magazine, London* 27: 9–15.

Stokes, George Gabriel. 1846. On Fresnel's theory of the aberration of light. *Philosophical Magazine, London* 28: 76–81.

Strnad, Janez. 1986a. Photons in introductory quantum physics. *American Journal of Physics* 54 (7): 650–652.

Strnad, Janez. 1986b. The Compton effect - Schrödinger's treatment. *European Journal of Physics* 7: 217–221.

Stuewer, Roger H. 1969. Was Newton's 'wave-particle duality' consistent with Newton's observations? *Isis* 60: 392–394.

Stuewer, Roger H. 1970. A critical analysis of Newton's work on diffraction. *Isis* 61: 188–205.

Stuewer, Roger H. 1971. William H. Bragg's corpuscular theory of x-rays and γ-rays. *British Journal for the History of Science, London* 5: 258–281.

Stuewer, Roger H. 1975a. *The compton effect: Turning point in physics*. New York: Science History Publ.

Stuewer, Roger H. 1975b. G.N. Lewis on detailed balancing, the symmetry of time, and the nature of light, *Historical Studies in the Physical (and Biological)/Natural Sciences*. 6: 469–511.

Stuewer, Roger H. 1998. History as myth and muse, (a lecture at the University of Amsterdam, 11 Nov. 1998) abstract: History and Physics, on-line at http://www.physics.ohio-state.edu/~jossem/ICPE/B3.html

Stuewer, Roger H. 2014. The experimental challenge of light quanta, in Janssen and Lehner, ed. 2014, 143–166.

Sudarshan, E.C.G. 1963. Equivalence of semiclassical and quantum-mechanical descriptions of statistical light-beam. *The Physical Review Letters, New Series* 10: 277–279.

Sulcs, Sue. 2003. The nature of light and twentieth-century experimental physics. *Foundations of Science* 8: 365–391.

Tango, W. 2006. Richard Quentin Twiss 1920–2005. *Astronomy and Geophysics* 47: 333.

Tarsitani, Carlo. 1983. L'effetto fotoelettrico e la natura duale della radiazione. In *Storia della Fisica*, ed. F. Bevilaqua, 165–190. Franco Angeli: Milano.

Taylor, Geoffrey Ingram. 1909. Interference fringes with feeble light. *Proceedings of the Cambridge Philosophical Society* 15: 114–115.

Tegmark, Max. 2007. Shut up and calculate, arXiv:0709.4024v1 [physics.pop-ph], 25 Sep. 2007.

Tegmark, M., and John Archibald Wheeler. 2001. 100 years of quantum mysteries, *Scientific American* (Feb.): 72–79.

Thagard, Paul. 2012. *The cognitive science of science: Explanation, Discovery, and conceptual change*. Cambridge, MA: MIT Press.

Thomas, Lewellyn Hilleth. 1926. The motion of the spinning electron. *Nature* 117: 514.

Thomson, Joseph John. 1893. *Notes on recent researches in electricity and magnetism*. Oxford: Clarendon.

Thomson, Joseph John. 1896. On the discharge of electricity produced by the Röntgen rays, and the effects produced by those rays on dielectrics through which they pass. *Philosophical Transactions of the Royal Society, London* 59: 274–276.

Thomson, Joseph John. 1897. Cathode rays, (a) *Notice of the Proceedings of the Royal Institution of Great Britain* 15: 419–432; (b) *Philosophical Magazine, London* (5) 44: 293–316; (c) summary in *Proceedings of the Cambridge Philosophical Society* 9: 243–244; (d) reprinted in *American Journal of Physics*, 15, 1948: 458–464.

Thomson, Joseph John. 1903. *Electricity and Matter, 1903*. Cambridge: Cambridge University Press (Silliman Lectures as guest lecturer at Yale University).

Thomson, Joseph John. 1907a. Röntgen, cathode, and positive rays, *The Electrician*, 5 Apr. 1907: 977–979.

Thomson, Joseph John. 1907b. Rays of positive electricity. *Philosophical Magazine, London* 6 (14): 359–364.

Thomson, Joseph John. 1908. *Die Korpuskulartheorie der Materie*. Braunschweig: Vieweg.

Thomson, Joseph John. 1908b. Positive rays. *Philosophical Magazine, London* 6 (16): 657–691.

Thomson, Joseph John. 1911. Röntgen rays, *Encyclopaedia Britannica*, 11th ed., 1910/11, vol. 23, 694–696.

Tolman, Richard C., and Sinclair Smith. 1926. On the nature of light, *Proceedings of the National Academy of Science* 12, no. 5: 343–347, 508–509 (positive comment on Lewis 1926a).

Tomonaga, Sin-Itiro. 1974/97. *Spin wa meguru*, (a) 1st Jap. ed., 1974; (b) Engl. transl.: *The story of spin*. Chicago: University of Chicago Press, 1997.

Trenn, Taddäus. 1976. Geiger-Müller-Zählrohre. *Abhandlungen und Berichte des Deutschen Museums München* 44: 54–64.

Troland, Leonard Thompson. 1916. Apparent brightness; its conditions and properties. *Transactions of the Illumination Engineering Society* 11: 947–975.

Troland, Leonard Thompson. 1917. On the measurement of visual stimulation intensities. *Journal of Experimental Psychology* 2: 1–33.

Troland, Leonard Thompson. 1922. The present status of visual science. *Bulletin of the National Research Council* 5: 1–120.

Turnbull, Herbert Westren, John F. Scott, A. Rupert Hall, and Laura Tilling (eds.). 1959–1977. *The Correspondence of Isaac Newton*. Cambridge: Cambridge University Press, 7 vols.

Turner, Mark. 2006. *The artful mind: Cognitive science and the riddle of human creativity*. Oxford: Oxford University Press.

Tutton, Alfred Edwin Howard. 1912. The crystal space-lattice revealed by Röntgen rays. *Nature* 90: 306–309, see also W.H. Bragg 1912.

Uhlenbeck, George Eugene, and Samuel Goudsmit. 1925. Ersetzung der Hypothese vom unmechanischen Zwang durch eine Forderung bez. des inneren Verhaltens jedes einzelnen Elektrons. *Die Naturwissenschaften. Wochenschrift für die Fortschritte der Naturwissenschaften, der Medizin und der Technik, Berlin* 47: 953–954.

Uhlenbeck, George Eugene, and Samuel Goudsmit. 1926. Spinning electrons and the structure of spectra. *Nature* 117: 264–265.

van Vleck, John, and Hasbrouck, 1924. The absorption of radiation by multiply periodic orbits, and its relation to the correspondence principle and the Rayleigh-Jeans law. *The Physical Review. A Journal of Experimental and Theoretical Physics, New Series* 2 (24): 330–346.

van der Waerden, Bartel L. 1960. Exclusion principle and spin. In *Theoretical Physics in the Twentieth Century*, ed. M. Fierz, and V.F. Weisskopf, 199–244. New York: Interscience.

Walker, Charles T., and Glen A. Slack. 1970. Who named the -ON's? *American Journal of Physics* 38: 1380–1389.

Wallace, Philip R. 1994. Comment on Louradour, et al. 1993. *American Journal of Physics* 62: 950.

Walmsley, Jan A. 2015. *Light: A very short introduction*. Oxford: Oxford University Press.

Wawilow, Sergei Iwanowitsch. 1954. *Die Mikrostruktur des Lichtes*. Berlin: Akademie-Verlag (Russian orig., 1950).

Wayne, Randy. 2009. Nature of light from the perspective of a biologist. What is a photon?, in: Mohammad Pessarakli, ed., *Handbook of Photosynthesis*. Boca Raton, Florida: CRC Press, 17–43, on-line at http://labs.plantbio.cornell.edu/wayne/pdfs/whatisaphoton.pdf (20 Sep. 2016).

Weidner, Richard T., and Michael E. Browne. 1997. *Physics*. Boston: Allyn & Bacon.

Weinberg, Steven. 1977. The search for unity: Notes for a history of quantum field theory. *Daedalus* 106 (4): 17–35.

Weingard, Robert. 1982. Do virtual particles exist? *PSA* 1: 235–242.

Weingard, Robert. 1988. Virtual particles and the interpretation of quantum field theory, in H.R. Brown and Rom Harré, eds., *Philosophical Foundations of Quantum Field theory*. Oxford: Oxford University Press, 43–58.

Weinmann, Karl Friedrich. 1980. *Die Natur des Lichts. Einbeziehung eines physikgeschichtlichen Themas in den Physikunterricht*, Darmstadt: Wissenschaftliche Buchgesellschaft.

Weizsäcker, Carl Friedrich, and von. 1931. Ortsbestimmung eines Elektrons durch ein Mikroskop. *Zeitschrift für Physik* 70: 114–130.

Weizsäcker, Carl Friedrich, and von. 1941. Zur Deutung der Quantenmechanik. *Zeitschrift für Physik* 118: 489–509.

Wentzel, Gregor. 1925. Die Theorien des Compton-Effektes. *Physikalische Zeitschrift* 26: 436–454.

Wentzel, Gregor. 1927. Zur Theorie des Comptoneffektes I-II. *Zeitschrift für Physik* 43 (1–8): 779–787.

Wentzel, Guido, and Guido Beck. 1926/27. Zur Theorie des photoelektrischen Effekts, *Zeitschrift für Physik* 40: 574–589, 1926, and 828–832, 1927.

Weyssenhoff, Jan. 1937. Natanson †, *Acta Physica Polonica* 11: 251–281 (with bibliography).

Wheaton, Bruce. 1983. *The Tiger and the Shark*. Empirical Roots of Wave-Particle Dualism, Cambridge: Cambridge University Press.

Wheaton, Bruce. 1987. Symmetries of matter and light. In *Symmetries in Physics*, ed. G. Manuel, 279–297. Bellaterra: Doncel.

Wheeler, John Archibald. 1978. The 'Past' and the 'Delayed-Choice Double-Slit Experiment'. In *Mathematical foundations of quantum theory*, ed. A.R. Marlow, 9–48. New York: Academic Press.

Wheeler, J.A., and Richard P. Feynman. 1945. Interaction with the absorber as the mechanism of radiation. *Reviews of Modern Physics* 17 (2–3): 157–161.

Wheeler, J.A., and Richard P. Feynman. 1949. Classical electrodynamics in terms of direct interparticle action. *Reviews of Modern Physics* 21 (3): 425–433.

Wheeler, J.A., and Kenneth Ford. 1998. *Geons, Black Holes and Quantum Foam - A life in physics*. New York-London: Norton.

Wheeler, J.A., and Wojciech Hubert Zurek (eds.). 1983. *Quantum theory and measurement*. Princeton: Princeton University Press.

Whitaker, Andrew. 2012. *The new quantum age*. From Bell's theorem to quantum computation and teleportation. Oxford: Oxford University Press.

Whitaker, Martin Andrew, and Bancroft. 1979. History and quasi-history in physics education. *Science Education* 14 (108–112): 239–242.

Whitaker, Martin Andrew, and Bancroft. 1985. Planck's first and second theories and the correspondence principle. *European Journal of Physics* 6: 266–270.

Widom, F., and T.D. Clark. 1982. Quantum electrodynamics uncertainty relations, *Physics Letters* 90 A: 280.

Wiederkehr, Karl-Heinrich. 2006. Photoeffekte. Einsteins Lichtquanten und die Geschichte ihrer Akzeptanz, *Sudhoffs Archiv* 90: 132–142.

Wien, Wilhelm. 1896. Über die Energieverteilung im Emissionsspektrum eines schwarzen Körpers, *Annalen der Physik, Leipzig* 299: 662–669; in English trans.: On the division of energy in the emission-spectrum of a black body. *Philosophical Magazine, London*, 1897 5th ser. 43 (262): 214–220.

Wien, Wilhelm. 1913. *Vorlesungen über neuere Probleme der theoretischen Physik*. Leipzig & Berlin: Teubner.

Wien, W., and Otto Lummer. 1897. Methode zur Prüfung des Strahlungsgesetzes absolut schwarzer Körper. *Annals of Physics* 292: 451–456.

Wilson, Patrick. 1782. An experiment proposed to determine, by the aberration of the fixed stars, whether the rays of light, in pervading different media, change their velocity according to the law which results from Sir Isaac Newton's idea concerning the cause of refraction and for ascertaining their velocity in every medium whose refractive density is known, *Philosophical Transactions of the Royal Society, London* 72: 58–70.

Wöhrle, Dieter. 2015/16. Photonen, Licht, Stoff- und Energieumwandlungen. Was ist Licht?, *Chemie in unserer Zeit* 49 (2015): 386–401 & 50 (2016): 244–259.

Wöhrle, D., W.W. Tausch, and W.D. Stohrer. 1998. *Photochemie: Konzepte. Methoden, Experimente*, Weinheim: VCH.

Wolfers, Frithiof. 1924. Interférence par diffusion. *Comptes Rendus hebdomadaires des Séances de l'Académie des Sciences, Paris* 179: 262–262.

Wolfers, Frithiof. 1925. Sur un nouveau phénomène en optique; interférence par diffusion. *Journal de Physique* 6: 354–368.

Wolfers, Frithiof. 1926. Une action probable de la matière sur les quanta de radiation. *Comptes Rendus hebdomadaires des Séances de l'Académie des Sciences, Paris* 183: 276–277.

Wolfers, Frithiof. 1928. *Éléments de la physique des rayons X, introduction à la radiologie médicale et à l'étude générale des rayonnements*. Paris: Hermann.

Wolfke, Mieczyslaw. 1913/14. Zur Quantentheorie, *Verhandlungen der deutschen Physikalischen Gesellschaft* 15 (1913): 1123–1129, 1215–1218 & 16 (1914): 4–6.

Wolfke, Mieczyslaw. 1914. Welche Strahlungsformel folgt aus der Annahme der Lichtatome, *Physikalische Zeitschrift* 15: 308–311, 463–464 (reply to Krutkow 1914).

Wolfke, Mieczyslaw. 1921. Einsteinsche Lichtquanten und räumliche Struktur der Strahlung. *Physikalische Zeitschrift* 22: 375–379.

Woodruff, A.E. 1966. William Crookes and the radiometer. *Isis* 57: 188–198.

Wootters, William Kent, and Wojciech H. Zurek. 1982. A single quantum cannot be cloned. *Nature* 299: 802–3.

Worrall, John. 1982. The pressure of light. *Studies in the History and Philosophy of Science, Cambridge* 13: 133–171.

Wright, James Remus. 1911. *The positive potential of aluminum as a function of the wave-length of the incident light*. Ph: D. Diss. in Physics, University of Chicago.

Wright, Winthrop R. 1937. The photoelectric determination of h as an undergraduate experiment. *American Journal of Physics* 5 (2): 65–67.

Wroblewski, A. 1985. De mora luminis: A spectacle in the acts with a prologue and an epilogue. *American Journal of Physics* 54: 620–630.

Wurmser, René. 1925a. La rendement énergétique de la photosynthèse chlorophylliene. *Annales de Physiologie et de Physicochimie Biologique* 1: 47–63.

Wurmser, René. 1925b. Sur l'activité des diverses radiations dans la photosynthèse. *Comptes Rendus hebdomadaires des Séances de l'Académie des Sciences, Paris* 181: 374–375.

Wurmser, René. 1987. Letter to the editor. *Photosynthesis Research* 13: 91–93.

Wüthrich, Adrian. 2011. *The genesis of feynman diagrams*. Heidelberg: Springer.

Yin, Juan. 2017. Satellite-based entanglement distribution over 1200 kilometers. *Science* 356: 1140–1144.

Zajonc, Arthur. 1993. *Catching the Light*. The entwined history of light and mind. New York: Bantam Books.

Zeh, H.Dieter. 1993. There are no quantum jumps, nor are there particles. *Physics Letters A* 172: 189–195.

Zeh, H.Dieter. 2012. *Physik ohne Realität: Tiefsinn oder Wahnsinn*. Heidelberg: Springer.

Zeilinger, Anton. 2003. *Einsteins Schleier. Die neue Welt der Quantenphysik*, Munich: Beck.

Zeilinger, Anton. 2005. *Einsteins Spuk - Teleportation und weitere Mysterien der Quantenphysik*, (a) Gütersloh: Bertelsmann, 2005; (b) Munich: Goldmann 2007; (c) English transl.: *Dance of the Photons: From Einstein to Quantum Teleportation*, New York: Ferrai, Strauss & Giroux. 2010. *(orig.* Random House): English.

Zeilinger, A., et al. 1997/98. Experimental quantum teleportation, (a) *Nature* 390. 1997): 575–579; (b) in *Philosophical Transactions of the Royal Society, London* A 356, 1998: 1733–1737.

Zeilinger, A., Gregor Weihs, Thomas Jonnewein, and Markus Aspelmeyer. 2005. Happy centenary, photon. *Nature* 433: 230–238.

Zeilinger, A., Xiao-Song Ma, et al. 2012a. Quantum teleportation over 143 kilometres using active feed-forward. *Nature* 489: 269–273.

Zeilinger, A., Xiao-Song Ma et al. 2012b. Experimental quantum teleportation over a high-loss free-space channel. *Optics Express* 20 (21): 23126, 8 Oct. 2012.

Index

A

Abraham, Max, 61
Absorption, 48
Accretion, *see* semantic accretion
Aether
    drag, 46
    Faradays lines of force, 109
    for Lorentz, 100
    Newton's, 94
Aether-drag
    experiment by Fizeau, 46
Al-Haytham, Ibn, 95, 167
Andrade, Edward Neville da Costa, 25
Angular momentum, 77
    interior, *see* spin
Annus mirabilis, 17
Anomalous magnetic moment, 90
Antibunching, 161
Anticorrelation, 159
Arago, Franois, 41
Armstrong, Harvey L., 171
Aryan physics, *see* Deutsche Physik
Aspect, Alain, 152, 155, 156
ATLAS collaboration, 167, 168
Attosecond laser, 176

B

Badino, Massimiliano, 133
Band, W., 25, 66
Barkla, Charles Glover, 49, 110
Bartoli, Adolfo, 52
Batho, Harold F., 173
Beams, Jesse, 51
Beck, Guido, 132
Begriff
    terminological definition, 5

Bell, John Stewart, 155
Bennet, Abraham, 52
Besso, Michele, 1, 21, 96, 98, 100, 101
Beth, Richard A., 81
Black bodies, 10, 65
    spectral density, 20
Black hole, 43
    as conceptual blend, 141
Blair-Michell effect, 43
Blair, Thomas, 41
Bloch, Ernst, 6
Bloch, Felix, 91
Blumenberg, Hans, 185
Bodenstein, Max, 50
Bohr–Kramers–Slater (BKS) theory, 128
Bohr, Niels, 32, 77, 128, 158
Bohr-Sommerfeld atomic model, 17
    quantum jumps, 50
    quantum theory, 31
    spin quantum number, 77
Bohr's frequency condition, 73
Boltzmann gas, 19
Boltzmann statistics, 84
Born, Max, 87
    correspondence with Einstein, 129
Bose-Einstein condensate, 85
Bose-Einstein statistics, 81, 175
    confirmation of, 161
Bose, Satyendra Nath, 85
Bothe, Walther, 119, 130, 132
Bragg, William Henry, 49
    binary model, 184
    mental model by, 110
    wave-particle duality, 68
Bragg, William Lawrence, 49
Brannen, Eric, 147
Bremsstrahlung, 20

© Springer International Publishing AG, part of Springer Nature 2018
K. Hentschel, *Photons*, https://doi.org/10.1007/978-3-319-95252-9

Printed in the United States
By Bookmasters